DATE DUE			
Apr 7 '75			

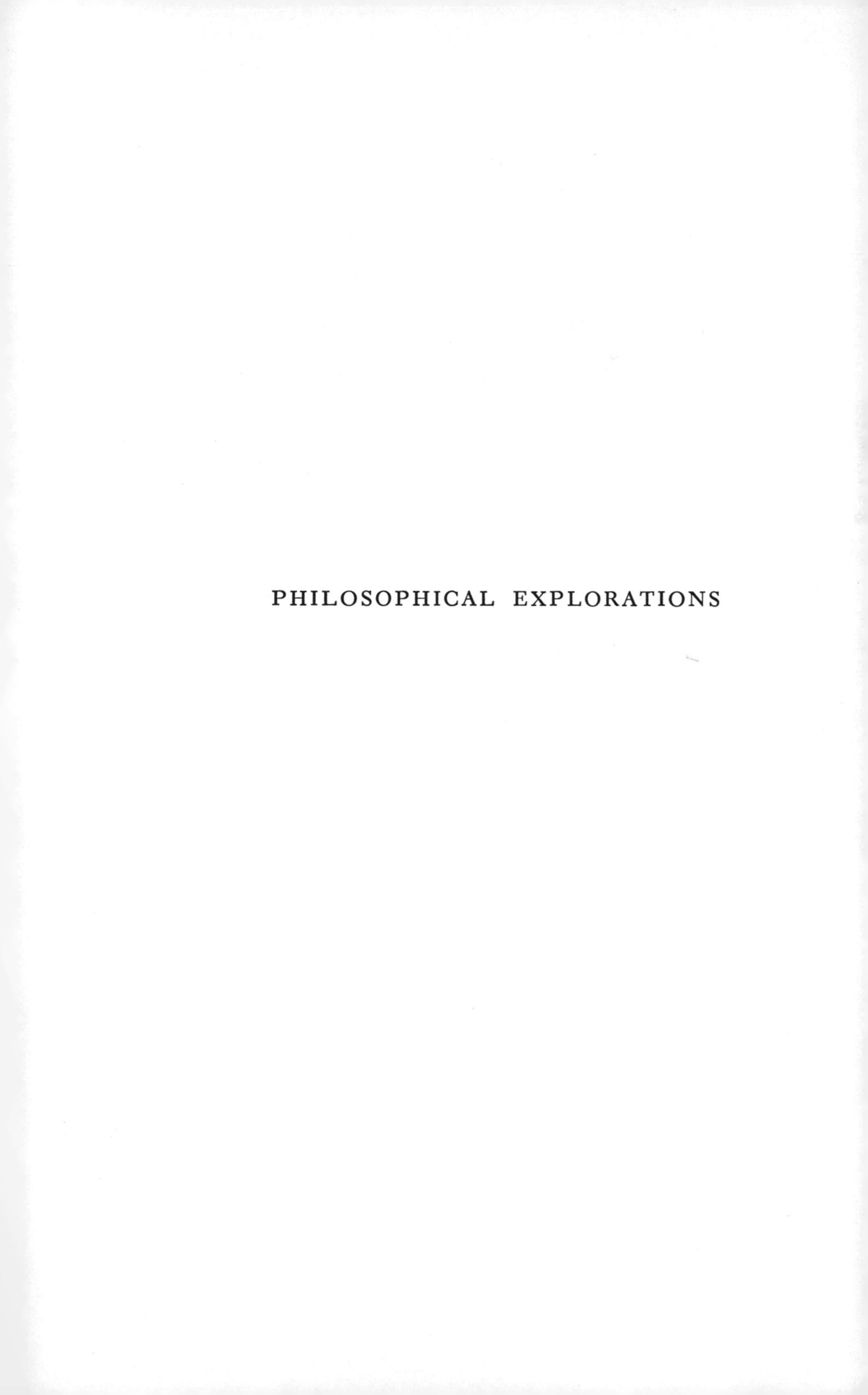

PHILOSOPHICAL EXPLORATIONS

PHILOSOPHICAL EXPLORATIONS
Φ

ROBERT S. HARTMAN

THE STRUCTURE OF VALUE: Foundations of Scientific Axiology

FOREWORD BY

George Kimball Plochmann

PREFACE BY

Paul Weiss

SOUTHERN ILLINOIS UNIVERSITY PRESS
Carbondale and Edwardsville

FEFFER AND SIMONS, INC.
London and Amsterdam

Library of Congress Catalog Card Number 65–12393
Printed in the United States of America
DESIGNED BY ANDOR BRAUN

To Rita

FOREWORD

MY COMMENTS about this remarkable book, which is an attempt to "introduce orderly thinking into moral subjects," ought not to cover the same ground as the excellent statement by Dr. Weiss on the chief doctrines of the work. Instead I should speak on certain aspects of Professor Hartman's methods and their mainsprings, suggesting along the way something of the scope and classic precision of this treatise.

The author posits a close parallel between the history of the knowledge of nature and that of the knowledge of value; but after a long account of the analogies between the two he carries the tale much further than mere analogies by presenting a tightly-knit logic of the two kinds of knowledge, part being common to both, the rest being broken into parts peculiar to each. The historical analogies might leave me, for one, unconvinced, mainly because I believe that the enormous complexity of the history of thought could scarcely find a place for any real similarities. Even in sciences so nearly allied as physics and chemistry, their respective early histories show significant differences between the types of experimentation, the uses made of mathematical schemata and formulae, and so forth. On the other hand, both do coincide in the fact that there *is* experimentation and that there *is* the use of mathematics. Mr. Hartman, generalizing from these facts, defines science as the application of *any* logical frame of reference to *any* subject matter. While this definition may include sciences so disparate as those of value and nature, the differences of disciplines so close as physics and chemistry seem to leave room for doubt concerning his generalization.

But Dr. Hartman has an ingenious logical point about the diremption between goodness and what may be called physical being, the being of fact: although the sciences studying them are separated as to subject matter, the basic *concepts* upon which these sciences depend are scarcely different at all: they have a definite logical relationship. To be a good *x* is not entirely other than being an *x* in the first place. It is almost as if Socrates in *Republic VI* were to have said that the illuminative sun is the idea of *being*, and that all vision depends upon it. At any rate, the near fusion of traditional metaphysics and axiology implied by this enables Professor Hartman, in his final chapter, to develop the theory of good as fundamental for a new organization of the sciences.

His logic of value is an intensional logic, though it has extensional applications; it is principally an analysis of predicate, not subject terms. The arithmetical approach makes use of well-known theorems in combinatory mathematics, dealing with the possible subsets of sets of descriptive properties of objects. The value of something is its *meaning*; it is, first of all, its being what it is. And since "what it is" can be exposited in terms of its properties, the theory of sets should—and indeed can—give an exact enumeration of the combinations of these properties. This shows to what degree and in what intensional configurations something really is what it is, i.e. what are the various forms of its being, and of its meaning—of its "whatness." One is reminded of the sign posted some years ago in the Bowery: "Coffee, 3 cents; Coffee that *is* coffee, 5 cents." But it should scarcely be assumed that the author is content to fractionate the individual properties of things, leaving the nature of the thing to take care of itself. Though the descriptive properties of each object may be ticked off on one's fingers, the thing itself as a whole is a unity that yet contains an infinity within its oneness. The unity of *this* man and of his infinity of properties is of a sort quite different from the unity of *a* man (or of mankind) and its denumerable set of properties. Cantor's theory of the infinite is invoked by Dr. Hartman here, and its connection with the theory of finite sets is interpreted.

Here again I incline to hang back, for the arrangement of

subsets of properties seems the kind of exactness on which Alfred North Whitehead was occasionally so hard—even though, in "Mathematics and the Good," he did stress the connection of goodness and arithmetical pattern. *My* world appears shot through with eagerness, fatigue, effort, distaste, and a hundred other sentiments and attitudes not so much interwoven as fused in a complex consciousness. Such a world would be troublesome to reduce to rigid frames; even Plato made a rough joke (*Statesman* 266a-d) which turns upon the carrying of mathematical applications too far. But again Hartman offers a rejoinder. We must not at first consider the human act of valuing; we consider instead the formal pattern of this act, the structure that lies behind our valuing. It is a thoroughly objective standpoint that the author adopts. His is the very opposite of Spinoza's theory that all good and evil are interior responses to things, not the measures of things themselves.

Although some of his procedures and constructions remind us of the recent Positivists, his theory of value takes us a long way from their insistence on the noncognitive character of valuation, and we find in this book that much the same types of logical devices do for nature and for value, though in slightly varied applications. Thus Hartman draws a distinction between the axiological treatment of some object, *x*, and its logical treatment. Although fact—what *x* is—remains the norm or determinant of value, and value is indeed a predicate, it is still possible to distinguish between the two propositions "*x* is a C" and "*x* is a good C." In the latter, "good" is not a predicate of *x* directly, but is instead a predicate of all those predicates enshrined in the descriptive name, "C." "Good" is thus a second-order concept, "*x* is a good C" a second-order function, and the logic of value is firmly based upon the power to reduce second-order functions to first-order functions. To show how this is done the author provides elaborate examples, one of them being that of the contrast between first and second sense of actuality (Aristotle, *On the Soul*, II, 1, 412a23 ff.), the having of a trait and the exercising of it. This problem (which an earlier book in the Philosophical Explorations Series happens also to have dealt with at some length —David Weissman's *Dispositional Properties*) is receiving attention again after a woeful lapse of many centuries. In Dr. Hartman's terminology, there can be functions of functions, and a predicate can be asserted and denied of a subject simultaneously, provided that the order of function be made clear.

It should be evident from this that the present approach to the good and to value generally is to treat these first of all as terms, not things. As with Moore, a definition is sought of each term that would make it unambiguously applicable to all instances properly falling under it, and inapplicable to everything else. This is not, then, a theory of moral sense or judgment, though once the main problem is disposed of, such further considerations fall into line. But Professor Hartman strikes me as outdoing those analysts who are satisfied once they have defined a constellation of individual words, "categories," because he shows also the interrelations between his terms through an "axiom," pointing out the way to locate all topics to which they are relevant, and to which they may unequivocally be attached.

Within this axiomatic framework, *The Structure of Value* uses a vast, intricate series of dichotomies: philosophy-science, analytic-synthetic, definition-exposition, category-axiom, logic-axiology, and so forth. Some of the flock of dichotomies remind us of distinctions long current or at any rate recently popular, while others are quite novel.

Having begun with literal divisions of this kind, based on his analysis of the notion of science, the author, on the basis of his "axiom" of the science of value, goes on to constructions which assort them in various patterns, as demanded by the axiomatic deduction. One of these patterns is tetrads: there are (to take an example from Chapter 5) pure logical propositions, mixed logical ones, pure axiological propositions, and fourthly those axiological propositions that are mixed. There is then no difficulty in passing on to other assortments involving more elements in new levels of combination; Dr. Hartman finds that great arrays are mathematically certain, axiologically suggestive, and dialectically defensible.

There is a second treatment of dichotomies, one by which the author offers a single pair of terms and then, to make his distinction stick, adds two more, parallel to the first not assorted with it: thus for Hartman the concepts of philosophy are analytic (and are called, moreover, categories), while those of science are synthetic (and are called axioms). In the former, which are material concepts, intension and extension vary indirectly; in the latter, which are formal, intension and extension vary directly. This distinction is fortified by another, showing the same procedure, where each successive pair narrows and strengthens whatever pairs have gone before, and manages dialectically to sum up its predecessors. Hartman divides concepts from propositions and each of them (following Kant) he subdivides into analytic and synthetic. It is the concepts in which he is interested. Those that are analytic are the categories which have danced their way through philosophy's long hour upon the stage. Among the synthetic, on the other hand, turn out to be the very axioms so badly needed for the founding of the sciences, including the science of value. Throughout the history of these sciences, both natural and axiological, synthetic statements regarding primary properties take the place of loose categories covering the secondary; and here progress begins.

A third treatment of paired concepts is to combine the two opposites. So Dr. Hartman differentiates formal value (the object of theoretic study), phenomenal value (the object of feeling), and finally axiological value, which merges the two. I say this to warn the reader against supposing that there is a total rigidity of terms. Which brings us to the last prominent logical device.

It is that one and the same expression can serve on two different levels of generality, and thus include itself as one of its species. For instance, Professor Hartman stresses that his is a formal science, formal axiology. (I have his Chapter 7 especially in mind here.) But value itself is not wholly amenable to formal treatment (as Professor Weiss makes clear in his Preface), and "value theory" has both a broad and a narrow, proper sense. So we may say that value theory both does and does not involve everyday feelings. (In Hartman's theory, these feelings appear in their primary properties: as configurations of descriptive properties). The reader should stay on guard, ready to distinguish separate senses of important terms, for otherwise paradoxes will arise.

As we said at the beginning, Dr. Hartman separates method from content, finding the difference between science and philosophy to be a methodological one, not a division of subject matter. Science for him is prior to philosophy in an order of precision though not of history; the philosophical axiologists, who came first, have been vague and have employed unclear categories; so it is high time, he believes, that scientifically exact thinkers supersede them, much as Galileo replaced the predominantly Aristotelian writers of the late Middle Ages. G. E. Moore has enunciated a rough prototype of the chief axiom of the new science of value (as well as giving expression to its leading paradox and warning against the perpetration of attendant fallacies), and it remains for a latter-day Newton to diagram the whole set of conclusions following from the reformulated axiom.

Sympathy and wit counterbalance Hartman's enormous load of erudition—a glance at his footnotes or a delighted sweep of the eye through his personal library will quickly convince one of his wide reading—and although he stresses the need for extricating scientific thinking about value from the act of valuing itself, we feel him to be directly engaged in a clear-headed reflection about the world and its goodness—and what he calls no-goodness. Sometimes in reading modern ethical philosophers one fears that the root problems of how to do well and fare well have passed by default into the hands of the psychoanalysts, the preachers, or maybe even the sociologists. Despite the forbidding logic and the chilly mathematical formalities of some parts of Hartman's treatment, we can rejoice, I think, that a scientifically stimulating, profound book can still be about the homeliest and most familiar of terms and the commonsense ways that these should be applied in our common world.

George Kimball Plochmann

Southern Illinois University
July 6, 1965

PREFACE

A GREAT idea—an idea in terms of which subsequent thought takes its start or uses as a pivot—strikes one at first glance as incredible, at second as incredibly simple, at third as simple-minded, then as obvious and true and eventually beyond refutation, until finally it gives way to another and is pushed aside, with its civilization, as unfortunately too crude and limited. I think Dr. Hartman has discovered such an idea and has expounded, applied, urged, and defended it with the devotion and passion of the true pioneer. It is a notion that belongs to the family of notions essential to what might be called the running broad jump advance to truth. Instead of, as in the standing broad jump, attempting to make a straight leap forward from some mark, the running broad jumper turns his back on the target, walks away from it, suddenly turns about, runs to his mark and then, and then only, leaps forward. By first moving away from his objective and preparing himself to jump, the running broad jumper is able to leap more expeditiously and effectively than otherwise.

Modern thought can be said to begin with the introduction by Galileo of something like the running broad jump method into science. Where his predecessors attempted to understand motion and change in terms of the obvious features and language of every day, Galileo offered a set of formulae couched in abstract, formal, nonobservable terms, defining an ideal, rational state of affairs. Combinations of these formulae, when subject to qualification and limitation so as to answer to some actual state of affairs, made it possible to measure precisely and predict a wide range of phenomena beyond the reach of any other method. Dr. Hartman has dealt with values in a spirit somewhat similar to Galileo's. He therefore speaks with justification of his own work as essentially scientific—though it is not experimental and though it does make use of a logic and deals with topics distinct from those which interest natural science. It is his claim that he has provided the first scientific account in terms of which value in all its dimensions can be analyzed, classified, interrelated, and calculated. This volume goes far to justify this striking claim.

There are many directions from which Dr. Hartman's view can be approached. One is through the idea of a formal system. A formal system is a set of relations, rules, and principles in terms of which ideal entities are constructed and interconnected. Mathematics is one such system, two-valued symbolic logic a second, theoretical physics, econometrics, others. One of Dr. Hartman's contributions is to lay the foundations for a systematic axiology, or formal science of values. This does not concern itself with appreciation or enjoyment, but with an abstract interrelating of values. It is a discipline related to such value-subjects as ethics, aesthetics, economics, and politics, as mathematics is related to physics, chemistry, biology, and geology. All its objects are constructions; they are as abstract, conceptual, and intelligible as a mathematical number or triangle is.

Axiology, for Dr. Hartman, has three distinct but interrelated branches dealing in formal ways with extrinsic, intrinsic, and systemic value.

Extrinsic value theory defines the nature and offers measures of the good and its cognates—bad, fair, ought, ought not, and the like. These terms have application to any and all the objects encountered in daily life. Those objects are usually dealt with by us not as unique entities, but as items in a class, sharing a number of features with the other items that belong to that class. Some of these features define what it is to be a member of the class. They are the minimal properties which a thing must have to be properly called by a common name. A watch, e.g., is a small instrument, designed to be carried by a person, and serving to mark the passage of time. What does not live up to that definition is not a watch. But there are good watches and poor ones. The poor ones obviously fall short of the nature of the good ones. The good ones have, in addition to the features which any watch must have, features which are lacking to the poor ones. They keep time, have at least two hands, need be wound only once in a while, and so on. Such features, for Dr. Hartman, make up the "exposition" of the watch. A good thing is one which has all the features which its exposition encompasses. What is not altogether good lacks some of these features. Since such a thing ought to be good, we can say of it that it should have some expository features which it now does not have, that it should conform more closely than it does to the full idea of such an object. A good thing completely conforms to the full idea of that kind of thing. This is one of Dr. Hartman's basic notions, in terms of which large areas of value theory are illuminated in fresh and sometimes startling ways.

Objects are not merely members of a class; they are unique, with a being, an interiority of their own. We recognize them to have an intrinsic value when we see any one of their features to be representative of all the others. Dr. Hartman sometimes speaks of a unique object as having an infinite continuum of features and as being beyond the reach of any other designation than that of a metaphor. More important, he takes uniqueness to have degrees; two individuals, though equally ultimate and irreducible, can therefore be compared as having more or less fulfilled their promise because they have more or less differentiated the continuum which is theirs natively. Dr. Hartman has not attended to this branch of his theory to the same extent that he has attended to the others, and much therefore remains obscure. But he has made it unmistakably clear that intrinsic and extrinsic value are distinct and ought never to be confounded, and that both differ from systemic value.

Good, fair, bad and the like, express degrees of extrinsic value. They can be interrelated, united, and quantified. This is done in a logic and calculus of value. The result is a series of theorems in a systemic treatment of extrinsic value.

Since one can deal with anything intrinsically by involving oneself with it, extrinsically by classifying it, and systemically by means of construction, one can apply any branch of value theory to any other. One can, for example, deal extrinsically with a systemic treatment of an intrinsic value. Dr. Hartman's value-calculus therefore is capable of endless expansion. Towards the end of the book, he has himself indicated some of the applications of his view.

Dr. Hartman is at home with the literature in the field of value, and he has examined his hypothesis in the light of other attempts. No one, I am convinced, who will allow himself to follow Dr. Hartman's relentless systematic inquiry into the nature of value will be able to free himself from the awareness that he has been engaged in a new intellectual adventure whose consequences and influences promise to be vast and varied.

Paul Weiss

New Haven, Connecticut
August, 1965

ACKNOWLEDGMENTS

I HAVE RECEIVED encouragement and inspiration for this work from colleagues and students in the United States, Mexico, and Europe. The late Professor Albert E. Avey has encouraged me throughout long and often difficult years; Professor Paul Weiss has clarified, in spirited conversations, many of the notions underlying my approach to value. He has read the entire manuscript, extensively commented on it, and honored it with his preface. Professor John W. Davis has contributed to the intension of the theory by acute essays and to its extension through his classes and his television lectures over the network of the University of Tennessee. In Mexico, the late Dr. Samuel Ramos, Coordinator of the Humanities, and Dr. Eduardo García Máynez, Director of the Center of Philosophical Research of this University, have, as administrators, contributed to the institution of research centers by which this University offers virtually ideal conditions for full-time research workers. As thinkers they have deepened the field of philosophy and created an atmosphere of scholarly dedication which it is a privilege to share.

Dr. García Máynez' work in the philosophy of law follows the same axiomatic lines as the present work in axiology. There are similar endeavors by individual thinkers in the various countries of Latin America, Europe, and North America. The book thus joins the ranks of an advance guard of works endeavoring to base social and humanistic disciplines on axiomatic foundations—or rather, a rear guard, for it covers the advance of traditional philosophy with modern resources toward the frontiers of ethical thinking.

My students, especially those trained in science, have been a constant source of inspiration. Their training has made them enthusiastic collaborators, their unhampered imagination original investigators. Their studies in formal axiology—which I hope to present at a later occasion—creatively extend the field in several directions; they helped to convince me of the fundamental correctness of the approach here presented.

My wife is my partner in this enterprise, which, although started some twenty years ago, has only begun. She typed the manuscript of this book and took care of the many details, necessary but often tedious, which are part of the writer's work. The book is dedicated to her in gratitude for her unflagging enthusiasm and unfailing intuition which saved me from many an impasse.

The present book is the considerably revised version of *La Estructura del Valor: Fundamentos de la axiología científica,* published in 1959 by Fondo de Cultura Económica, Mexico and Buenos Aires. In spite of the long process of gestation of this book, no material published previously in English was incorporated in the Spanish version. However, such material, as well as materials published in the interval between the Spanish and the English version, is included in this book. It is from the following articles: "The Analytic, the Synthetic, and the Good: Kant and the Paradoxes of G. E. Moore," *Kant-Studien,* 1953/54; "Value Theory as a Formal System," *Kant-Studien,* 1958/59; "The Logic of Description and Valuation," *Review of Metaphysics,* 1960; "The Logic of Value," *Review of Metaphysics,* 1961; "The Logical Difference Between Philosophy and Science," *Philosophy and Phenomenological Research,* 1963; "Four Axiological Proofs of the Infinite Value of Man," *Kant-Studien,* 1964/65. Material from this book has been used in a discussion article "Axiology as a Science: Rejoinder Note to Professor Neri Castañeda's Review of *La Estructura del Valor,*" in *Philosophy of Science,* 1961. Finally, some material has been used from my contributions to *The Language of Value,* Ray Lepley, ed. (New York: Columbia University Press, 1957) and *Knowledge in Human Values,* Abraham Maslow, ed. (New York: Harper and Brothers, 1958). Acknowledgment for permission to print this material is gratefully made to the journals and publishers mentioned.

This book has immeasurably benefited by the interest and acuteness of the editor of this series, Professor George Kimball Plochmann. His drive for perfection and conciseness, coupled with his deep understanding of the issues and sympathy for the author, have inspired me to rethink and often reformulate the argument. The result is, I believe, a better book than the original.

Robert S. Hartman

Centro de Estudios Filosóficos
Universidad Nacional Autónoma de México
October, 1965

CONTENTS

FOREWORD vii

PREFACE xiii

ACKNOWLEDGMENTS xvii

INTRODUCTION 3

PART ONE THE STRUCTURE OF SCIENCE

1 Philosophical Ethics and Scientific Ethics 25

2 The Structure of Science 64

PART TWO THE FOUNDATIONS OF VALUE SCIENCE

3 The Concept of Axiological Science 95

4 The Axiological Reinterpretation of Moore's Ethical Theory 121

PART THREE THE STRUCTURE OF VALUE

5 The Elements of the Axiological System 153

6 The Systematic Import of Formal Axiology 193

7 The Empirical Import of Formal Axiology 249

Summary and Outlook 303

NOTES 315

INDEX 361

The Structure of Value
Foundations of Scientific Axiology

Spirit can understand itself as spirit, but feeling cannot understand itself as feeling. Science must clarify the judgments of value which grow out of reflective feeling and transform them into cognition. Thus they will cease to be the unrestricted property of individual sentiments and become sublime truths beyond the changes of moods and attitudes.

— HERMANN LOTZE

We are entering upon an age of reconstruction, in religion, in science, and in political thought. Such ages, if they are to avoid mere ignorant oscillation between extremes, must seek truth in its ultimate depth. There can be no vision of this depth of truth apart from a philosophy which takes full account of those ultimate abstractions, whose inter-connections it is the business of mathematics to explore. . . . The paradox is now fully established that the utmost abstractions are the true weapons with which to control our thought of concrete fact.

— ALFRED NORTH WHITEHEAD

But what abstract discipline can ethics, or aesthetics, or any non-physical interest of philosophers fall back upon? Mathematics does not apply to all phenomena. Certainly the "values" of ethics, as they are conceived at present, defy mathematical treatment. . . . How, then, can we hope for the sort of abstract analysis of values which might lead to the discovery of really fundamental ideas? The indispensable tool of all philosophy is logic. Pure logic, of which pure mathematics is a branch, can deal with any subject matter whatever; it is not limited to just such concepts and premises as will define the sciences of quantity. But its possibilities in application to ethics, aesthetics, history, etc., are as yet entirely unexplored. Analyses of such concepts as "historic fact," "value," "life," and many others have never been undertaken. We have never seriously tried to find the conceptual elements out of which a science of ethics, for instance, could be made.

— SUSANNE LANGER

INTRODUCTION

> Some students begin by forming an opinion . . . and it is not till afterwards that they begin to read the texts. They run a great risk of not understanding them at all, or of understanding them wrongly. What happens is that a kind of tacit contest goes on between the text and the preconceived opinions of the reader; the mind refuses to grasp what is contrary to its idea, and the issue of the contest commonly is, not that the mind surrenders to the evidence of the text but that the text yields, bends, and accommodates itself to the preconceived opinion.
>
> — FUSTEL DE COULANGES

> He who confuses the study of the *object value* with the study of the concrete *valuable objects* or even with the study of the acts of valuational attitudes, is in the position of a man who assigns the study of arithmetic to the botanists because he learned to count with apples and nuts, or who confuses higher mathematics with the psychology of counting because there would be no theory of numbers without people who know how to count.
>
> — THEODOR LESSING [1]

This book is an undertaking in traditional philosophy which, unfortunately, will appear to many revolutionary. It attempts to introduce orderly thinking into moral subjects. By "orderly thinking" it understands a procedure which accounts with a minimum of concepts for a maximum of objects. It thus conceives of a theory of values as, or analogous to, a science which from a minimum of axiomatic assumptions draws a multitude of conclusions, in a pattern so varied and detailed that its features mirror the multitude of those found in the realm of values.

Value theory is thus conceived of as a pattern isomorphous with the value realm. The *structure of value* is the structure of the pattern as relevant to, or explicatory of, this realm. This conception presupposes that there are value phenomena, that they form an orderly pattern, and that this pattern can be mirrored in a theoretical structure, the theory of value or axiology. Value thus appears on three levels: that of the axiological pattern, *formal value* or "*value*"; that of the value realm, *phenomenal value* or *value;* and that of the combination of both, *axiological value*. All three levels together form the *science of value*.

The present book does not present this science; it gives only preliminary and fundamental considerations concerning it. It does not discuss specific values such as economic, aesthetic, or religious, and its reference to moral value, although frequent, is illustrative and provisional rather than substantial and definitory. The system of theoretical axiology is sketched out only in sufficient detail to establish its possibility and make plausible its theoretical and practical import. Eventually, however, the theoretical pattern presented is expected to cover the realm of value in all its details, as the theoretical pattern of natural science covers its own subject matter, nature. Orderly thinking in the field of value is thus regarded as being not different from orderly thinking in the field of fact: it means construction of patterns isomorphous with the subject matter.

This ideal of orderly thinking is, of course, the ideal of knowledge of most of traditional philosophy. It has been proclaimed for both natural and moral philosophy from Plato to Whitehead, and has been the ideal of thinkers on thinking from Ockham to Mach. While in natural philosophy it has been well realized and the corresponding method has celebrated triumphs, it is largely out of fashion today in moral philosophy; and there exists, of course, no corresponding method. The reasons for this lag in moral science are many, both historical and systematic, all coming down to a very recent insight: that the borderline between fact and value is extremely thin yet runs extremely deep, an invisible crack in our understanding. Philosophy has either overlooked or exaggerated it. It is our task to follow this line, both in length and in depth, and to show how fact and value run along the two edges of this minuscule crack, one fitting tightly into the other and following its convolutions, no matter how intricate — yet both sharply and forever divided.

To follow this trek we need a sharp intellectual instrument: a science of value. And the conception of such a science presupposes several distinctions that have never been clearly made.

In natural science they did not have to be made because they were obvious. But they have to be made explicit in moral philosophy where they are anything but obvious, owing to the oversight and exaggeration mentioned; for to make them shows up sharply the parallelism of natural and moral science; and such parallelism is too much of a division for those who see no difference between natural and moral science, and too little of a division for those who see too much of a difference. Such oversight does not make the gap disappear, nor does such exaggeration make it any deeper or sharper, it only makes it cruder.

Failure to make the distinctions in question has been one of the main obstacles, if not *the* main obstacle, to the development of a science of value and, indeed, of orderly thinking in the value field.

A botanist dissecting a rose is in no danger — or temptation — of thinking that in dissecting it he is dissecting himself, or that he smells like a rose. He cannot possibly confuse himself with his subject matter; but a psychologist is in a more precarious situation in this respect. He has to identify himself with his patient, and it is not impossible, at least theoretically, that he may do so to such an extent as to confuse himself with his subject. Needless to say that, if he did so completely and to the point of losing his rational detachment, his usefulness as an analyst would cease, and he would become a patient himself.

Value theoreticians have thought of themselves as closer to the psychologist than to the botanist. And, rather than being aware, and being wary, of the danger of involvement in their subject matter to the point of losing their rational detachment, have made such loss of detachment, vaguely, a condition of their axiological activity. Actually no systematic investigation of the relation between the value theoretician and his subject matter, value, has so far been made. This relation, therefore, has remained obscure. In particular the danger of losing one's usefulness as a *theoretician of* value by being involved *in valuing* has not been examined or even realized. As a result, the two — value theory and its subject matter, value — are constantly being confused: valuers believe they analyze value, and value analysts believe they value; philosophers of value believe they must be involved, committed, etc., and valuers who are committed, involved, etc., believe they must philosophize. Thus philosophy and value, value and ideology, commitment and rationalization are all mixed together. These confusions play havoc, not only in moral philosophy but also in social science, politics, etc.

Thinking in the value field presupposes distinctions never explicitly made. Once made, these distinctions are difficult to challenge; and their intellectual obviousness, coupled with the fertility of their application, makes them presuppositions of any philosophy that pretends to be scientific. At the very least, they help to draw the line between those who want to be experts in a field and those who want to be subject to the analysis of experts.

The fundamental distinction in question is the self-evident one between thinking and the object thought. This implies divisions between thinking and doing, between content and form, between subject matter and method, between practice and theory, between use and meaning. These distinctions, in turn, determine those between order and disorder, between clarity and confusion, between coherence and fragmentation, and between relevance and triviality. In other words, the fundamental distinction between thinking and its object is the condition *sine qua non* for the order, clarity, coherence, and relevance of a theory, while the fusion of thinking with its object leads, in different degrees, to the disorder, confusion, fragmentation, and triviality of a theory. All this, of course, is too obvious for natural science and traditional philosophy even to be mentioned. But for value theory this clarification is of importance. Much of that theory has been based on the confusions in question: the three levels of value — formal value, phenomenal value, and the combination of both, axiological value — are not distinguished. This means, simply, that value has not been made an object of orderly thinking. It has not been made really the subject of a *theory*; and "value theory" is a euphemism.

For us, value is an object of knowledge like any other, not different in this respect from the rose for the botanist or the electric current for the physicist. Botanists, as we said, do not smell, and physicists do not spark. In the same way, *the axiologist does not value but analyzes value.* When the botanist hands a rose to his fiancee or the physicist pulls the bread out of the electric toaster, they do not act as either botanist or physicist, but as human beings in everyday situations. These situations happen to exemplify certain features of their professional subject matter. But *as* botanist and physicist their task is to be professionals and experts in roses and currents, that is to say, to be familiar with the *funda-*

mental principles and general laws that underlie *all* roses and *all* currents.

The axiologist is the expert in value. When he values he does not act as an axiologist but as a human being in an everyday situation which happens to exemplify certain features of his professional subject matter. But *as* axiologist his task is not to value but to be professional and expert in value, to be familiar with the *fundamental principles and general laws* that underlie *all* valuing.

This book, then, is to value as a treatise on pulmonary physiology is to breathing. The lung specialist is not a yogi or a fakir specializing in breathing exercises, he has no commitment to a particular way of life of which breathing exercises are a part: he breathes like anybody else; and if there is something wrong with his breathing he will probably go to consult a colleague. The value specialist is neither a saint nor a fiend specializing in valuing exercises; he has no commitment to a particular way of life of which value exercises are a part; he values like anybody else. And it may well be that, when the science of valuation is as fully developed as is the science of medicine, the axiologist, if there is something wrong with his valuing, will need to consult a colleague.

This book, then, gives not instances of valuation but the *principle of value.*

The botanist is not afraid that he would fail to enjoy roses by dissecting them; the value expert does not fail to enjoy the *experience* of value merely because he knows the *principle* of value. On the contrary, it may be said that he enjoys more, in a certain subtle way, just as knowledge of the botanist, of the electrical physicist and of the lung specialist gives a certain subtle hue, a piquant new dimension to their human activities — a fact exploited by the hostess who asked a famous surgeon to carve the steak. In general, theoretical knowledge of a field does not destroy the human involvement in the field, but deepens it by rational penetration. To think otherwise is fundamentally to misunderstand the human use of reason. The distinction between thought and its object does not separate the two; on the contrary, it fuses them. For thought cannot be thought unless it is distinct from its object, and only then can it penetrate into the latter. Thinking thus must be separated from doing in order to fuse with it. To fuse the two without distinction is to confuse them.

Such confusion is that of the dilettante, not the expert. The dilettante musician enjoys the music but he does not know the score. The expert musician knows the score, which does not destroy his enjoyment but deepens it — his feeling for music is rationally structured. *Theoretical axiology is the score of the value realm: through it the feeling for value becomes rationally structured.*

If there were no experts in roses there would be no expertly grown roses. If there were no theorists in electric currents there would be no toasters, no power stations, telegraphs, X rays or light bulbs. If there were no lung specialists, there would be no help in breathing for those who need it. If there were no experts in musical harmony there would be no composition, no musical instruments, no orchestras. There would, however, still be roses, electrical phenomena, breathing, and sound. But all these would be in a state of nature. There would be wild roses, natural currents, natural breathing — none for swimming, singing or iron lungs — and the music of the birds. In the same way there will always be valuing in the wild or natural state, whether there are experts of it or not. There will always be everyday knowledge of value by the valuers, just as there is knowledge of bird song among the primitives and of breathing among the witch doctors. But there will be no expert knowledge in the way we are used to, no *scientific knowledge of value.*

It so happens that value knowledge is still in the natural state. Valuing is just going on; there are no experts. There are observers and samplers of "wild" values and their natural occurrences. There are "naturalists" who collect values, go out into the woods and gather samples. Much of value theory today is a result of such hiking expeditions into valuation, with values sampled, classified, stuck on needles, dried, pickled and preserved. Treatises on value are the rucksacks full of such samples, reports on the expeditions of collecting, the situations of catching, the contexts of trapping, and the like. There are a few who feel that there is a pattern to the whole field and many who feel that its beauty lies in having no pattern. There are professional hikers who have methodological aspirations, write Baedekers of value contexts, and raise the process of collecting to a method, indeed, the ultimate method of knowledge. Thus, the Darwin of value theory has not yet appeared, and should he appear he would be greeted with expressions of horror by the boy scouts of valuation. There are, however, metaphysicians and phenomenologists, who admire the beauty and variety of the value realm, describe it in glowing colors, often of their own imagination, and ascribe its order to the wisdom of the creator or to the insight of the observer.

Thus, on the one hand, there are the collectors of value items, some of whom commend the lack of method as a method; on the other hand there are the metaphysicians and phenomenologists who spin patterns out of themselves. But there are no *scientists of value;* and few even who believe that a science of value is possible: that there is an *axiological pattern isomorphous with, and formally structuring, the whole value realm.* There is no Darwin, no Newton, no Lavoisier of value; and there are powerful obstructions, especially from the side of the samplers, to any systematic endeavor.

Sampling can become a science only if there is at least one sampler who transcends that method, steeping himself in the subject matter to such an extent that he sees the totality of it at one glance; the characteristic feature that is its essence. Thus Goethe saw the *Urpflanze,* while strolling under the palms of Palermo, Darwin the pattern of evolution while reading Malthus, and Newton the law of gravitation while feeling the impact of the famous apple. There is much literature on this kind of experience — sometimes called the "Aha" experience — and the ecstasy or shock of recognition. Sometimes this shock is a physical one, as in the case of Franklin when he pulled the lightning down through his kite, sometimes it is a vision so overpowering that the visionary is unable to fulfil its promise, or able to fulfil it only in part, as with Kepler. Usually, the first visionary receives only the initial shock; there is a whole series of shocks following, each widening and deepening the vista, each followed by an army of secondary investigators who learn to handle safely and routinely what consumed the life of the pioneer. Thus Franklin led to Faraday, Faraday to Maxwell, Maxwell to Marconi and Edison, and these last to the hi-fi shop around the corner. Kepler led to Newton, Newton to Laplace, Laplace to Einstein, and Einstein to Fermi and Admiral Rickover, to the atomic pile and its technicians.

In the process of science the initial vision is continuously deepened and widened; and its ecstasy, its shock, is owing to the sudden recognition of the *infinite unity of the whole field.* Such discoverers, then, see nothing *new;* they only see the integrative core of the totality of all the phenomena in question.

In value theory this initial vision has already occurred. The

Benjamin Franklin, the Kepler, has already appeared. But as with many profound discoverers, he has not been understood. He caught value and stated that there was an essence to it, that there was a unity to the whole field, and that samples could exhibit but never exhaust it. He stated that there is an essence of value unlike anything else and unlike any of its samples. To confuse it with either some other essence or one of its samples constitutes a fallacy. A science of value is not only possible but necessary and will begin a new era in value theory.

This discoverer was G. E. Moore. He even came close to stating the structure of this essence, and, for any careful reader, made amply clear its axiomatic nature.[2] Unfortunately, neither the samplers nor the metaphysicians of value took him seriously enough to follow him. Granting that he might have captured value and brought it down from the sky, they were unable to follow his experiment. Moore never told them how his kite was constructed or how the information was obtained. The experiment was unrepeatable. Even though he did detail the composition of the current, his words were so dark as to be paradoxical, not only to his readers but to himself; so that all he effected was a general albeit admiring headshaking. The science of value, though its axiom was suggested as early as 1903 and proposed in 1922 in a manner neither more nor less darkly than were Kepler's laws of planetary motion, is still unwritten.

The present book is a first attempt at writing it, or rather, at stating the conditions for writing it. It proposes to take Moore seriously and to take what he says about value as the axiom of a science. It believes that he *has* written, as he claimed, the prolegomena to a science of value, and that what he says not only makes sense but contains the essential truth about value: it constitutes the axiom of scientific axiology. We try to clarify this axiom, develop it into a theoretical structure, and show how this structure can be used to explain value phenomena. The result, we expect, will be a *nonnaturalistic science of value*, which in its final form will probably be more like the science of music than the science of nature.

This attempt runs counter to much that is done today in moral philosophy. In particular, it is not afraid of system, for it regards the wholesale condemnation of systems as one of the results of the confusions mentioned. Value philosophy, indeed Western philosophy in general, insofar as it is not metaphysics or phenomenology, is haunted by the fear of systems. The existentialist revolt on the European continent, the pragmatic revolt in the United States, the positivist revolt in England — especially in its "therapeutic" form — all have conspired to make systematic thinking in philosophy unpopular. As a result "doing" has fused with thinking, content with form, subject matter with method, practice with theory, use with meaning, in short, all the confusions mentioned have made their appearance. Articles have largely replaced books, flashes of insight the sustained fire of argument, all kinds of intellectual devices the clear light of reason. Optatives, imperatives, approvals, proposals, arguments, persuasions, rituals, ceremonials, warrants, vindications, commendations, dispositions, etc., taken from miscellaneous disciplines such as rhetoric, jurisprudence, economics, psychology, sociology, anthropology, have all been brought forward as substitutes for "formal" thinking — and this at a time when experts in most of these disciplines are bending all efforts toward advancing to the kind of thinking their imitators in value theory eschew. We need only think of the great work of Kelsen in jurisprudence.

What these devices have in common is that they fuse discursive with nondiscursive elements, words — or thoughts — with actions, so that the latter appear as the former and the former as the latter: actions appear as words and words as actions. These devices, then, serve as miniature prototypes for supposedly possible new "logics" in which the confusions we mentioned become principle: where thinking *about* a subject is fused — and confused — with the subject itself. In value theory, especially, such devices are designed to merge the activity of valuation with the thinking about it. Every little value action — or type of it — is supposed to carry its own little reason. Thus, the value theoretician is relieved from seeking the Reason for the whole and may content himself with observing and classifying all the little reasons in their little contexts, hoping against hope that one day, in a far off future, it will all make sense. This hope is futile, for only knowledge systematized within a unifying framework is cumulative. Problems treated by themselves with their own "solutions" do not grow together by themselves, as the history of alchemy has shown — and as the history of philosophy is still showing. The cumulative process of science is the *differentiation of the unitary vision* of which we spoke. Without such a vision the "connected but separable problems" will remain forever separated and their "connection" an illusion. Unless order is made the primary object of thought, disorder sprouts like a weed. It is the natural state of things. The value realm today is in this natural state, a state of high entropy: split up into sample valuings, "prototypal contexts," and "manageable fragments." A confusion castigated already by Plato is celebrated as the discovery of a new method. Plato himself has become unpopular. And what we call orderly thinking has become anathema. Disorder has become the order of the day, an order militantly given by a large school of philosophers. On the other hand, the metaphysical and phenomenological schools confuse value and theory of value, value instances and value in itself, in their own way; not, indeed, by identifying contextual mechanisms with value, but by identifying conceptual reifications with value.

For one who, to quote Einstein,[3] has the "passion of comprehending," the situation in value theory is a sorry one. The passion seems to have gone and tinkering taken its place. "What is it," asks Einstein, "that impels us to devise theory after theory?" It is the passion for unification, for simplification, for comprehending a totality in terms of most simple premisses. The true theorist is neither an empiricist nor a metaphysician but a combination of both.

> Every true theorist is a kind of tamed metaphysicist. The tamed metaphysicist believes that not all that is logically simple is embodied in experienced reality, but that the totality of all sensory experience can be "comprehended" on the basis of a conceptual system built on premisses of great simplicity. The skeptic will say that this is a "miracle creed." Admittedly so, but it is a miracle creed which has been borne out to an amazing extent by the development of science.

This heaven-and-earth-moving passion not only is missing in today's value philosophy, it is suspect. The situation was no different in natural science:

> Every theory is speculative. When the basic concepts of a theory are comparatively "close to experience" its speculative character is not so easily discernible. If, however, a theory is such as to require the application of complicated logical processes in order to reach conclusions from prem-

> isses that can be confronted with observation, everybody becomes conscious of the speculative nature of the theory. In such a case *an almost irresistible feeling of aversion* arises in people . . . who are unaware of the precarious nature of theoretical thnking in those fields with which they are familiar.

Value theory is still "close to experience," either in its naive empirical or its hypostatized metaphysical form. In both cases, precise logical formulation gives rise to suspicion and "an almost irresistible feeling of aversion": in the former case because such formulation is regarded as high-flown and authoritarian, in the latter because it is regarded as pedestrian and constraining. Thus, value philosophers either lie too low or fly too high — they either lack or they overdo passion, valuing too pettily to rise to systematic unity, or too sweepingly to stoop to systematic detail. In both cases they value rather than analyze value. In the first case, they describe value practices, in the second case they hypostatize them. They are practitioners not theoreticians. They value both value and value judgments rather than analyzing them. Thus, they value in the second and third power.

If they were told that they were valuing rather than analyzing value, most value philosophers would think themselves complimented. However, if the subject matter of value *theory* is value, then most of today's value "theories" — being valuations — are subject matters *of* value theory rather than value theories. In this respect pragmatism, positivism, and phenomenology are alike. To all of them — as is particularly clear in Dewey, Heidegger, and some of the ethicists of the Wittgensteinian school — thinking is a kind of doing. In all of them, words, prepared either etymologically, psychologically, or in some other way for a contextual role, take the place of exactly defined terms; the unity of cognition becomes that of action, use becomes meaning, implication becomes "contextual implication" and, as a German philosopher has recently said, the place of insight is taken by tracing word senses, determined half-consciously by the mind in the twilight of uses.[4] Discursive reason for such philosophers becomes a nuisance, if not a "malignancy," a "divorce from practice," and order, which is unity in variety, dissolves into pure variety indistinguishable from chaos, its unity being confused with constraint or regimentation. What these philosophers fear — and rightly so — are not systems but pseudo-systems, structures that force the mind into patterns which do not correspond to reality. They fear heteromorphous, not isomorphous systems: strait jackets, not maps. The latter are the systems of science, based on phenomenal exploration, the former are the pseudo-systems of philosophy, based on conceptual speculation. The cure for them is not lack of any system but isomorphous system: accounting with the minimum of concepts for the maximum of phenomena. It is orderly thinking.[5]

This is all that is meant by *science*. It is in this sense that the scientific method is to be introduced into moral philosophy. Once this is done, the philosophies of value will reappear in moral science as *ways of valuing* — rather than as alternative axiologies.

This book then examines *the transition from philosophy to science in the field of value*. In the field of natural fact this transition took place three hundred years ago, and may serve as a model for the process in moral philosophy. Those not familiar with the revolutionary changes wrought by Galileo in natural philosophy will be somewhat surprised — though, it is hoped, not shocked, as were Galileo's contemporaries — by the radical changes in moral philosophy which the transition to moral science will be shown to entail. Those on the other hand familiar with the Galilean revolution will regard the development here examined as no more than a matter of course. The ideal reader of this book, therefore, will be both thoroughly familiar with the history of science and profoundly disturbed about the moral chaos of our time. He who knows the history of science but is not disturbed by the moral chaos will immediately perceive, by the analogy of the present situation in moral philosophy with the Galilean in natural philosophy, the seriousness of our crisis, and the necessity and soundness of the new departure. He who *is* profoundly disturbed by the moral crisis but is unfamiliar with the transition from natural philosophy to natural science will be emotionally involved in the situation we describe; and our solution, that of *effective involvement by scientific detachment*, will appear to him paradoxical. He will value, or rather disvalue this book, instead of regarding it detachedly as a rational argument; and he will thus illustrate its presupposition that no *understanding of* valuation is possible without *detachment from* valuation. He will be like the reader of a treatise on the physiology of breathing, afraid to lose his breathing power, or like a lay reader of pathology involved with the text and illustrations, afraid of catching any of the sicknesses so graphically described. The reader who will have the greatest difficulty in following our argument will be he whose involvement in valuation has taken the form of theories about being involved, which on theoretical grounds forbid him to be detached. These theories are, from the point of view of value science, rationalizations for a refusal to think about value. Unfortunately they abound in value theory. For this reason many a value philosopher will be the one least sympathetic to value science.

The transition from moral philosophy to moral science, therefore, has much of the excitement today that the corresponding transition had in natural philosophy — strenuous opposition of philosophers to the adventurous spirit of new scientists and laymen. For (something we have forgotten in the three hundred years of triumphant natural science) the transition from natural philosophy to natural science not only meant the triumph of science but also the tragedy of philosophy.

It may be true that natural science is the offspring of natural philosophy, but this offspring is so radically different, and its birth so painful, that the parenthood is all but forgotten. Little remains of astrology in today's astronomy, little of alchemy in today's chemistry, little of Aristotle's mechanics in today's mechanical science. Almost every fundamental feature of modern science is explicable in opposition to the corresponding feature in the mother philosophy. If, therefore, we speak seriously of a transition from moral philosophy to moral science we cannot help but get into opposition to value philosophy. Its features will have to appear as value alchemies and astrologies compared with the future value chemistries and astronomies. The value philosopher, therefore, who reads this book must be asked, for the duration, to set aside all he has learned, to perform a phenomenological reduction, an *epoché,* start with a clean slate, and follow the argument from the beginning. To the scientist what we shall say will be obvious; and the layman will understand it without too much difficulty except for some technicalities.

As was said before, the present book represents a first attempt, the laying of the foundations of scientific axiology; in particular in its relationship to natural science. The three parts of the present book, "The Structure of Science," "The Foundations of Value Science," and "The Structure of Value" make clear that the book is concerned, as its subtitle says, with laying the foundations of a scientific theory of value. To do so, it has to first explain the nature of science in general (Part I); then, in the light of this examination, analyze the nature of value science (Part II); and, finally, present the actual structure of value itself in the light of this science (Part III).

The following theses summarize the argument of this book:

I. There is a difference between philosophy and science, which consists in the method and not in the content of the two disciplines.

II. This difference is a *logical difference,* based on the nature of the concepts used in philosophy and those used in science. The former are categories (abstracted, implicative, and material concepts), the latter are axioms (constructed, deductive, and formal concepts). In the former, intension and extension vary inversely; in the latter, intension and extension vary directly. The properties of things corresponding to this distinction are called, respectively, "secondary" and "primary" properties.

III. Natural philosophy has changed, historically, from the use of philosophical concepts to the use of scientific concepts, in the sense defined above. The result has been the transition from natural philosophy to natural science, which is based on the distinction between "facts" of secondary properties, and "facts" of primary properties.

IV. "Fact" appeared on three levels, that of pure theory (formal fact), that of observation (phenomenal or situational fact), and that of the combination of both (scientific fact).

V. These distinctions showed up methodological fallacies within natural philosophy, as in alchemy and astrology, in particular the confusion between analysis and the analyzed (the fallacy of method), and the confusion between different natural sciences (the naturalistic fallacy).

VI. Value philosophy — the traditional moral philosophy — has to pass through the same development as natural philosophy.

VII. This development would give for value results analogous to those which the corresponding development gave to fact, namely, the distinction between "values" of secondary value properties and "values" of primary value properties.

VIII. "Value" has to appear on three levels, that of pure theory (formal value), that of the value situation (phenomenal or situational value), and the combination of both (axiological or scientific value).

IX. These distinctions have to clarify methodological fallacies contained in moral philosophy, in particular the confusion between analysis and the analyzed, that is to say, between formal value and phenomenal or situational value (the fallacy of method), and the confusion between different moral sciences such as ethics, aesthetics, economics, psychology, sociology, theology, etc. (the naturalistic fallacy).

X. The principal task of today's scientific axiology is the development of the first level, that of the pure theory of value, and the determination of formal value.

XI. In the book, this task is approached in four ways: first, through the logic of Bertrand Russell; second, through the axiology of G. E. Moore; third, through the method of Galileo; fourth, through the polyguity or homonymity of the term "good." A fifth approach, through Husserl's phenomenology, is suggested.

1] According to Russell, a number is the class (extension) of classes (extensions) similar to a given class (extension). There also ought to be a concept which is defined as "the class of intensions similar to a given intension." This is the concept of a value. If there is given an intension of n predicates (a set of predicates) determining a certain extension (class) then the things belonging to the class (the class members) having n properties are *good* such things (good class members), and all those having less than n predicates are *less than good* (fair, bad etc.) such things. The value "good" then is the totality of intensions similar to the intension n, and the values "less than good" are the totalities of intensions similar to intensions less than n. The set of all sets of predicates (intensions) which contain *all* their predicates (n) is the value "good," and the set of all sets of predicates (intensions) not containing all their predicates is the value "less-than-good." The set of all these sets *of* intensions similar-to-a-given-intension, then, is "Value" — rather than this or that value — just as the class of all classes *of* extensions-similar-to-a-given-extension is "Number" — rather than this or that number. In a word, both number and value are common properties of sets, number of extensional sets and value of intensional sets.

2] The paradox of Moore of "the two different propositions [that] are both true of *goodness,* namely: 1] that it does depend *only* on the intrinsic nature of what possesses it . . . and 2] that, *though* this is so, it is yet not itself an intrinsic property" is solved as all paradoxes must be solved, namely, by showing the different logical levels in question: the negative proposition — concerning what goodness is *not* — refers to the thing itself; and the positive proposition — concerning what goodness *is* — refers to the concept of the thing. If "good" is a property then it is *not* a natural property of the thing itself but *is* a property of the concept of the thing, namely, its intension's being fulfilled by the thing.

3] The third deduction compares Galileo's approach to the problem of motion with a possible axiological approach to the problem of value. In both cases the problem is one of finding a standard of measurement. Galileo found the standard of measurement of motion by disregarding the secondary qualities of the phenomenon and concentrating on its primary qualities, that is, qualities amenable to measurement; so that what was measured was not the sense phenomenon of ordinary life with secondary properties but a construct consisting of primary properties. What is to be measured in value measurement is the ordinary sense object not only *as* possessing its secondary properties, but this very possession is what measures its value. Hence for value measurement *the secondary properties must be used as primary properties*. The *meaning* of a thing becomes the *measure* of its value.

4] The fourth deduction leads to the same result, and demonstrates, moreover, the purely formal and nonnaturalistic character of the theory. "Good" is conceivable, as axiological writers from Aristotle to Edwards have suggested, as a homonym applicable in many different contexts, with a different set of criteria for its employment in each. This is an exact description of the logical nature of the variable. "Good" is a variable and its values are actually fulfilled intensions. A similar thing is true of other axiological values. The values "less than good" are variables and their values are actually not fulfilled intensions. In general, *value is that variable the logical values of which are axiological values.*

5] The fifth approach, through Husserl's phenomenology, would regard the fulfillment of intentionality as that of intrinsic value. This would convert phenomenology into axiology.

XII. The use of the logical concept as axiological measure presupposes the elaboration of intensional logic, which has been neglected during the development of extensional logic in the last hundred years. Such elaboration leads to the notion of intension as a set of predicates and to the structuralization of this set according to the logic of sets. Any set of descriptive qualities defines a fact, any subset of such a set defines a value.

XIII. The result of such elaboration is the inversion of the relation between fact and value. Fact is that set of predicates, p, in terms of which the totality of subsets of predicates (that is to say of values) V_t, is ordered, according to the combinatorial formula, $V_t = 2^p - 1$. Fact appears thus as the ordering norm of value.

XIV. This means that in valuation one leaves out of account the normal set of secondary properties (of fact or of the thing in question) and freely combines and recombines the elements of this set. Such combinations and recombinations of secondary properties are *values*. As *elements of* values the secondary properties of fact become the primary properties of value.

XV. The combinatorial calculus applied allows the exact measurement of values. Finite intensional sets measure systemic value, denumerably infinite intensional sets measure extrinsic value, non-denumerably infinite intensional sets measure intrinsic value.

Although the book is a development of G. E. Moore's "Prolegomena to any future Ethics that can possibly pretend to be scientific," [6] and owes more to Moore than to any other philosopher, to none is it less congenial. Where Moore revels in questions, this book composes an answer; where Moore loses himself in analysis, this book finds its structure in synthesis. Where Moore stresses search, this book emphasizes discovery — and indeed Moore's own discovery. But this discovery, for Moore, was almost accidental and certainly incidental. He never expected to find anything; and whenever told he had found something he was surprised and defiant. His search for truth was much like the Jews's search for the Messiah — it was acceptable only as long as not found. Moore was interested in the pursuit of truth, not in finding it. Analysis was for him an end in itself. His stimulation came not from the state of the world but from what philosophers had to say about it.[7] His definition of 'good,' though he labored over it off and on for forty years, was only a by-product of what alone mattered for him, doing philosophy. His *science of Ethics* was never taken seriously by him — nor, for that matter, by anyone else. But it *is* being taken seriously in this book.

Moore did not *answer* the quest for the nature of good discussed in this book but he *confirmed* an answer I had found independently. "Such general and yet definite principles," says Kant with respect to Hume, "are not easily learned from other men, who have had them only obscurely in their minds. One must hit on them first by one's own reflection."[8] In this sense my answer may be said to be based on Moore's; and hence Moore appears in these pages as an answerer rather than a questioner, a role alien to him. Yet, it is this role, I believe, and no other in which he will survive in philosophy — or rather in science. For his foundation of the science of Ethics is his immortal contribution to human thought. In this alone he surpasses, and overcomes, philosophy. In all the rest, the overwhelming rest of his life's work, he was a philosophers' philosopher, a master of analyses which never came to an end and petered out, after endless involutions, for sheer lack of time.[9] Only in his analysis of 'good' was he driven to ever clearer determination of the problem, ever closer precision. Indeed, as Keynes said, in *Principia Ethica* Moore "carried the use of ordinary speech as far as it would ever be possible to carry it, in conveying clear meaning. *For still greater precision one would have to proceed by mathematical symbols.*"[10] This kind of procedure I have been driven to employ in this book.

Thus, the book points beyond Moore and his immediate influence on the present generation of philosophers. This influence was incalculable. Not, indeed, through his doctrine or method, of which he was never aware, but through his person.

"I never learned anything from him," one of Socrates' disciples once said, "but every time I spoke with him I became a better man." Moore, says Morton White, "did not persuade me, I am bound to say, of the validity of a single one of his main philosophical doctrines, but he was living proof of the importance of honesty, clarity, integrity, and careful thinking in philosophy. . . . These qualities of Moore meant more to me when I began to stand on my own philosophical legs than all of the machinery of *Principia Mathematica*, than all of the learning of the learned, than all of the wisdom of the ancients." [11] Moore was the great inspirer of the young: "Do your philosophy yourself."

I have taken his advice. If Moore had seen this book, he would undoubtedly have gone into his famous act, described so well by Keynes, greeting it "with a gasp of incredulity — *Do* you *really* think *that* — . . . wagging his head in the negative so violently that his hair shook. *Oh!* he would say, goggling at you as if either you or he must be mad; and no reply was possible." [12] Yet, philosophy must go on even beyond Moore, indeed, even beyond itself. The ethics Moore had in mind is not philosophy, it is science. As such it is *method*. The reader of this book, therefore, ought to learn not only what is value but how to value. The reply to Moore, in the last resort, ought to be moral practice. It ought to be, through his thought to follow his example.

PART ONE

The Structure of Science

I

PHILOSOPHICAL ETHICS AND SCIENTIFIC ETHICS

True "thinking" is not analytic thinking, which persists in mere identity; but thinking which unfolds synthetically in patterns of variegated, mutually constitutive formal elements. To demonstrate the mutual dependence between the individual elements is not "deducing" one from the other by its meaning, so that both vanish in one undifferentiated genus. Rather such reduction of thought to abstract and hence empty generic units is opposed by the functional view which stresses the "concrete" multiformity of intellectual procedure: a multiformity which is rational not in the sense of being a concealed identity — at bottom a tautology — but in the sense of variety as variety manifesting a certain structure. By virtue of this structure variety becomes understandable to us, as a comprehensive pattern. — ERNST CASSIRER

Synthesis is achieved when we *begin from* principles and run through truth in good order, thus discovering certain progressions and setting up tables, or sometimes general formulas, in which the answers to emerging questions can *later* be discovered. Analysis *goes back to* the principles in order to solve the *given* problems only. . . . It is more important to establish syntheses, because this work is of permanent value.

— GOTTFRIED WILHELM VON LEIBNIZ [1]

1. *The Definition of Ethics*

There are two ways in which to define any term, the intensional and the extensional. The term "ethics" could be defined intensionally by genus and differentia: as "the theory of human conduct," "the discipline of goodness," "the doctrine of choice," and the like. Extensionally it could be defined as the things that are called "ethics": the writings of the ethical philosophers. Both definitions seem unsatisfactory. The first defines ethics by its subject matter which serves as the differentia of the genus, "theory," "discipline," or the like. But because there are so many different ethical philosophies each using a distinct differentia it is difficult, if not impossible, to know which is *the* differentia of ethics, and hence which "ethics" is ethics. One ethical theory seems as "good" as another; but nobody seems to know the nature of this "goodness." What are the criteria of "a good ethics"? Obviously, to determine this there is needed not an ethics but a metaethics, a theory about ethics rather than an ethics. But if "good" is the differentia of an ethics — as in the definition "the discipline of goodness" — what would be the differentia of a metaethics? These questions are at present unanswered. Ethical theories stand side by side, without systematic connection or, for that matter, systematic distinction. The first way to determine ethics — the intensional — leads into a jungle where trees, weeds, and creepers are intertwined and the light of systematic reason is unable to penetrate.

What "ethics" is remains unanswerable if we approach it from the side of ethical subject matter, for nobody really knows what that is. "Goodness," "the good," "human conduct," "utility," "pleasure," "choice," "the will," "the will of God," "the good will," — these and many others have been proposed. Men who determine ethics in this way know neither how to justify their selection nor how to define what they have selected. To say that the subject matter of ethics is human conduct presupposes, first, a justification of *why* "human conduct" has been chosen as the fundamental ethical term, rather than, say, "the good," or "choice"; second, it presupposes a definition of "human conduct" delimiting it clearly against all the other fundamental terms that have been and are being proposed as the essence of ethics. Since it is unknown by which method such a justification or definition could be achieved, to determine ethics by its subject matter would mean either to disregard the question of which subject matter to choose — a question which may well be *the* fundamental question of ethics — or else to become entangled in discussions which for lack of a method could not be resolved and of which the relevancy to ethics would always remain doubtful.

The only way left, then, is to determine ethics by the second, the extensional method: to understand by ethics the writings of the ethical philosophers, from Plato and Aristotle to Scheler and Nowell-Smith. Ethics would then be the sum total of philosophical writings under the heading "Ethics." It is this definition, unlikely though it may sound, which must provide us with a map for the jungle of ethical subject matter. This definition is at least exact. Any philosophical or other text whose title is, or refers to, "ethics," "morals," and the like is a member of the class of "ethics." By determining ethics in this way we are enabled to discuss these texts as a class and find a common property of all of them. This result can be reached without reference to the subject matter of ethics, yet it has far-reaching consequences for this subject matter.

2. *Philosophical "System" and Scientific System*

In defining "ethics" as the sum total of ethics texts, we do reach immediately one important result. There is an obvious and immediate difference between texts in ethics and texts in, say, atomic physics. When we read a text in the former, Aristotle's or John Dewey's, we understand almost everything they say. Their language is more or less that of ordinary discourse. With atomic physics it is different: everyday language and intelligence avail us nothing. There are few everyday nouns in it, and the very letters of the language are interspersed with symbols unrecognizable unless we have learned them by special effort. The physics text seems more complex, and more *precise*.

We ought to ask the reason for this difference. Are the phenomena of human life so much less structured than those of nature that they can be understood through language simpler and less structured? Or is a physics text complex and structured beyond necessity? Aristotle's physics was written in the same kind of language as his ethics; Galileo introduced the mathematical language found, in highly developed form, in the atomic physics text. Does, then, the whole development of modern physics rest on a mistake? Or is ethics simply behind in its development? In view of the fact that situations of human life seem to be at least as complex — they are actually much more so — as those of nature, should not the language of ethics be equally structured? The fact that it is not, and that in an age of Einsteinian physics we must fall back on Aristotelian ethics might well be the reason for our incapacity to control life morally as we control nature physically.

There are, of course, ethics texts, such as Kant's or Spinoza's, more complex and structured than those of Aristotle and Dewey. But even these do not play the role that textbooks in physics do. We do not live in accordance with such ethics, the way we live in accordance with the physics of Einstein and Heisenberg when we fire an atomic rocket, the concepts of Maxwell and Edison when we switch on the light, or, for that matter, the concepts of medicine when we call in a doctor. Who, in a moral emergency, has consulted any of these texts and found in them the solution of his difficulties? Who has called in an ethicist? Ethical philosophers evidently live in an ivory tower, removed from and without the slightest influence on the course of human events which they supposedly guide. Ethics is still a philosophy, not a method. It lacks the precision, hence the complexity, needed to be applicable to human situations. Even the ethics of Kant and Spinoza, although more complex and consistent — hence more difficult to understand — than those of Aristotle and Dewey, are still much too simple and too inexact. It seems a paradox that a text to be efficient in everyday reality should be incomprehensible to everyday minds. But this is the case. The core of the most concrete is found in the most abstract.

We have, then, different degrees of complexity, of easiness or difficulty of texts. The ethics of Aristotle is easier than that of Kant, and the ethics of Kant is easier than the physics of Einstein. Since all these texts are designed to explain to us large regions of the world, moral or natural, these differences signify something of profound importance for understanding the moral and the natural world.

Though the moral philosophies in question are "systematic" to some degree, there is a wide chasm between *their* systematic character and that of a text by Einstein or Dirac. To understand Kant's ethics we must know the Kantian system, and knowledge of Spinoza's system does not essentially help us to understand the ethics of Kant. But to understand the texts of Einstein or Dirac we do not need to know their respective systems. What we do have to know is the *system of mathematics*. And knowing *that* system we can understand any other text in physics as well.

There is, then, a fundamental difference between a philosophical "system" and a scientific system. The philosophic "system" is a unique one constituting the philosophy of an individual thinker. There is no objective, universal philosophical system which helps us understand any text of ethics or of philosophy in general. But there is a universal system in natural science by means of which one can understand any scientific text whatsoever: mathematics.

A system comprehending the texts in a large region of knowledge is more comprehensive and significant than a "system" which comprehends only a few texts or arguments in a limited region of knowledge. The more comprehensive system is more general, and at the same time more specific, more abstract and at the same time more concrete, than a less comprehensive system: to account for phenomena in *a larger region* it must be more general, to *account* for these phenomena it must be more specific.

It is characteristic of mathematics to be both highly abstract and profoundly concrete; and it is this that gives it its efficiency in actual life. Philosophy, and ethics in particular, lack such a universal and precise instrument — or language — and hence lack actual efficiency. They are "easy," hence inefficient. Even the "technicalities" of Kant and Spinoza cannot be compared with those of an exact science.[2] In a physics text everyday language is almost entirely missing, or does not signify what it does in everyday use. There is a greater difference between the Kantian system and Einstein's than between the Aristotelian system and Kant's. The latter both use everyday words — "happiness," "duty," etc. — and even though the meanings of these words are both more general and more precise than, say, in a newspaper editorial, they are not separated from their commonsense basis. Mathematics, on the other hand, uses symbols so separated that one must learn it like a foreign language. The texts of physical science, even if written in words, are suffused with symbols, so that the words have symbolic rather than everyday meaning. Physical terms such as "force" or "mass" signify something entirely different from what they do in everyday language or in philosophy. In the last instance they mean nothing more than certain operationally-defined mathematical formulae. These same words in a philosophical text — unless it be a text in the philosophy of science — have more or less their everyday meanings.[3] Indeed, there is a philosophical fashion which makes such meaning the basis of its investigations; and in spite of some effort to distinguish "use" from "usage" it is not clear in which sense these meanings are distinguished from those in the dictionary.[4]

There are, then, two classes of systems, the philosophical and the scientific. The first are individual and particular "systems"; there is no one system that embraces them all. The second are all parts of one overarching system, a superstructure or universal pattern which, in the natural sciences, is mathematics. But mathematics is not the only such system possible, the only superstructure that determines exact sciences. The musical sciences are structured by the theory of harmony, its keys, chords, scales, and so on. What determines the structure of electronic computers is symbolic logic. There is the possibility of an infinity of such *super*structures; without them intellectual disciplines cannot be exact sciences. Thus, if ethics is to advance from philosophy to science, a formal structure has to be found for it. This structure is the *logic of value* or *formal axiology*.

The *systematic* or *axiologic* ethics of which we shall speak — and this is fundamental — is not "systematic" in the philosophical sense but systematic in the scientific sense. It is an ethics itself structured by a formal superstructure. This latter, formal axiology, is to moral philosophy as mathematics is to natural philosophy or as the theory of harmony is to music.[5]

Systematic ethics, then, is not just another philosophical "system" like Kant's or Aristotle's. It is a genuine system *by which these other "systems" can be analyzed*. It is a *metaethics* — an ethics of ethics, a system which enables us to evaluate ethical philosophies, to comprehend them methodically, and to compare one with another by objective and logical standards. It is a *norm for ethics*, as mathematics is a norm for physics; or historically, as mathematical physics is for other forms of physics, such as Aristotle's, Plato's, Hobbes's, and so on. While these philosophical physics are today history, the presystematic ethics are still with us, still philosophy. *Systematic* ethics, on the other hand, is not yet actuality; it is still of the future: the kind of "future ethics that pretends to be scientific" and to which G. E. Moore has written the "prolegomena." We are examining in this book *the transition from philosophical ethics to scientific ethics*.

So we must understand in detail the difference between philosophy and science, that is, between the two kinds of system of which we have spoken.

3. *Analytic Concepts and Synthetic Concepts*

This difference can be determined with logical precision: philosophical "systems" use *analytic* concepts while scientific systems use *synthetic* ones.[6] Analytic concepts are those whose intensions consist of *predicates*, while synthetic concepts are those whose intensions consist of *formal relations.* Philosophy and science thus stand in an exact logical relation — and the transition from a philosophy to a science is a definite logical step. It was performed in natural philosophy by Galileo, who made the transition to natural science by a far-reaching invention: he substituted formal relations for the material predicates of Aristotelian philosophy — he transformed Aristotle's analytic concept of *change* into the synthetic concept of *motion.* An analogous procedure must be followed in moral philosophy if it is to be transformed into science: the analytic concepts *good, value,* and the like must be transformed into synthetic concepts.

We shall first examine the differences between these two kinds of concepts.

A concept is a mental content having a "double aperture." [7] On the one hand it has meaning or intension, on the other, reference or extension. The intension is a set of words or symbols, the extension a set of items, concrete or abstract. To take the example of an analytic concept, the concept "chair" has for its intension the group of words: "knee-high structure with a seat and a back," and for its extension the set (or class) of all chairs that are, have been, or will be.

Definitions of concepts, as we have seen, can be either intensional or extensional. In Webster's *New Collegiate Dictionary* the definition of "chair" is: "A seat, usually movable, for one person. It usually has four legs and a back, and may have arms." This is an intensional definition. If we want to know in greater detail what a chair is, we look up "seat" — but here we are disappointed for we read: "A chair, stool, bench, pew, etc., or the part of it on which one sits," — which is not the intension but the extension of the concept "seat." The definitions of the dictionary thus jump back and forth between intension and extension. The *intension* of "seat" would be, for example: "A plane or curved surface destined for locating the inferior prolongation of the human spine." And this in turn could be determined by the intension of "plane," "curved," "surface," "destined," "locate," etc. Thus, in the last instance, all intensions are connected.[8] In a coherently intensional dictionary we could "systematically" go from one word to the next until we had exhausted the whole dictionary — and it would make no difference with which word we started. All analytic intensions imply each other.

It is this which makes possible analytic "systems," which are nothing but chains of implications of more or less abstract concepts. The chain originates in the intensive definition of the concept with which we begin, proceeds to the intensional definitions of the concepts contained in the first definition — let us call them the secondary definitions — and hence to the definitions of the concepts contained in the secondary definitions — let us call them the tertiary definitions — and so on. Thus the primary definition in our example is that of "chair"; "a knee-high structure with a seat and a back." The secondary definitions are those of "knee-high," "structure," "seat," and "back," respectively. The tertiary definitions are those of the concepts contained in the secondary definitions, e.g., those of "seat": "plane," "curved," "surface," "destined," etc. This closer and closer determination of an analytic concept may, of course, go on ad infinitum — or rather, to be exact, as many different times as the dictionary can be exhausted. For, once we have exhausted it the first time, we can begin all over again with another word and make the whole journey from word to word in a different order.

This game of stringing together series of definitions or intensions, and of definitions and intensions *of* definitions and intensions, and of definitions and intensions of definitions and intensions *of* definitions and intensions, and so on, need never end for there are enough words in every language to continue throughout all the time astronomers have calculated life on earth and the planets to continue. At present the game has gone on for some five thousand years, in ordinary discourse and in philosophy — for the process of following an intension to its implications, and these implications to their implications,[9] and so on, is at the core of analytic thinking, as we shall see in greater detail — and there are enough permutations of the words of a language to make possible original analytic thought for another ten thousand million years. There are, in Webster's dictionary, for example, some 50,000 words, most of which signify analytic concepts. We can arrange these words in series of implications 50,000 times; and by taking series of less universal implications, that is, of implications that do not exhaust the entire dictionary but only parts of it, we would arrive at all the permutations of the 50,000 words possible, which are 50,000 factorial, an astronomical number ($1 \times 2 \times 3 \times 4 \ldots \times 50{,}000$); and we could do this in $2^{50,000}$ groups of words — a supra-astronomical number, considering that 2^{64}, the number of grains of wheat which King Shirham of India "promised" to his Grand Vizier for the invention of chess, is already 18,446,744,073,709,511,615, and the number of particles in the universe is of the order of only 2^{263}. Thus, practically even though not mathematically,[10] there is possible an infinity of analytic "systems."

These "systems" may be referred to an infinity — again practically, not mathematically — of groups of things because the *extensions* of the totality of analytic concepts include all the knowable things possible. As the extension of the concept "chair" is all the chairs that are, have been, or will be, so the extension of any analytic concept — and of any of its generalizations — is the totality of all the things thus named. The totality of *all* extensions, then, includes the totality of all possible knowable things. These things are, by their intensions, grouped in classes, in sets, as are the intensions themselves, but the things are not ordered within these sets. The analytic intensions determine the properties which each member of the extension or class has but they do not determine the order among these members. The concept "chair" determines the class of chairs, but not the order among chairs. A class thus is nothing more than a grouping, a set of things which are intensionally similar; but it does not constitute any extensional order.[11] The only order connected with it is the intensional order of the implications — and this order, as we have seen, is vague and arbitrary. If there is a law that orders such series of implications it has not yet been found, in spite of energetic efforts to discover it made ever since antiquity.[12] Neither the exact number of concepts which can be implied by one intension, nor their order, nor the law of synthesis formed by all the implications of one intension can at present be determined. Therefore we said that these implications follow each other in series. Even though the order is a little more complicated than that — it consists of series of series — it is not for that matter more determined.[13] It is a vague order, like layers of clouds, one above and beside the other — formations which still await their meteorologist. The molecules of these clouds of intensions — which surround, as "meaning," the analytic concepts — are the *predicates*, that is to say, the words which constitute the content of each intension.

These predicates, the "members" of the intension, as we may say, lack exact interrelationship, just as do the members of the extension. They have their meanings, but these meanings in turn are vague, and so are the meanings of these meanings, and so on. Hence, the vagueness of everyday language, and even more (because of its higher abstraction) the vagueness of philosophical language. Analytic language lacks the precision of scientific language. The latter uses concepts of nonpredicative and relational intensions. The intensions of the scientific or synthetic concepts, in other words, do not contain predicates but formal relations.

We shall see in detail what this means, but a provisional example will do for the moment. The Aristotelian concept of movement — or rather change — was an analytic or philosophical concept. It consisted of analytic but not formal relations, that is, of relations between predicates: "fulfillment (*entelecheia*) of the potential *qua* potential," "change of potentiality into actuality," "fulfillment of the movable *qua* movable," and the like. From these predicates Aristotle derived a multitude of consequences which, much to the detriment of the intellectual development of mankind, were as vague as the original concepts; and their occasional numerical form only concealed the basic vagueness.[14] They were derived more or less in the manner in which one lexical definition is derived from another — a procedure whose arbitrariness appears still two thousand years after Aristotle in what was taken for science — natural philosophy — throughout the Middle Ages, in alchemy,[15] in the Aristotelian physics of Galileo's time, and, later still, in the eighteenth century, in the phlogiston theory of early chemistry.[16] A classic example of analytic implicative reasoning is the famous argument of Francesco Sizzi, chief Aristotelian at Padua, against Galileo's discovery of the moons of Jupiter:

> There are seven windows given to animals in the domicile of the head, through which the air is admitted to the tabernacle of the body, to enlighten, to warm, and to nourish it. What are these parts of the microcosmos? Two nostrils, two eyes, two ears, and a mouth. So in the heavens,

> as in a microcosmos, there are two favorable stars, two unpropitious, two luminaries, and Mercury undecided and indifferent. From this and many other similarities in nature, such as the seven metals, etc., which it were tedious to enumerate, we gather that the number of planets is necessarily seven. Moreover, these satellites of Jupiter are invisible to the naked eye, and therefore can exercise no influence on the earth, and therefore would be useless, and therefore do not exist. Besides, the Jews and other ancient nations, as well as modern Europeans, have adopted the division of the week into seven days, and have named them after the seven planets. Now, if we increase the number of the planets, this whole and beautiful system falls to the ground.[17]

Arguments like these appear peculiar to us today only because Galileo introduced a very different manner of reasoning into natural philosophy. The number seven today has lost its implicative power, and arguments like Sizzi's seem to us collections of *non sequiturs*. But in moral philosophy, as we shall see, this kind of reasoning still abounds; and we do not find it peculiar.

Galileo defines the concept "motion" — which for him, as for Aristotle the corresponding analytic concept, was the central concept of natural philosophy — in a manner very different from Aristotle: not analytically but synthetically, not by predicative but by formal relations. The relation which defines motion — rather than movement or change — is $\frac{s}{t}$, a relation between the space traversed by a body and the time required for this traversal. For us today this is a matter of course — if a body moves a hundred miles in five hours, its "velocity" is twenty miles an hour — but for Galileo's age this reduction of something which was a cosmic principle inherent in both the supralunar and the sublunar sphere, the world of God and all its creatures, to a common arithmetical division, was so revolutionary as to be heretical.[18] Galileo's contemporaries did not understand what mathematical abstractions could have to do with physical reality — as we today have difficulty in understanding what logical abstractions could have to do with value reality. Galileo was the first to discover and to proclaim the power of formulae over reality. He stepped, like Alice, through a looking glass into a world of pure forms, where the world of concrete reality appeared transfigured, transformed — and in its essence.[19] His simple relation between *s* and *t* could be elaborated, augmented; a whole new world could be created out of it — the world surrounding us today.[20] Galileo's formula led to a multitude of consequences; but instead of random implications, as contained in Aristotle's teleological principle, these consequences were precisely determined, exactly defined, logically deduced and demonstrated, and empirically verified. They led to the systems of Newton and Einstein — systems in the strict sense, based not on material analytic but on formal synthetic concepts, not on predicative but on relational intensions.

The fundamental logical difference between analytic and synthetic concepts corresponds to a basic historical change. Our present historical situation demands the same change in moral philosophy that Galileo brought about in natural philosophy: *we must transform moral philosophy into moral science.* We have to replace the analytic concepts, "good," "beautiful," etc. by synthetic ones. The results ought to be scientific ethics as against philosophical, scientific aesthetics as against philosophical, scientific metaphysics as against philosophical; in short, moral science as against moral philosophy.

This science ought to have all the properties of natural science, in particular universal and precise applicability to reality — of being *a method rather than a philosophy.*

The reason that science is and philosophy is not such a method is, precisely the difference between analytic and synthetic intension. For these intensions determine in their turn different extensions, or referents: the "things" to which analytic intensions refer are different from the "things" to which synthetic intensions refer. Thus, the *worlds* to which these two classes of concepts refer are different. The world of analytic concepts is, as we have seen, without interrelationship between its individuals, and exhibits only vague interrelationships between its classes. On the other hand, the world to which synthetic intensions refer is a world of interrelated individuals, a world of order, complexity, and precision, a world incomparably richer and fuller. For the time being this fullness of a world of synthetic concepts is only to be found in the world of natural science and of music. Once synthetic concepts are used in moral science the world of morality will be as much richer than our present world, as our present world is richer than that of the Middle Ages. This is no utopian wish but a deduction from our logical point of departure, the extensional determination of ethics. This definition thus has already at this point yielded a morally relevant result. It will be the task of formal axiology to define the kind of "individuals" to which the synthetic concepts of moral science are to be applied and between which these interrelationships are to be constituted.

The mutual relation between intension and extension in analytic and synthetic concepts, respectively, determines the fundamental *logical* difference between philosophy and science: *In the analytic concepts of philosophy, intension and extension vary inversely whereas in the synthetic concepts of science they vary directly.*

While the inverse variation of intension and extension of analytic concepts is discussed in every textbook of logic the rule of direct variation of synthetic concepts is not. Yet, it is equally obvious and conspicuous in the development of science. The progress from Galileo to Newton was an increase of intension and a corresponding increase in extension of the Galilean concept "motion." Newton increased the intension of this concept by adding to the Galilean relations — $\frac{s}{t}$ for uniform motion and $\frac{1}{2}gt^2$ for uniformly accelerated motion — a whole network of additional, more comprehensive relations — between force and momentum, momentum and velocity, velocity and mass, mass and density, density and volume, etc. The Galilean relations were thus incorporated in a network of relations applicable to many more things than Galilean mechanics. Newton thus not only increased the intension but also the extension of the original Galilean concepts. Moreover, he increased this extension both quantitatively and qualitatively. Quantitatively, by including the whole inventory of the natural world, from planets to tides, and not only individual bodies but also combinations of bodies; qualitatively, by extending his application from the "ideal cases" of Galileo to real cases — by defining the deviations of the actual phenomena with respect to the mathematical formulae of the system. Thus the intensional increase also increased the extension both quantitatively and qualitatively: it increased the scope and the exactness of the application.

The reason for the greater power of the synthetic over the analytic concept is that the latter may be and usually is being considered in isolation, whereas the former makes sense only within the totality of a system of which it forms a part. The synthetic concept, in the sense used by the early scientists, is deductive while the analytic concept is inductive. There is no synthetic concept outside a deductive system, while the analytic concept is usually in search of a system. It is the system that gives the synthetic concept its power. The system itself is the *synthetic intension.*

Thus, the definition of a polygon as "a closed figure of *n* sides" may serve as intension of the *analytic* concept "polygon," within a first approximation to a system, based on sensorial apprehension, such as the Euclidian. But it may not serve as the intension of the *synthetic* concept in a synthetic system — as a *term* which serves as variable one of whose possible values is "polygon." In such a geometrical system the intension of "polygon" is the totality of the system in question. As we have seen, the analytic intension of the concept "motion" is the Aristotelian "the fulfillment of the potential *qua* potential"; but its synthetic and formal intension is the whole system of post-Galilean mechanics.[21]

That the intension of a term of physical science is not any specific equation but the system itself in its totality, is confirmed among others by Pierre Duhem, who dedicates Chapters V and VI of his famous treatise to the demonstration of the presence and unity of the entire physical system in each particular experiment: "Physical science is a system that must be taken as a whole; it is an organism in which one part cannot be made to function except when the parts that are most remote from it are called into play, some more so than others, but all to some degree." Duhem also clarifies the difference between abstractions, such as "All men are mortal," and the *constructions* which form the symbolic apparatus of science: "The symbolic terms connected by a law of physics are . . . not the sort of abstractions that emerge spontaneously from concrete reality." [22]

The law of direct proportionality between extension and intension means that the wider the applicability or extension of a scientific system the richer (more "complicated" according to Duhem [23]) will be the system in question, that is, precisely, the synthetic intension. A complex system is applicable to a more numerous and complicated set of phenomena; on the other hand, a less complex system has a reduced field of application. The system of Newton, as intension or content of the concept "material world," is more complex than the intension of the same concept in Galileo, and its application to phenomena of the material world is far greater than Galileo's. The system of Einstein, in turn, as intension of the same concept "material world," is more complex than the Newtonian and can be applied, precisely therefore, to even more phenomena of the material world. The reason is the very obvious one that there are more interrelations between more things than between fewer things, and that, if the interrelations are contained in an intension, the more interrelations there are in an intension the more are the interrelated things to which the intension can be applied. Conversely, the more interrelated things there are the more relations will be contained in the intension which applies to them.

As we have mentioned, few current textbooks of logic discuss the variation in direct proportion of synthetic concepts. On the other hand, there are a number of classics which clarify this proportion, from the *Regulae* of Descartes to the writings of Cassirer and the *Logic* of Garcia Bacca.[24] Descartes indicates clearly that it is always by "subtraction" that the notes "yielded by analysis become progressively more simple and it is by 'addition,' in the syntheses, that the successive terms become more complex."[25] The same is made clear by Kant in his *Logic*. While the logic of analytic abstraction is the Aristotelian, the logic of synthetic construction is the Cartesian, elaborated by Kant. The essential characteristic of the synthetic process is that "each sequence or deductive series consists of terms growing in complexity as the sequence advances" and that in the same degree grows the applicability of the series, until the science covers "exhaustively" its object or, in Descartes' view, the science becomes "coextensive with, and exhaustive of nature."[26]

Cassirer makes the difference between the inverse proportion of analytic intension and extension, and the direct proportion of synthetic intension and extension the basis of his *Substance and Function*.[27] "The increasing extension of the [analytic] concept corresponds to a progressive diminution of the content; so that finally, the most general concepts we can reach no longer possess any definite content." While this is true of the analytic concept, or material abstraction, the opposite is true of formal construction, such as mathematics. "When a mathematician makes his formula more general, this means not only that he is *to retain* all the more special cases, but also be able *to deduce* them from the universal formula. . . . Here the more universal concept shows itself also the more rich in content." For the philosopher, abstraction "is very easy," but "the determination of the particular from the universal so much the more difficult. For in the process of abstraction he leaves behind all the particulars in such a way that he cannot recover them." The mathematician, on the other hand, and the scientist in general, does not abstract from the manifold before him but creates for its members a definite relation by thinking of them as bound together by an inclusive law. The more formalization proceeds the more the concrete particulars stand out. Therefore Cassirer entitles this paragraph The Mathematical Concept and Its Concrete Universality. This means that the particular forms of Euclidian geometry, for example, such as "polygon," obtain their complete concretion only in the nonEuclidian geometry, in their formalization rather than in their description. "The intuition of our Euclidian three-dimensional space only gains in clear comprehension when, in modern geometry, we ascend to the 'higher' forms of space; for in this way the total axiomatic structure of our space is first revealed in full distinctness."

The same fact is shown with great clarity by Huntington in "The Duplicity of Logic." The "duplicity of logic" is precisely that between the logic of analytic abstraction and the logic of synthetic construction. Huntington's illustrative example is that of the difference between descriptive and constructive geometry. The first, he says, is based on "premathematical observation of space," the latter on formulae of variables whose possible applications are the bodies described by description. Once the system is attained "from the point of view of abstract geometry the particular system of observational geometry may now be laid aside, as a scaffolding is laid aside after the finished building is completed. The concrete system of observational geometry is only one of a multitude of systems (*K*, *R*) which satisfy the postulates — an important one to be sure, and the one which provided all our motivation in the selection of the postulates, but not a logical part of the structure of the final discipline. The set of postulates itself now stands on its own feet and is quite independent of any particular interpretation which may be given to the symbols *K* and *R*." [28] The system is very much more complex than the corresponding Euclidian geometry and it is of correspondingly wider application.[29]

In summary, synthetic concepts arise and develop within the body of a system. The latter, in turn, arises from an axiom, as the structure of a web from a spider. As the spider is not part of the web, so the axiom is not an element of the system. The axiom is *not a synthetic concept,* but that concept which contains in germ the whole of the system. The richness of the system will depend on the fecundity of the generative axiom, or of the group of such axioms.

The axiom is neither an analytic nor a synthetic but a *singular* concept of a specific kind. It is the *symbolic form of the core of a phenomenal field.* Although the system is developed and grows out of a deductive process, starting from and based upon the axiom, the axiom itself is captured only through another and very distinct form of knowledge: a direct and immediate intuition.[30]

Let us now apply all this to moral philosophy, ethics in particular. To know in an orderly fashion, that is, to account with a minimum of concepts for a maximum of phenomena, presupposes a definite kind of concept: concepts, namely, whose *precision increases with their generality.* It is the sign of prescientific concepts to *lose* precision with generality. It is therefore wise for workers in prescientific fields to stick closely to the phenomena, as was the custom of the wiser alchemists and as is that of the contextualists in today's value theory. Value theory is still in the prescientific stage. If it is to advance to the scientific stage of large-scale order rather than small-scale orders, new concepts must be introduced; for in the degree that the old ones become more general they fizzle into complete vagueness. They are categories whose very wideness makes them empty. Such categories may do for philosophy, but not for science. Here we need *axioms* rather than categories.

The difference between these two kinds of concepts has been precisely determined: precise and exact concepts, as those of mathematics and physics, were called *synthetic* and are constructions (*syn-thesis* = con-struct) of the human mind. They have no direct connection with sense reality but essential relevance for it. They are posited in such a way that they contain "all that is worth thinking" ("*axiomata*") about a field of phenomena: they contain its essence; and the wider their applicability the more precise their meaning. Vague and general concepts, as those of philosophy, were called *analytic* and are abstractions from sense reality, directly connected with but of no essential relevance for it — a kind of mental calcomania, dis-solving (*ana-lyon*) and drawing off (*abs-trahens*) sensory qualities. While synthetic concepts arise from thinking relevantly about a subject, penetrating to its essence, analytic concepts arise from talking about a subject, arguing (*kat-agoreuein*) about it in the market place (*agora*). The "solution-by-argument" school of value theory thus goes back to the pre-Socratic beginnings of categorial analysis, the talk of the people in the market place. Analytic concepts contain what is at the surface of the phenomena, sense properties which are open to anybody's inspection.

If this is so, and if value properties are nonsensory, nonnatural properties, as it has been supposed — and as we shall assume — then it is obvious that they cannot be rendered by analytic concepts, which lack relevance for value reality by virtue of the very essence of this reality: they lack, we might say, relevance a priori. The only concepts left to account for value reality, then, are the synthetic. While analytic concepts are heteromorphous with respect to both fact- and value-reality, synthetic concepts have been proved isomorphous with fact-reality and must now be proved to be isomorphous with value-reality. The concept "value," then, must be changed from an analytic to a synthetic concept. Value as an analytic, philosophical concept is an empirical abstraction; as a synthetic or scientific concept it is a formal construction and its meaning becomes more precise the more general is this construction. We move from a category of philosophy to an axiom of science.

The philosopher who has given a formal construction to value within a philosophy he regarded as "Prolegomena to any future Ethics that can possibly pretend to be scientific," is G. E. Moore. His construction of value was as follows: *The value of a thing is a nondescriptive property which depends only on the descriptive properties of the thing.* This construction was not precise enough to serve as axiom for a formal system of value science; but we can make it precise enough by a simple addition.

Here emerges a fundamental *logical* difference between traditional ethics and axiological or scientific ethics. The unitary vision of the scientific creator has a strictly logical basis: *he sees the field synthetically rather than analytically.* Galileo changed the concept "movement," Lavoisier that of "element," Newton that of "force," all analytic concepts long used in natural philosophy, to precise synthetic ones. In the same way Moore changed the analytic concept "goodness" from an analytic to a synthetic one: from "pleasure," "satisfaction," etc. to "*nondescriptive property depending on descriptive properties.*"

The resulting ethics, which he did not succeed in formulating but whose nature he clearly discerned, is fundamentally different from any traditional ethics. Traditional ethics is philosophy, the new ethics is science; traditional ethics is categorial, the new ethics is axiomatic. The first is substantial, the second, functional; the first, abstractive, the second, constructive; the first, material, the second, formal; the first, naive, the second, critical; and the first is predicative, the second, relational; the first implies at random, the second deduces with precision; the first loses specification in the degree that it gains generality, the second gains specification in the degree that it gains generality; and while the first is an aggregate of general assertions, the second is a method — the method of conducting life worthy of man's rationality.

Scientific ethics is more difficult than a merely philosophical ethics, and requires greater intellectual effort. Eventually, to understand ethics scientifically the same effort will have to be made as in the understanding of an exact science. Only then will ethics have the excellence it ought to have. For *omnia praeclara tam difficilia quam rara* — everything excellent is as difficult as it is rare. These final words of Spinoza's *Ethics* apply as much to ethics itself as to the moral life.

4. *The Logical Concept and the Philosophical Concept*

Since we have determined the difference between philosophy and science *logically,* and since logic is a part of philosophy, we must now determine the particular role that logic plays in the transformation of moral philosophy to moral science.

In the curricula of colleges and universities logic is one of the traditional disciplines of philosophy, along with metaphysics (or ontology), epistemology, ethics, and aesthetics. Actually, however, logic has a very special position in philosophy. In some respects it is already a science, in other respects it is still philosophy. It is instructive, even indispensable, to examine the difference between the two kinds of concepts in the light of the curiously divided nature of logic and, closely connected with this, to examine the relationship of logic to the other philosophical disciplines.

There is a jocular saying that the philosopher knows less and less about more and more until he knows nothing about everything, while the scientist knows more and more about less and less until he knows everything about nothing. There is a grain of truth in this saying, and it concerns the difference between analytic and synthetic concepts. Philosophy deals with the most general and abstract concepts; and of the meaning of these concepts, as we have seen, not much can be said. One can say something definite of the concept "apple" and even of that of "fruit"; but what can one say of the meaning of the concept "thing" or that of "being"? If this concept is the most abstract possible then, according to the rule of inverse variation between analytic extension and intension, this concept refers to "everything" — and this implies that it means "nothing." Since the most general concepts, such as "being," "good," "beautiful," etc., are precisely the subject matter of philosophy there is a grain of truth in saying that philosophy "knows nothing about everything." Aristotle suggested this more technically by saying that the highest genera, such as the categories, have no definition, Kant suggested it by showing that the dialectical use of reason in traditional philosophy is illegitimate, and Hegel by identifying pure being and pure nothing. Ontologists, therefore, from Plato to Heidegger, if they wanted to say anything, had to leave analytic language; and, refusing to use synthetic language, they had to have recourse to the language of singular concepts, myth and metaphors. Rather than becoming scientists they thus became poets. But they could not remain philosophers — in the sense defined — and say anything significant of Being.

Science does not deal with the whole of Being but with very specific beings. Not only with apples in general but with species of apples; not only with species of apples but with individual apples; not only with individual apples but with the organic composition of individual apples; not only with the organic composition of individual apples but with the physical elements of such composition; not only with these physical elements but with the elements of these elements, to the point of discovering new particles having no reality in the actual world but only within the system of physics; only ideal reality or ideality. Thus, it may be said that science makes reality disappear and knows everything of that which is not, of "nothing."

We know, however, that this knowledge of science is extremely powerful. The "nothing" of which science "knows everything" is then no "nothing"; it is reality itself, even though in a very different form from that which is treated in ontology. It is reality seen from behind the scientific looking glass, reality seen in its *logical* core. When science deals with the most general concepts, such as "motion," "matter," etc., these concepts instead of losing their specific meaning, as they do in philosophy, on the contrary, acquire such meaning. Instead of losing meaning in the degree of their generalization, they fill themselves with meaning in this degree. So instead of saying that the scientist knows more and more of less and less until in the end he knows everything of nothing, one should say that he knows more and more of the more and more significant until in the end he knows everything of the most significant. This "most significant," to be sure, although qualitatively "everything," is quantitatively so small as to be practically nothing — *it is the symbol in proportion to that which it symbolizes.*[31]

Science, then, deals with general concepts as does philosophy, but by a logic, a method which, far from weakening the meaning of these concepts, strengthens it. Historically, as we have seen, this method was first used by Galileo, who filled the abstract form of a philosophical concept, which meant "nothing" to him — the Aristotelian concept of "movement" — with a new meaning of mathematical relations which to him meant "everything." This transformation of natural philosophy into natural science was not immediately recognized in all its revolutionary significance, even though Galileo himself recognized very clearly the scope of his new method. But the "scientific revolution" took many years to become aware of its own method, in particular, its radical difference from philosophy. In the last century the words "scientist," "physicist," etc. were still anathema to "natural philosophers" such as Faraday. Galileo himself, although he took the fundamental step from philosophical to scientific logic, was so much more philosopher than scientist that quantitatively more of his work is philosophical justification rather than scientific elaboration of his new method — he had spent, he said, more years in the study of philosophy than months in mathematics. Even today, few textbooks in logic state the *logical* difference between science and philosophy as that between kinds of concepts, being content instead with superficial descriptions of what they call the hypothetical method — in a sense which, as Kant himself observed,[32] is more psychological than logical. It is no exaggeration to say that the fundamental *logical* difference between philosophy and science is largely unkown even today.

The general concepts used in philosophy, then, are relatively unstructured, in a definite logical way, while the corresponding concepts of science are precisely structured. A most general concept in philosophy being a *category,* we shall call the philosophical method that of *categorial analysis.* Here "analysis" not only means the relative lack of structure of the philosophical concept, due to its abstractive nature, but also its relative lack of relevance to actuality.

As yet, there is no general agreement regarding the nature of the most general concept in science, the concept that is structured to a degree that it gives rise not only to the whole coherent system of the science but also mirrors faithfully the total variety of the corresponding actuality — the concept which manifests variety *as variety* (to speak with Cassirer), both theoretically and practically. We may call it the *isomorphous concept,* and thus show both its structure and its relevance to actuality. But there is a better name, more useful in axiology. This concept is the *axiom.* So the scientific method is one of axiomatic synthesis. The term "synthesis" here means not only that the scientific method is constructive and hence precise, but also that it is relevant to actuality. We include, in other words, the epistemological Kantian meaning of the terms "analytic" and "synthetic" in their logical meaning.

It would be a long step forward in moral philosophy if one could use the same method of axiomatic synthesis, of *general specification* or *specific generalization,* that is used in science. If, for example, one could apply this method to the concept "good" or "value," one would convert value theory from philosophy to science. Such a transformation of the central concept of moral philosophy would change the whole of this philosophy into science, in the same way that the transformation of the central concept of natural philosophy, that of "movement" by Galileo, transformed the whole of natural philosophy into natural science. Thus, one could transform not only ethics, but all the traditional branches of moral philosophy — metaphysics, logic, epistemology, and aesthetics — into sciences. Moreover, this would affect the large fields of knowledge which today are neither philosophy nor science but in between, the humanities and so-called social sciences. In the empirical sense these disciplines have ceased to be philosophy, but in the conceptual they still are philosophy, using philosophical concepts but applying them to empirical material.

This hybrid method is exactly the same as that of the prescientific natural philosophers. Alchemists and astrologists applied analytic concepts of natural philosophy to material that was to be dealt with by the synthetic concepts of natural science. Social and political "scientists" and ethical philosophers of today apply analytic concepts of moral philosophy to material that is to be dealt with by the synthetic concept of moral science. Our social and moral knowledge, thus, is in the alchemistic stage, and our social and moral life bears witness to it. The split in our intellectual requirements for natural science on the one hand, and social science on the other, may be illustrated by the unique example of the philosopher who applied philosophical concepts to the materials of *both* natural *and* moral science, Hegel. The different acceptance accorded, on the one hand, his natural and, on the other, his moral (political) philosophy is characteristic for our age. His natural philosophy is regarded as nonsense, but his political philosophy, either directly or through the inverting mirror of Marx, is regarded by some as the acme of *science*.

An ideology may be defined as the application of analytic concepts to empirical material. The alchemists and astrologists thus were no "objective scientists"; they were *ideologists;* and it was this that Galileo had to overcome. Today social and political "scientists" as well as ethical philosophers are no "objective scientists" either but apply ideologies. These ideologies, today fighting against each other, will have to be overcome by the moral scientist (as was pointed out in the remarkable speech by West-German President Theodor Heuss on New Year, 1958). Ideologies thus are the pseudosystems of which we spoke above, the straitjackets rather than the maps. Unless there is a synthetic system for comparison, the insanity of analytic systems is not easily recognized. Quacks in *natural* philosophy are easily recognized today but were not in the Middle Ages where they occupied respected positions. Quacks in *moral* philosophy are not easily recognized today; they occupy high positions in the academic, political, and religious hierarchy.[33]

What is needed, then, are the Galileos and Lavoisiers in the humanities and the social sciences — sociology, anthropology, political science, etc. The analytic concepts in these fields must be replaced by precise synthetic ones: in sociology concepts such as "group," "class"; in anthropology concepts such as "man," "race"; in political science concepts such as "state," "democracy," "ideology," and so on. The same holds for the concepts of philosophy proper; in metaphysics, "being," in ethics, "good," in aesthetics, "beautiful," in logic, "concept," in epistemology, "knowledge." All these concepts must be given the same precision possessed by the concepts of natural science.

We are primarily interested in the concept of value, "good." But we cannot define this concept if we do not at least mention the specifications of the other philosophical concepts. In particular, we need to specify the concept of logic, "concept." For, obviously, since all specific concepts are species of the genus "concept" all specifications of concepts imply — in the sense examined — specifications of the genus "concept." The *logical* concept, thus, is of supreme importance for the precision of the other philosophical concepts, which are its species. From this follows an inescapable conclusion. If logic is the theory of the genus "concept," and the other philosophical disciplines are theories about species of concepts, these disciplines are species of logic. Logic is the genus within which the other philosophical disciplines are species.

One might suppose from this that it would be simple to determine the concepts of moral philosophy with scientific precision, for all that would be necessary would be to determine the differentiae by which the specific concepts, "being," "good," "beautiful," etc. are distinguished from the logical concept "concept." Actually, however, this task is most difficult; for the concept "concept" is as vague as the other philosophical concepts. It belongs precisely to that part of logic that is still philosophy. One might have thought that so fundamental a concept as that of "concept" would have been scientifically elaborated in the renaissance of logic in the last fifty years, not to mention the long history of logic since Plato and Aristotle. Unfortunately, this is not the case. The concept "concept" has never been treated in all the history of logic with the dedication it deserves, so that it is very much in the same state that it was in with Aristotle.[34] There occurred in logic what occurred in philosophy in general: only one half of the subject matter was determined with precision, the other half was neglected — and the two events, one in logic and the other in philosophy, are intimately related. The development from natural philosophy to natural science is based on the development of that part of logic which has been determined with precision, namely, extensional logic, and later mathematical logic and mathematics itself; while the lack of development of moral philosophy is based on the lack of development of that part of logic which has been neglected, namely intensional logic.

The concept "concept" has been determined with scientific precision in logic only in its extensional but not in its intensional aspect. Logic has investigated what the application of concepts to classes means but it has not investigated what concepts mean. The concept "concept" has this peculiarity, that "extension" and "intension" are the characteristics of its Intension. While the Extension of the concept "concept" is the class of concepts, the Intension of the concept "concept" is the set of properties which each concept has in common with every other concept. But this is the very structure of extension and intension we have described earlier. Thus, conceptual extension and intension are the content of the Intension of the concept "concept." The Intension of "concept" is relational — it contains the formal relation between "intension" and "extension" which is either direct or inverse — and hence this concept is a synthetic one, and the branch of knowledge that deals with it, namely logic, is a science. But the intension of "concept" can also be represented analytically, when no formal relations are given between the intensional characteristics, "intension" and "extension," and the "relation" between them is left undefined. In this case, the Intension is predicative rather than relational. Here logic is philosophy rather than science.

The extensional characteristic of "concept" has been developed in modern logic — up to a point [35] — but not the intensional.[36] On the contrary, the extension of concepts has been explicitly identified with the Intension of "concept"; and it has been held that all that there was to the meaning of "concept" was "extension." As we have said, only one half of logic has been converted into a science and the other half has been neglected.[37] A logic of intension is missing, a logic which investigates and structures conceptual intension; and which examines and structures the interrelationships between intensions, up to and including the totality of all intensions. Such a logic would give us *Meaning* as such — rather than what logic gives us today, an inventory of the world. And if, as Plato said so emphatically, the meaning of the world is its *value*, what is missing is a logic of value.

Actually, the logic of intension is today less developed and less known than it was some one hundred and fifty or two hundred years ago. The logics of that time, such as Wolff's and Kant's, are intensionally more "modern" than today's logic texts which have practically no intensional part (and in whose indexes the word "concept" rarely appears.) [38] The excursions of modern logicians into intension are faltering.[39] By identifying conceptual extension with the intension of "concept" they amputate, at one stroke, one half of logic — and it may well be the better half. It is not strange, then, that today's logic lacks philosophical significance and is largely a manipulation of symbols which are empty rather than formal; and whose significance for metaphysics [40] is specious and for ethics and aesthetics zero. Thus, although logicians from Whitehead [41] to Fitch [42] have made us splendid promises for a logic of ethics, no such ethics, let alone axiology, has so far been forthcoming from this quarter, in spite of half a century of concentrated and brilliant technical development. To judge from past performances of logicians in the field of ethics, notably Russell, who frankly admits his failure,[43] not much can be expected in this respect either. If this is the empirical evidence of the case, the a priori evidence is that extensional logic is in principle inapplicable to the value realm.

Extensional logic can as little be applied to the phenomena of moral philosophy as scholastic logic could be applied to the phenomena of natural philosophy. Modern logicians are, with respect to value phenomena, in the same situation as were scholastic philosophers with respect to natural phenomena. In their positivistic school which, negatively, applies extensional logic to value phenomena and thus, magically, causes them to disappear, they act in the same way as "the professor of philosophy of Pisa laboring before the Grand Duke with logical arguments, as if with magical incantations to charm the new planets out of the sky." [44] Just as the medieval schoolmen preferred to look through their theological glasses rather than through Galileo's telescope, so today's positivists and naturalists prefer to look through the telescopes and microscopes of science rather than through the lenses of inner vision discovered by G. E. Moore. No wonder their argument against Moore is the same as that of the astronomers against Galileo: "We don't see what you pretend to be showing us."[45] The reason is, of course, that Moore's lenses, as those of Galileo's telescope, are crude and the possibility of hallucination cannot be discounted, as little as could that of optical illusion in the case of Galileo. Thus, what is needed is a more precise instrument. In the case of Moore, this means to sharpen his insight. This sharpening, in turn, requires finer tools of grinding and polishing, and these must be supplied by a new logic.

As we shall see, the scientific precision of moral philosophy is based on the precision of the intensional characteristic of the concept "concept," just as the scientific precision of natural philosophy is based on the precision of the extensional characteristic of this concept.

This latter precision is that of mathematics. The fundamental concept of mathematics, that of "Number," is defined in modern logic by the extensional characteristic of the concept "concept." This characteristic is the concept "extension" or "class." Mathematical number is defined by the concept "similarity of classes" or "similarity of extensions." A number is the class (extension) of classes (extensions) similar to a given class (extension). If the given class is a couple, then the number 2 is the class of all couples. "Number" itself is the class of all such numbers of couples, trios, etc.

There also ought to be a concept which is defined by "the class of all intensions similar to a given intension." This concept happens to be the concept "a value," as it will be elaborated from the determination given by G. E. Moore. If there is given an intension containing a number n of predicates (a set of predicates) determining a certain extension (class) then the things belonging to the class (the class members) which have n properties are good such things (good class members); and all those which have less than n predicates are "less than good" ("fair"), "not good" ("bad"), or the like, such things or class members. Thus, if there is given an intension of four predicates, like that of "chair" — "a knee-high structure with a seat and a back" — then a thing which is a chair and has all these properties is "a good chair" while a chair which lacks any of these — be it a seat or a back or the correct height or structure — is "not a good chair" or "a bad chair." Thus, the value "good" is any intension similar to the intension n and the values "fair," "bad," and the like are any intension similar to intensions $<n$. The set of sets of predicates (intensions) which contain *all* their predicates (n) is the value "good," and the set of the sets

of predicates (intensions) which do not contain all their predicates is the value "less than good," "fair," "bad," "no good," and the like. The *set of all these sets of intensions similar to given intensions, then, is Value* — rather than this or that value — just as the class of all classes of extensions similar to given extensions is Number — rather than this or that number.[46]

The definition of Value, thus, is the intensional analogue of the logical definition of Number. Inversely, the definition of Number is the extensional analogue of the definition of Value. Just as number is defined by similarity of extensions, so value is defined by similarity of intensions.[47] *Value* then is that variable, the logical values of which are axiological values.

The analogy between value and number appears in the parallelism of the "naturalistic fallacy" in the philosophy of both Moore and Russell. For Moore, the naturalistic fallacy was the confusion of generic value with specific values as well as that of different specific values with each other, such as the natural, metaphysical or psychological with the moral.

This fallacy is contained in the very nature of the category. It inheres in categorial — rather than axiomatic — thinking. It was also found in mathematics as long as number was regarded philosophically rather than scientifically (as Russell says, "mathematically"). As late as 1884 Frege had to make clear that number is "as little an object of psychology or an outcome of psychical processes as the North Sea." [48]

> The question "What is number?" [says Russell] is one which, until quite recent times, was never considered in the kind of way that is capable of yielding a precise answer. Philosophers were content with some vague dictum such as "Number is unity in plurality." A typical definition of the kind that contented philosophers is the following from Sigwart's *Logic* (§ 66, section 3): "Every number is not merely a *plurality*, but a plurality thought *as held together and closed, and to that extent as a unity.*" Now there is in such definitions a very elementary blunder, of the same kind that would be committed if we said "yellow is a flower" because some flowers are yellow. Take, for example, the number 3. . . . The number 3 is something which all collections of three things have in common, but is not itself a collection of three things.[49]

Exactly analogous is the argument of G. E. Moore about Value. To define Value as pleasure or the like would be to define a kind of value as Value itself. It would be the same as to hold, when we say "an orange is yellow," that "orange" means nothing else but "yellow" or that nothing can be yellow but an orange. Such definitions will not do for the *science* of ethics that Moore has in mind. "We should not get very far with our science, if we were bound to hold that everything which was yellow, *meant* exactly the same thing as yellow. We should find we had to hold that an orange was exactly the same thing as a stool, a piece of paper, a lemon, anything you like. We could prove any number of absurdities; but should we be the nearer to the truth? Why, then, should it be different with 'good'? . . . There is no meaning in saying that pleasure is good, unless good is something different from pleasure." [50]

What Russell and Moore have to say applies not merely to the fields of number and value, but to the relation between the generic and the specific in *any philosophy*. Before there can be a *science* of mathematics, of axiology, or any other, the confusion between generic and specific in the corresponding philosophy must be eliminated. In natural philosophy, the same fallacy was committed in alchemy, where colors were confused with substances and it was thought that the color of thing A could be given to thing B by mixing A and B; or when species of disciplines themselves were confused, such as religion and chemistry — when a prayer was needed to transmute an element.

Our definition of value as "similarity of intensions" does away with the confusions in value theory which led Moore to the formulation of the naturalistic fallacy. It is seen immediately that this definition is not an analytic but a synthetic one; not one by predicates, such as "pleasure," "intelligence," "preference," "self-realization," but one by logical relations — that, in a word, it is a scientific and not a philosophical definition.[51] It is a construction rather than an inference. It does not assume simple particulars that *are* values; it posits *a logical structure that has all the properties required for* values.

Since this definition is, as will be seen, the postulate of *formal axiology*, formal axiology is with respect to intension what mathematics is with respect to extension. And *what mathematics is to natural philosophy, formal axiology is to moral philosophy.* It follows that the analytic concepts of moral philosophy can be converted into synthetic concepts by the use of axiological relations, just as the analytic concepts of natural philosophy have been converted into synthetic concepts by the use of mathematical relations. There are no a priori reasons, then, for the frequent assertions, made especially by positivists, that a science of value is impossible.

5. *Logical Positivism and Axiological Positivism*

Failure to develop intensional logic is the reason that the logical positivists, far from succeeding in establishing a *moral science* do not even believe a *moral philosophy* to be possible. One may agree with positivism in its critical view of philosophy, but disagree with its negative conclusion. The traditional philosophical disciplines are indeed vague and meaningless and ought to be regarded as species of (traditional) logic. But this does not mean that they ought to disappear. On the contrary, they ought to be reconstituted on the basis of a new logic. The counterpart in physics to the positivistic attitude in ethics would be Einstein's decision, since it was proved that the ether did not exist, to lean back and pronounce all natural science as "nonsense." Instead, he designed a new frame of reference for the old one that had rendered things nonsensical. Not so the radical logical positivists. Observing that traditional philosophy appears nonsensical in the frame of reference they apply, they lean back and pronounce all such philosophy nonsensical. They never for a minute doubt their frame of reference. Like a man wearing blue glasses they swear the world is blue. Although they supposedly apply the scientific method they never once apply the truly scientific procedure of changing the frame of reference when it renders the subject matter absurd. Old nonsense, the history of science has amply demonstrated, makes sense in new frames of reference.[52] In the fashion of naive scholastic realism they condemn the subject matter rather than critically examine their instrument of inquiry. This is a typical example of dogmatism, methodologically identical with that of the schoolmen. In the struggle for a new ethics we have the Galilean controversy in a new key. That the dogmatism to be overcome appears in the name of the scientific method introduced by Galileo makes the situation especially piquant.

We thus differ from the positivists on one decisive point: our conception of logic. For the logical positivists there is only one logic, extensional or class logic, and as they see very clearly that the concepts of metaphysics, ethics, aesthetics, etc. do not fit into this logic, they conclude that therefore these disciplines are not "logical," that they are not disciplines at all, and "literally nonsense." Thus, as the modern logicians amputated one-half of logic, the logical positivists amputated one-half of philosophy itself, moral philosophy. Since this is all that is left of philosophy — natural philosophy has already been converted into natural science — they are killing philosophy altogether.

For the scientific ethicist there is not only one logic but two: he presupposes by the side of extensional or class logic, intensional or value logic (or whatever other kind of logic will yield formal axiology). The positivistic point of departure, which is that the philosophical disciplines are species of the genus "logic," leads him to the concept of a moral science. He could call himself an *axiological positivist* for he considers moral philosophy the same way the logical positivists consider natural philosophy. But he cannot afford the dogmatism of the logical positivist; he cannot declare that since his is the only true logic and natural philosophy appears in it as nonsense, natural philosophy is a psychosomatic phenomenon, a kind of grunt, like "hm" or an ejaculation like "aha" and natural science an hallucination.[53] He conceives that there are two different kinds of logic, both interrelated, applicable to two realities, that of nature and that of morals, and he concedes to the logicians of the first their right to deal with natural philosophy, reserving for himself the right to deal with moral philosophy. He takes his *right* to deal with moral science — and thus do what the logical positivist regards as impossible — from the *fact* that he is doing it (if and when he does). But the axiological positivist reserves his right to answer the assertion of the logical positivists that what he is doing is "literally nonsense" by stating that from *his* point of view what *they* are doing is nonsensical, and is so not only in metaphysics, ethics, and aesthetics, but even in epistemology, logic, and theory of science.[54]

The efforts of the logical positivists to amputate half of philosophy because, as for Procrustes, it does not fit into the bed which for them is the only one there is, is against all philosophical and scientific tradition. All philosophy has aspired to specify its general concepts even though this effort up to the present has succeeded only in regard to natural philosophy. But the moral philosophers, although less successful, were no less industrious and methodologically correct. On the contrary, this tradition in the field of moral philosophy is as old as philosophy itself — and older than in the field of natural philosophy. As late as 1695 Locke believed natural *science* to be impossible — in spite of the works of "the incomparable Mr. Newton" — but he never doubted the possibility of a scientific *ethics* as obvious and precise as mathematics.[55] He only repeated a tradition which goes back to Socrates and Plato, that of the maieutic method and the elaboration of the good as measure in *diairesis*, the division, or organization, of the ideas.[56] The concepts which these philosophers wanted to determine with scientific precision, a precision which they perceived to be the very core of the philosophical method and only justification of the philosopher, were the concepts of moral philosophy, in particular "goodness."

The positivistic direction in the field of morals is not only against all philosophical but also against all scientific tradition. It is, in particular, against what the positivists call empiricism. They simply do not apply this method to moral phenomena. They thus deny their own fundamental principle. The empirical scientist penetrates to the core of the phenomenon, and with all his powers seeks to discover the rational law which rules it. All the effort of science is bent on making order out of chaos. If the positivists were interested in the phenomenon of value they would try to penetrate to the core of this phenomenon and bend every effort to search for the law which rules it. Instead, they are content with the kind of superficial observations Moore has exposed over half a century ago, or with criticisms of the assertions of moral philosophers. Thus, instead of penetrating to the core of the phenomenon, they pit words without sense — the analytic concepts of traditional ethics — against words equally without sense, because they belong to the wrong frame of reference. Thus, they convert the discussion about values into a battle of words — precisely the kind of scholastic procedure that provoked the scorn of Galileo and Descartes. Nothing, said Galileo, is more "revolting" than the reference to texts when it is a matter of life; "our discourse must relate to the concrete world and not to one of paper. . . . If, indeed, you wish to continue in this method of study, then put aside the name of philosophers and call yourselves historians or memory experts; for it is not proper that those who never philosophize should usurp the honorable title of philosopher." [57]

As the natural philosopher should observe nature and draw his conclusions from nature, not from analyzing textbooks, so the moral philosopher should observe the moral phenomena, drawing his philosophy from these rather than from textbooks on ethics. "What would you say," Galileo wrote to Kepler, "of the leading philosophers here to whom I have offered a thousand times of my own accord to show my studies, but who, with the lazy obstinacy of a serpent who has eaten his fill, have never consented to look at the planets, or moon, or telescope? Verily, just as serpents close their ears, so do men close their eyes to the light of truth. To such people philosophy is a kind of book, like the Aeneid or the Odyssey, where the truth is to be sought, not in the universe or in nature, but (I use their own words) by comparing texts!" [58] Naturally, ethics texts are nonsensical if they do not thoroughly, that is, synthetically rather than analytically, account for the moral phenomena. "True philosophy," wrote Galileo, "expounds nature to us, but she can be understood only by him who has learned the speech and symbols in which she speaks to us." The same ought to be true of moral philosophy.

Thus, the discussion between axiologists and positivists should be about value; but usually it is about words concerning value. The positivists counter the empty words of the traditional axiologists by denying the object of the inquiry altogether and are in no way different from Galileo's positivistic contemporaries. Like them, our contemporary positivists hold fast to an obsolete frame of reference which distorts the reality to be discovered. Unlike them, however, in their attitude toward value, they deny their own principles: supposedly positivists, they are in truth negativists, if not nihilists; standard-bearers of science, they are really obscurantists; champions of empiricism, they speak with a priori certainty about what is not; advocates of precision, they are imprecise; upholders of expertness, they are inexpert; pioneers of analysis, they refuse to analyze; and paragons of criticism, they are uncritical. They neglect to analyze the basis of all their thinking, the supreme value of empirical logic; for such analysis, of course, presupposes a theory of value. Thus, in the last instance, they are irrational and take things on faith, as Kecskemeti rightly observed. Theirs is, as it has been said recently, "the most tyrannical of all dogmatisms: the dogmatism of prohibitions, which philosophy and science must abolish in order to continue their advance." [59]

Hence, the old quest must be taken up anew. Natural science has taught us to think *with precision in generality* and *with generality in precision.* The transformation of natural philosophy into natural science has not changed the generality of the object of knowledge; physics deals with the most general problems of space, time, and matter. But it does so with precision, scientifically, not philosophically. What has changed is the method, not the content of natural philosophy. To know with precision it is unnecessary to limit the object one wants to examine. One can think in the largest possible contexts yet with precision. The sciences typical of the most general and at the same time the most precise thinking are mathematics and logic. By means of an axiologic, an intensional logic, one could think with equal precision in ethics, aesthetics, metaphysics, and epistemology as extensional logic does in physics, chemistry, and astronomy. But before thinking with precision in moral philosophy one must ask oneself what it *means* to think with precision in these disciplines. What is ethical precision? This question we should answer if we are to understand the fundamental difference between philosophical and scientific ethics.

6. *Ethics and Metaethics*

Scientific ethics determines its terms with synthetic precision, or rather, first determines how ethical terms *can* be determined with precision. When traditional ethics speaks of "norms," "values," "duties," etc., systematic or scientific ethics analyzes these terms, determines their meaning, their interrelationships, and establishes a network of *formal relations* which can be used in ethics as mathematics can be used in physics or astronomy. *Systematic ethics is a formal discipline dealing with variables some of whose values are the terms of ethics.* Thus it is a propaedeutic, anterior to ethics; it is not ethics but the determination of ethics and even of the possibility of ethics. It is metaethics, and deals *with* ethics as its subject matter, just as ethics has as its subject matter human conduct. Systematic ethics, thus, deals with the knowledge of human conduct and makes this knowledge more precise. It is, in a word, the precision of ethics by synthetic formulation.

This being so, systematic ethics is on a higher logical level than ethics. It is "meta" — "beyond" — ethics. The neglect of the differences in logical levels, although eradicated in logic, is still common in ethics. It appears over and over as the fallacy of confusing method and content which, as we have seen, is the bane of moral philosophy. We call it the *fallacy of method.* The method of dealing with something is on a higher logical level than the content dealt with.[60] These levels have been overlooked in philosophical ethics, and fallacies have resulted, only one of which was seen by G. E. Moore.

Because the metadiscipline is more general, it has a wider extension. If, moreover, it is synthetic rather than analytic it also has a richer content or intension. Metaethics is not only more extensive but also more intensive, more precise than ethics. It applies to the totality of all ethical philosophies and is an instrument for their analysis and interrelation.

As there is metaethics so there are — and ought to be — metaaesthetics, metametaphysics, metaepistemology, and other philosophical metadisciplines to analyze and interrelate the diverse philosophical disciplines, aesthetics, metaphysics, epistemology, etc. And "beyond" these metadisciplines there ought to be a meta-metadiscipline which analyzes and interrelates the metadisciplines. This discipline is metaphilosophy, that is to say, *metaphilosophia-moralis,* which is on the same logical level as *metaphilosophia-naturalis,* or mathematics. It is, what we called formal axiology. As mathematics is the natural metaphilosophy containing the patterns for natural philosophies, so formal axiology is the moral metaphilosophy containing the patterns for moral philosophies. As the mathematical patterns can be applied to the various disciplines of natural philosophy — physics, chemistry, astronomy, etc. — so the axiological patterns can be applied to the various disciplines of moral philosophy — ethics, metaphysics, aesthetics, etc. And as the mathematical applications by Galileo, Lavoisier, and Newton constituted, respectively, physics as a metadiscipline "beyond" Aristotelian physics, chemistry as a metadiscipline "beyond" alchemy, astronomy as a metadiscipline "beyond" the earlier astronomies and astrologies, so the axiological applications constitute metaethics beyond traditional ethics, metaaesthetics beyond traditional aesthetics, metametaphysics, beyond traditional metaphysics. These metadisciplines differ from the traditional disciplines in being more exact as well as more comprehensive — in possessing greater generality joined with greater precision: in being sciences. They are more *systematic.* "Systematic," "scientific," and "meta" discipline are thus synonymous. *Systematic ethics* is *scientific ethics* or *metaethics*. It is a part of formal axiology, as are metaaesthetics, metametaphysics and metaepistemology.

In the historico-logical scheme of the development of science — a scheme which confirms the assertions of philosophers of science, such as Duhem, that *historical* analysis of science is at the same time *logical* analysis — we have, then, logic at the highest level, with its two divisions, extensional logic and intensional logic. Extensional logic is applied to mathematics, mathematics to the natural sciences, and the latter to their predecessors, the corresponding natural philosophies — chemistry to alchemy, astronomy to astrology, etc. Intensional logic is applied to formal axiology, formal axiology to the moral sciences, and the latter to their predecessors, the corresponding moral philosophies — metaethics to ethics, metaaesthetics to aesthetics, etc.

The traditional moral philosophies would thus transform themselves, through the establishment of moral science, into historical disciplines with the same methodological position as the traditional natural philosophies. This does not mean, however, that they would become obsolete. While for the practicing scientist of today the natural philosophies of the past have little interest, for the philosopher trying to understand, rather than apply, these sciences, they are indispensable. The moral philosophies are even more important to the moral scientist — the axiologist — for in morality there will not be the external transformation of the world which natural science has effected through technology. The moral sciences will bring about an internal transformation of man, relating intensions rather than extensions, meanings rather than objects; and for this transformation the insight of moral philosophers into the nature of man is going to be more useful than that of the natural philosophers of the past into the nature of matter. But there is another, more significant reason why the new moral science will not change moral philosophy so radically as natural science has changed natural philosophy: *natural philosophy was to a large degree moral philosophy*.

7. *Physics and Ethics — Metaphysics and Metaethics*

Because natural philosophy was to a large degree moral philosophy the scientists of the Renaissance had to proceed drastically. Galileo, for example, had to invert the relation between form and matter of Aristotelian physics, between the method of analyzing nature and the object of this analysis. Aristotle used nature to explain logic, Galileo used logic to explain nature. The fundamental Aristotelian categories were teleological categories of the physical process: potentiality and actuality, unified through movement, as the cause which converts potentiality into actuality. Through these categories Aristotle not only solved physical problems but also logical ones, as that of the unity of the definition (*Metaphysics*, H, vi). Galileo, on the other hand, used logic — taken as mathematics — to solve physical problems. To do this he had first to demote the categories of Aristotle from their high metaphysical rank, and convert "movement" from a metaphysical concept to a physical phenomenon. He then had to find new explicatory categories for this *phenomenon*, and these he found in mathematical relations. Thus, the demotion of "movement" meant at the same time the elevation of mathematics and of sense observation — both again anathema to the medieval mind.[61] Since mathematical relations are specifically logical, we may say that he used logic to explain nature, thereby inverting the Aristotelian method. Thus, Galileo wrote the true metaphysics (in the logical sense explained above) to the physics of Aristotle. The latter then appeared as a collection of methodological, consequently factual, errors. It is, of course, methodologically speaking more correct to explain nature by logic than logic by nature. But this was by no means obvious to the Middle Ages, nor is it obvious to the value philosophers of today. If there is very little literature about the value of logic, there is almost none about the logic of value.[62]

Although we are going to discover similar methodological errors in moral philosophy and will have to invert some old ethical arguments in the same way that Galileo had to invert the Aristotelian arguments — in the fallacy of method — this inversion will be less radical than the Galilean inversion of Aristotle.[63] Suppose that traditional axiology has used the *category* of value in order to explain features of logic, and that we have to invert this relation, using logic to explain the *phenomena* of value.[64] This inversion, although logically as radical as the Galilean, appears less so, for value is not something obvious and sensory like movement. We do not know, really, what value is, whether it is a category or a phenomenon. Hence what axiological science is going to determine about value will scarcely shock our senses — and our world view — as Galileo shocked his contemporaries with his new definition of motion. The new position of value within the frame of reference of axiology will not be felt as a "degradation," at least not in the same degree that the new position of motion within the frame of reference of Galilean mechanics was felt to be. In some instances, the fallacy of method actually causes a feeling of demotion of value which makes it difficult to accept formal axiology. But the methodological shock to our knowledge is milder than was the theological shock to the emotions of Galileo's contemporaries. Value and logic are closer than motion and logic. Thus, the substitution of logic for value, though offensive to some degree, is not comparable in severity to the shock suffered by medieval man. What he had to give up he lived with all his soul. What we have to give up are some superficial errors which, moreover, concern mostly professional philosophers. We are not *aware* of value, we do not *think* about it, as medieval man was aware of, and thought about, God. What he had to give up was irretrievably lost. We neither have to give up nor are we going to lose. We are only going to gain: a new world of meaning, a new awareness of the values we normally live *by* though we do not yet consciously live *them*. The new science will enrich our world, not destroy it. The reason is that we are already living one-half of this new world, the extensional part, and only need to add the other half, the intensional. The medieval man lived, from the point of view of the modern age, in the *wrong kind of world*.[65] It had to be destroyed before the new age could become reality. Galileo had to be a destroyer as much as a builder; we can afford, above all, to build.

The rearrangement and reorganization of moral philosophies by moral science, therefore, will not disarrange and disorganize these philosophies to the point of making them antiquated the way the natural philosophies have become through natural science. For, as we have mentioned, these natural philosophies contained fundamental elements of *moral* philosophy. The natural *sciences* had to eliminate these elements with determination. In doing so they encountered *exactly the same argument against the use of mathematics in natural science that is raised today against this use in moral science:* that there is no place in mathematics for the Good. Since the core of nature was motion, and the core of motion the Good — as the end of a process — the world order, so ran the argument, could be understood only in terms of the Good. It is obviously more gratifying to insist on the use of mathematics in order to bring the Good relevantly into the world than to insist on this same use in order to extirpate it from the world — as Galileo was forced to do.[66] He only began the process of the secularization of science. Newton still had to have God in his famous scholium; and it took another 150 years for Laplace to observe that he did not need this hypothesis. Between moral philosophy and moral science there are no such radical confusions as there were between theology and mathematical physics. The corresponding situation in moral philosophy would be the existence in *moral* philosophy of elements of *natural* philosophy which *moral science* would have to eliminate, the same way that natural science had to eliminate the elements of moral philosophy. Although there does exist this same confusion in axiology — as Moore has shown in what he calls the "naturalistic fallacy" — this confusion is very much more subtle than it was in natural philosophy; so subtle indeed that the majority of today's ethicists do not recognize it.

The difference between moral philosophy and the new moral science is then very much less radical than was the rupture between natural philosophy and natural science. While this has practical advantages — a modern axiologist is in no danger of being burned alive — it has certain intellectual disadvantages. Since the differences in question are so subtle, and merely on the logical and perhaps the phenomenological but not on the sensory plane, there is a danger that these differences may be underestimated and looked upon as irrelevant hairsplitting. But the differences between moral philosophy and moral science are *logically* of the same importance as those between natural philosophy and natural science. This logical character is what determines the nature and efficiency of the new moral science, as it does that of any science.

2

THE STRUCTURE OF SCIENCE

> As in Mathematicks, so in Natural Philosophy, the Investigation of difficult Things by the Method of Analysis, ought ever to precede the Method of Composition. This Analysis consists in making Experiments and Observations, and in drawing general Conclusions from them by Induction. . . . This is the Method of Analysis: And the Synthesis consists in assuming the Causes discover'd, and establish'd as Principles, and by them explaining the Phaenomena proceeding from them, and proving the Explanations. . . . And if natural Philosophy in all its Parts, by pursuing this Method, shall at length be perfected, the Bounds of Moral Philosophy will also be enlarged.
>
> — ISAAC NEWTON

> Of all Galileo's achievements this was his greatest gift to posterity: the inductive method, the core of all exact science; extended in the years that followed it proved to be the key to the mysteries of being, opening up ever new strata, ever new depths. . . . But he did far more. He changed man himself. The man who knows what he can wrest from nature is not the same as the man who resigns himself to nature in passive contemplation. Once discovered, the dynamic of cosmic being awakened the dynamic of the human heart.
>
> — FRIEDRICH DESSAUER [1]

1. *Philosophical Reality and Scientific Reality*

The difference between analytic concepts and synthetic concepts, which defines the difference between philosophy and science, implies different characteristics of philosophy and science, respectively. Philosophy, as we have seen, does not have any structure fundamentally different from that of everyday discourse. Concept is linked to concept, each concept, by implication and association, gives rise to other concepts, and carries within its intension the couplings, as it were, to hook on to the rest of the language and its meanings. The whole concatenation of concepts mirrors the world as it appears to us; for analytic concepts are abstractions from this world and refer to it. Philosophical discourse is based on sensible intuition; and its only difference from everyday discourse is that the latter describes concretely what the former describes abstractly. But what both thus describe is only a vague likeness of the world, for analytic concepts lack specificity. It is a reflection of the world, but a blurred, misleading one. Yet it is the kind of reflection we live by. We understand immediately what analytic concepts mean to say. It is relatively easy to follow an analytic argument but difficult to follow a synthetic one.

Science as a network of formal relations rather than a succession of predications does not refer to things and properties. Relations are not matters of the senses. To follow a scientific argument, it is not enough to know the world and recall it to one's mind. It is necessary to *think*, formally and systematically, and to master the vocabulary and symbolism of the system. The scientific argument is directed to the intellect, not to sensible intuition. It demands seeing relations in themselves. The things and properties of the world are nothing but symbolic knots tying up these relations. The relations do not *refer to them* but form a pattern to which the world, by intricate procedures of application, is made to conform. The pattern does not mirror the world nor does the world reflect it; it orders the world, and the world is that of science insofar as it follows this order. The scientific pattern was not abstracted from the world but creatively produced by the scientist, after penetrating to the core of the phenomenon. It was produced out of him like a bee's honeycomb — to use the Baconian simile [2] — after his impregnation with the phenomenal essence. Therefore it is much more difficult to apply science to the world than to understand the world in terms of everyday or philosophical discourse. Science does not speak of the things and properties of the world but of their capacity of entering into formal relations. Thus, the art of the engineer is not only to know the mathematical system but to find in concrete situations the relations which correspond to those of the mathematical formulae. To find this correspondence, to see the isomorphism between concrete reality and a formal system, is a true art and can only be learned as an art, and like any art, by practice. This is the only meaning of "practical reason" that makes axiological sense.

Everyday discourse, then, refers to the world immediately, every word, in its context, says what it means and names what it refers to. Philosophical discourse refers to the world mediately, through the classification of its abstractions. It is everyday discourse generalized; and by specifying the generalization, that is, by retracing the steps which led to the generalization, one can understand philosophical discourse. Also, many philosophical generalizations are the same as those which common sense itself arrives at in its effort to understand the world and in its own work of abstraction. Concepts such as "immortality," "justice," etc., are understandable without exemplification. But this understanding is vague. There is not yet, as we have seen, any law structuring generalization. Therefore philosophical discourse, although close to everyday talk, is both in reference and in meaning, in extension and intension, much more vague than either everyday or scientific discourse. If it is "easier" to understand than the latter it is because this "understanding" is specious. Actually one cannot "understand" a philosophy, one can only guess oneself more or less closely toward it. The positivists are right in this respect; there is much in philosophy that is closer to poetry than to science. And if it is the task of philosophy to understand the world it would be better for it to be simply descriptions of the world. But the task of philosophy is not to understand the world. Its task is speculation, to carry analytic arguments wherever they will lead. If we want to understand the world we must either abolish philosophy, as some positivists suggest, or transcend it, as the scientists do.

Scientific discourse has no *reference*, either directly or indirectly, to the sensible world. It is *applicable*; and application is not reference. It is, at best, a very special kind of reference, a creative enterprise, the use of a tool, as is all art. A formal system is like a chisel that sculptures the world in the image of the artist.[3] If the chisel's — and the artist's — application to the marble is "reference" to the statue, then application of a system is reference to the world. But understanding the system in no way guarantees the ability to apply it to the world, nor does knowledge of the world guarantee the ability to interpret it by a system. On the contrary, the system often presupposes a way of looking at the world contrary to common sense, as with Copernicus and Einstein. A science thus has a complex structure consisting of *theory* and *practice*, of formal relations and material observations. The coordination of the two is, strictly speaking, the science.

A reality to which a science applies is, then, different from that to which a philosophy refers. The first is *part of the science*, the practical part: the totality of all the material observations which constitute the applied science. It is, as we have said, a world of interrelationships, of *order*, a world *created by* the science rather than *abstracted from* reality.[4] The science uses *elements* of the world according to its own laws, and combines them to form new structures, both theoretical and practical, as does natural science in technology. Our environment is largely a scientific creation. The world to which philosophy refers, on the other hand, is the world of everyday observations — of secondary rather than primary qualities. It is what Husserl calls the *Lebenswelt.*[5] This world is not so firmly bound to philosophy as the scientific world is to science. It does not have the same kind of rigorous order. It is looser, more arbitrary, just as is philosophy itself. The world, from antiquity to the middle ages, was a world of philosophy — and we all know the difference between it and the modern world. It was a world with few empirical interrelations. Not only its extension but also its intension was loose, its meaning crude. The Middle Ages were not what the romantic imagination paints, but instead a rugged age much more insensitive than ours today. Tortures, burnings, hangings were public spectacles, their thrill was like today's bullfights or prizefights. The ladies of high society went to the place of execution the night before and slept in their coaches in order not to miss the entertainment.[6]

> However ghastly and shocking these tortures were to any normal person, what is far more shocking is the fact that so many people of fashion found pleasure and excitement in them. In this crucial respect that extra-refined Parisian society was on the same level as the Iroquois Indians whose delight it was to prolong the suffering of their victims — on the same level as those untutored savages but with no excuse. The admirable elegance of the eighteenth century was indeed, as measured by later standards, only a veneer, concealing the most disgusting license and brutality, not only in the underworld but in the upper one, in the very highest spheres, even in the sphere of royalty which was generally supposed to be almost divine.[7]

As science forced the human mind to minuteness of perception and conception, it also forced a refinement of the senses. Our modern world is one of such refinement — in spite of its defects which are defects of refinement, either of individual perversion or of technological efficiency, or of both, as in the Nazi regime. The historical names of this refinement are *Humanism* and *Enlightenment*.[8]

Philosophy and science have each its own world, reflecting the different structures of the two kinds of knowledge. From this follows a most important consequence for the new moral science: if moral philosophy converts itself into science it must not only assume the structure of science but, through this very structure, change the moral world into a new one. Today's moral reality is still philosophical; it is not fundamentally different from that of antiquity or the Middle Ages. We have the same fundamental values and disvalues, even though we practice them with greater refinement, including torture. The new moral science ought to revolutionize our moral understanding itself and hence our moral practice, in the same way that natural science has revolutionized our understanding of nature and our sensitivity to it. The precise knowledge of the axiological relations ought to make us more sensitive to moral reality. It ought to teach us more profoundly the art of living.

This new moral world is difficult to imagine. For the moment we are only taking the first step in its direction, that of establishing the new science. But the analogy between natural science and moral science can give us an idea of this new world. As the science of nature details the experience of spatio-temporal events, so the science of value details the *meanings* of these events. As the science of nature refined, to a point heretofore inconceivable, the physical sensitivity of measuring instruments, so moral science will refine, to a degree inconceivable today, man's own moral sensitivity. Compared to this new world of meaning the present world of facts will appear as dull and crude as the world of the Middle Ages appears to us today.[9] The new world will be one of qualitative richness and not just, as our present Western world, one of quantitative wealth. Owing to the nature of axiological science, which refines man's own sensitivities rather than those of machines, man's own senses will become the finest detectors of value meanings. The whole tenor of life will be quickened by a new spirit similar to the age of the Renaissance. The new science will structure our sensitivity as physical theory has structured our intellect. It will open our understanding to the measure of value as the old science has opened our intellect to the measure of things. It will refine our inner environment as natural science has refined our external environment.

Thus, the new science will add spirit to technology, value to energy, human sensitivity to the sensitivity of instruments. It will develop *man* as natural science has developed *matter*. The world of natural science will be succeeded by the world of moral science. The world of value will follow the course of the world of fact, to ever subtler refinement of moral sensitivity. For value must follow the structure of moral science as fact followed that of natural science. As the structure of each science is refined, so is its subject matter.[10]

The relationship between these two structures, then, ought to clarify for us the relationships between *fact* and *value*.

2. *The Threefold Structure of Science*

Value, we said, must be seen on three levels: that of theory, formal value or "value"; that of practice, phenomenal value or value; and that of their combination, axiological value or, as we may now say, scientific value.[11] Fact, of course, exists on the same three levels: the formal one of theory, formal fact or "fact"; the commonsense level of practice, phenomenal fact or fact; and the combination of both, scientific fact. Thus, the physical-science equations for "energy" constitute a formal fact: it is "energy" defined as these equations. The puffing of a steam engine is a phenomenal fact, a *fact;* and the understanding of the engine in terms of the equations, or the exemplification of the equations by the engine, is a scientific fact. The formal fact is a set of formulae constituting a synthetic concept; the phenomenal fact is a set of sense observations combined with an analytic concept; and the scientific fact is a different set of sense observations, a schema (a set of primary rather than secondary properties), combined with a synthetic concept. In value theory we have so far only the second level, phenomenal value and its analytic concepts. What we need is the first level, formal value, which will teach us the third, scientific value. The difference between fact and value, then, is vague as long as both are seen phenomenally; and for this reason value theory finds itself in the difficulties already described, of either confusing value with fact or else separating them so completely that they show no relation at all. The distinction becomes precise as soon as the formal level is added in the field of valuation. Then *fact*, scientific fact, *is what is subject to natural science*, and *value*, scientific value, *is what is subject to moral science*. We must be clear about the three levels on which both fact and value can appear.[12] These levels constitute the structure of science.

As we have seen, the "world of fact" is the result of an historical development. What in the modern world is "fact" was not fact in the ancient and medieval world and vice versa; and what even today is "fact" in some advanced western societies, such as the technological environment of Western Europe and North America, is not fact in the underdeveloped countries of Africa and Asia. The same is true of value. Happiness, for example, was entirely different for the medieval man from what it was for Aristotle, or what it is for us. The frequently-made distinction between fact and value as the realms of objectivity and subjectivity, respectively, is invalid; it is not in accordance with the "facts," that is, the actual observations of people reacting in different cultures to the "same" data. What to an American appears as one thing, appears to an African as an entirely different thing. If an American and a North African were to draw the battleship Missouri — the experiment has been made — the American's drawing would look like a photograph and the African's like a dream, for the latter lacks a frame of reference to make it understandable to his intellect, even though his eyes see it — just as the schoolmen did not see what Galileo showed them. We see what we have a mind to see.

The so-called world of the senses, thus, is dependent on the higher rational faculties, not to mention the complexities of the perceptional apparatus itself.[13]

The modern world picture came into being in the Renaissance, when the mathematical frame of reference was applied to natural phenomena. It is, again, largely a creation of Galileo, whose empirico-mathematical method directed man's eyes toward nature.[14] Before Galileo another frame of reference was used, the theological, but it did not apply to "natural" phenomena. The application of the theological frame of reference to the world of the senses produced what, again, we would call today a dream.[15] What we call facts today emerged together with, and by application of, the mathematical framework. Galileo himself only gradually understood this when he recognized that appeal to the senses was an inadequate means for convincing his learned adversaries. What was needed was an entirely new way of looking at the world, "a different kind of thinking cap, a transposition in the mind of the scientist himself," [16] a new use of the mind and the senses: the welding of sense actuality to the mathematical rather than the theological pattern. Thus Galileo created the empirico-mathematical world picture: the analysis of the sense world — which thus turned "empirical" — in terms of mathematics, and the interpretation of mathematics in terms of the sense world. There emerged both a new sense world and a new pattern; and both combined in a new *natural philosophy* which today is called *natural science.*

Facts are complex, not easy to know; what were facts before Galileo and Lavoisier were no facts after them. Before them "facts" were the empircal counterparts to philosophies, to analytic concepts. After them, "facts" were the empirical counterparts to sciences, to synthetic concepts. Both kinds of "facts" were empirical but in different ways. The former "facts" lacked empirical and systematic import, the latter "facts" do have such import. Empirical and systematic import is the capacity of synthetic concepts to explain in minute detail the widest extension of reality. Analytic concepts lack this import. A *philosophy, then, is distinguished from a science by its lack of empirical and systematic import.*

a) EMPIRICAL AND SYSTEMATIC IMPORT

Hempel compares a scientific theory to

> a complex spatial network: Its terms are represented by the knots, while the threads connecting the latter correspond, in part, to the definitions and, in part, to the fundamental and derivative hypotheses included in the theory. The whole system floats, as it were, above the plane of observation and is anchored to it by the rules of interpretation. These might be viewed as strings which are not part of the network but link certain points of the latter with specific places in the plane of observations. By virtue of those interpretive connections, the network can function as a scientific theory: From certain observational data, we

> may ascend, via an interpretive string, to some point in the theoretical network, thence proceed, via definitions and hypotheses, to other points, from which another interpretive string permits a descent to the plane of observation. In this manner an interpreted theory makes it possible to infer the occurrence of certain phenomena which can be described in observational terms.[17]

Hempel's analogy brings to mind a simile: the whole of a science is like a floating rig used for the drilling of tidelands oil. The platform is the theory, the subaquatic soil the data, the drill the interpretive connection. The drills probe the subsoil at unconnected points and, through the machinery on the platform, these points are meaningfully connected. The theoretical system, in other words, determines through empirical interpretation the interrelationship between the observations, and the total set of the observational data receives its unity through the interrelationships within the theoretical system.

In order to be applicable, the formal system must be precise and detailed enough to account for the interrelationships of the empirical data. It must have both systematic and empirical import. A set of merely general principles does not constitute an applicable system; the key term of neo-vitalism, "entelechy," for example, lacks empirical import. It is a merely analytic concept of a high level, a category; and as such it even lacks systematic import. On the other hand, a term such as "universal gravitation" does have both systematic and empirical import. *Theoretical* or *systematic* import permits "the establishment of explanatory . . . principles in the form of general laws or theories" and is the center of a network of formal relations. To frame concepts with *merely empirical* import — low-level analytic concepts — is relatively easy, they "can be readily defined in any number, but most of them will be of no use for systematic purposes. . . . It is . . . the discovery of concept systems with *theoretical* import which advances scientific understanding."[18]

There is a fundamental difference between the empirical and the theoretical part of a science. The empirical part *describes and abstracts from phenomena;* it gives at best general definitions, but no universal connections. The theoretical part, on the other hand, *constitutes a pattern* of universal connections "to which the individual phenomena conform."[19] Whenever a definition is found which constitutes a pattern of this kind, it is so general that it seems to be entirely removed from the empirical and unconnected with it. Actually, through the process of axiomatic identification, it represents the very essence of the reality in question.

The development of science from empirical analysis to meaning analysis of analytic concepts to theoretical constructs (synthetic concepts) [20] is connected with a progress from everyday language through kinds of technical language to theoretical or systematic language. The latter, in the natural sciences, is mathematical. "The initial stages of scientific inquiry are stated in the vocabulary of everyday language. The growth of a scientific discipline, however, always brings with it the development of a system of specialized, more or less abstract, concepts of a corresponding technical terminology." [21] The technical terms "bear little resemblance to the concrete concepts we use to describe the phenomena of our everyday experience." [22] In a theoretical system the concepts used are "fictitious" and without any obvious connection with empirical reality. A theoretical system is "an uninterpreted theory in axiomatic form which is characterized by 1] a specific set of primitive terms . . . 2] a set of postulates, primitive, or basic, hypotheses; other sentences of the theory are obtained from them by logical deduction . . . Euclidean geometry . . . as 'pure geometry,' i.e. as an uninterpreted axiomatic system, is logically quite independent of its interpretation in physics and its use in navigation, surveying, etc." [23] Interpretation gives the system empirical import.

We shall call this kind of interpretation "axiomatic interpretation" to distinguish it from the "empirical interpretation" in terms of observation of a system mentioned above. *In axiomatic interpretation a whole field of phenomena is the interpretation of a system.* Thus, for example, "pure geometry does not express any assertion about the spatial properties and relations of objects in the physical world. A physical geometry, i.e. a theory which deals with the spatial aspects of physical phenomena, is obtained from a system of pure geometry by giving the primitives a specific interpretation in physical terms. Thus, e.g. to obtain the physical counterpart of pure Euclidean geometry, points may be interpreted as approximated by small physical objects . . . ; a straight line may be construed as the path of a light ray in a homogeneous medium; congruence of intervals as a physical relation characterizable in terms of coincidences of rigid rods; etc. This interpretation turns the postulates and theorems of pure geometry into statements of physics." [24]

Axiomatic interpretation always means that *something logical is identified with something nonlogical.* Every formal system is, in the end, a specification of logic. The identification of an element of a logical system with a term signifying a nonlogical phenomenon produces the axiom of a science — that science which studies the field of which the phenomenon is the essential element. A ray of light is the essential element of the field studied by optics, and optics arose from the identification of the term "ray of light" with an element in a formal system, namely "straight line," a term of pure geometry. Pure geometry, in turn, as any formal system, is in the last resort a specification of logic.[25]

The axiom of a science thus arises from the *identification of something logical with something nonlogical.* This means, of course, that the nonlogical, the phenomenon in question, must have been analyzed down to its essential structure in such a way that it is seen to correspond to a logical element. Such analysis is not possible through observation, it is a matter of conceptualization and ever subtler refinement of the conceptual structure. This refinement is both a matter of thought and of symbolism. In other words, the production of a science through axiomatic interpretation — and there does not seem to be any other way for producing a science — is largely a matter of the refinement of *language.* This is strikingly shown in a classic passage from a science as solid and concrete as chemistry. This science arose through Lavoisier's analysis of *chemical language* rather than chemical compounds:

> When I began the following work, my only object was to extend and explain more fully the memoir which I read at the public meeting of the Academy of Sciences in the month of April, 1787, on the necessity of reforming and completing the nomenclature of chemistry. While engaged in this employment, I perceived, better than I had ever done before, the justice of the following maxim of the Abbé de Condillac, in his *Logic,* and some other of his works.
>
> "We think only through the medium of words. — Languages are true analytical methods. — Algebra, which

> is adapted to its purpose in every species of expression, in the most simple, most exact, and best manner possible, is at the same time a language and an analytical [we would say "synthetical"] method. — The art of reasoning is nothing more than a language well arranged."
>
> Thus, while I thought myself employed only in forming a nomenclature, and while I proposed to myself nothing more than to improve chemical language, my work transformed itself by degrees, without my being able to prevent it, into a treatise upon the elements of chemistry.[26]

The procedure of axiomatic interpretation, in the sense of axiomatic identification of a logical with a nonlogical element, is of fundamental importance for the creation of the science of value. Here the nonlogical element is "value." The axiom of the science of value is established when the *logical* term is found which is to be identified with "value" — in our case the term "similarity of intensions." With this identification the system of *logic* is at the disposal of the language of *value*. This identification presupposes that value language itself has been analyzed to such a point that value appears in its own logical structure. Only then can the logical term in question be recognized as corresponding to the term "value." Just as the term "ray of light" could not be identified with the geometrical term "straight line" until and unless the essence of a ray of light was recognized as its *being* a straight line, so "value" could not be identified with "intension" (or rather "one-to-one-correspondence of intensions") until and unless the essence of value was recognized as *being* intension. In the case of the ray of light, the axiomatic recognition was in part empirical and in part, actually a far larger part, conceptual. In value analysis no empirical observation is at all possible. Thus, it is the analysis of the concept, and the term, "value" that must lead us to our goal. We must analyze value *language*. To quote Lavoisier again, "*We cannot improve the language of any science without at the same time improving the science itself; neither can we, on the other hand, improve a science without improving the language or nomenclature which belongs to it.* However, certain the facts of any science may be and however just the ideas we may have formed of these facts, we can only communicate false impressions to others while we want words by which these may be properly expressed."[27]

The scientific analysis of value language must follow analysis of fact language. The latter, we have seen, has three levels, 1] empirical language, which describes situations in everyday terms; 2] technical language, which is of two different kinds, (*a*) concept analysis, (*b*) interpretation of concepts in terms of a theoretical system (it is never "facts" that are subsumed but special sets of — primary — properties, "ideal cases"); [28] and 3] the theoretical system itself, systematic language which does not describe any situation but is *applied to* situations and orders them autonomously and normatively into a whole, thus producing the total empirico-theoretical structure which constitutes the science.

It is usually not made sufficiently clear that the distinction between these levels (and thus that between the various "imports") is based on the distinction between, and the logical structure of, synthetic and analytic concepts. Theoretical import is nothing but the intension of a fundamental-synthetic concept, of the axiom or postulate of a science; and its empirical import is the corresponding extension. A *science, thus, is adequately defined as the combination of the intension and extension of an axiom,* where by "axiom" is meant a fundamental-synthetic concept arrived at by axiomatic interpretation: a formula giving rise to a system applicable to reality. "Universal gravitation" is such a concept. Its intension is a system, its extension a universe. "Entelechy," on the other hand, is an analytic concept, the intension of which is a string of other analytic concepts and the extension of which are vague phenomena, of "merely empirical import, of no use for systematic purposes." Analytic concepts rule the lower levels, (1) and (2,*a*), of scientific language, whereas synthetic concepts rule the higher levels, (2,*b*) and (3). The creation of an exact science, thus, is the transition from level (2*a*) to (2*b*), from analytic to synthetic concepts. The creator of a science has before him an analytic concept, as Galileo had the analytic concept of the Aristotelian movement. The extension of this concept, of merely empirical import, were the phenomena of movement, from snakes to the soul. To these phenomena Galileo had to apply his new conception, the intension of a synthetic concept. Through such application the phenomena were not changed — things went on moving after Galileo as they had before him — but they were newly understood, systematically ordered, and relevantly rearranged. Thus, many of the things subject to motion in Aristotle became irrelevant in

Galileo's mechanics. On the other hand, new phenomena were included, not thought of in Aristotle. Thus, analytic extension changed to synthetic extension. Implied in the precision of the synthetic concept there appeared an operational procedure, measurement. This is the logical reason why the concept of science includes that of method, hence of action. Analytic concepts, on the other hand, are vague, hence inapplicable, and can be thought forever without any subsequent action. Thus, a science is the combination of the intension and extension of a fundamental-synthetic concept, an axiom, which constitutes a *method* and necessarily leads to action; while a philosophy is the combined intension and extension of a fundamental analytic concept, a category or principle, which lacks method and remains speculative. The lower, analytic levels of science are philosophy. The creation of an exact science is the application of a synthetic intension to an analytic extension — which cuts up this extension into primary elements — and the subsequent substitution of the synthetic extension for the analytic.

Applied to value theory, this means that its present philosophical stage may be regarded as the analytic levels of a possible science, and that to create this science a synthetic concept of value must be applied to the present analytic extension of the vague concept "value" — the welter of phenomena we regard, vaguely, as valuational. Through this application these phenomena will not be changed — people will go on loving and hating, being thrilled and bored, happy and unhappy — but these phenomena will be newly understood, ordered, and relevantly rearranged. Thus, many of the things subject to value theory today will become irrelevant. On the other hand, new phenomena will be included, not thought of in today's value philosophy.[29] Thus, analytic extension will change into synthetic. Implied in the precision of the synthetic concept there will appear an operational procedure, measurement. This will bring about enrichment and refinement of moral life. The present analytic concepts of value, on the other hand, will go on being thought, without any subsequent action, just as laymen today think vaguely and analytically about psychiatry, medicine, and technology, in spite of the experts in the field, whom they call in when needed. When the axiologist becomes a scientist, he too will be called in.

To summarize, we would have to find in moral science the levels of discourse we have discussed. To take the example of physical science, these levels are (1) empirical description of physical situations, real definition of physical events in terms of some feature(s) of the physical situation, culminating in empirical concepts; (2,*a*) analysis of such concepts, their testing in practical and experimental physics; (2,*b*) theoretical physics, subsumption of the results of (2,*a*) under (3); (3) autonomous (nominal, synthetic) definition of physical events in terms of a purely formal pattern, "mathematics for physicists." This latter language, through the process (2,*b*), patterns and remodels the empirico-analytic language (2,*a*) as well as the empirico-real language (1).

For value language, we would have correspondingly, (1) empirical description of value situations, real definition of value in terms of some feature(s) of the value situation, culminating in empirical value concepts; (2,*a*) analysis of such concepts, their testing in practical and experimental ethics (casuistry); (2,*b*) theoretical ethics, subsumption of the results of (2,*a*) under (3); (3) autonomous (nominal, synthetic) definition of value in terms of a purely formal pattern, "axiology for ethicists." This latter language, through the process (2,*b*), patterns and remodels the empirico-analytic language (2,a) and the empirico-real language (1).

Neither value language (3) nor (2,*b*) as yet exist to any significant degree. Thus value theory in general, and ethics in particular, consists largely of levels (1) and (2,*a*); its concepts lack empirical and systematic import — they are more like "entelechy" than like "universal gravitation." On the other hand, it follows from what has been said that value theory and ethics would be sciences as soon as level (3) was supplied. The present arbitrariness of their empirical value determinations — pleasure, purpose, interest, preference, etc. — would disappear and a definition of value possessing both systematic and empirical import would take its place.

Whereas the creation of a new science appears perfectly well defined as the transition from analytic to synthetic concepts, the very heart of the matter remains obscure: the way to make this transition. How does the creator of a new science proceed from the analytic concept given him to the synthetic which he is to define? What is the essential relation between these two sorts of concepts?

This is the logical formulation of the problem of axiomatic

interpretation: how does the creator of a science find the logical structure of the phenomenon given to him in an analytic concept, in such a way that this structure appears isomorphous with a logical element and thus becomes, by identification with this element, a synthetic concept? The problem may be stated in psychological, phenomenological, and other *material* terms. Our simple *formal* distinction of analytic and synthetic concepts states it in logical terms and gives us a clue to its solution.

b) THE RELATION BETWEEN ANALYTIC AND SYNTHETIC CONCEPTS

We have based the difference between philosophy and science on the Kantian notion of analytic and synthetic concepts because Kant has elaborated this difference most clearly and has done so in connection with the difference between philosophy and science. These differences have been lost in subsequent logic to the point where some modern logicians, as we have seen, believe that Kant's terminology is only metaphor. Also, his distinction between analytic and synthetic *concepts,* on the one hand, and analytic and synthetic *judgments,* on the other, has been confused, so that what is called "analytic" propositions in the logic books of today are propositions containing synthetic concepts in the sense of the Kantian *Logic.* These propositions are synthetic in the sense of being part of a synthetic system. Kant himself does not clearly distinguish the analyticity and syntheticity of judgments formed by analytic concepts from the analyticity and syntheticity of judgments formed by synthetic concepts, even though the latter lies at the basis of his fundamental distinction between analytic and synthetic unity in the *Critique of Pure Reason.* Primarily, the distinction between analytic and synthetic *judgments* refers *only* to judgments with *analytic concepts,* such as "body," "water," and the like. Synthetic concepts, or constructs, do not form analytic and synthetic judgments in the same sense. Hence, systems such as logic or mathematics, consisting of synthetic concepts, ought not to be said to give rise to analytic — or synthetic — judgments. There is a fundamental difference between "All bodies are extended" and "$2 + 2 = 4$." The first is a genuine analytic judgment with analytic concepts, the second is a relation between constructs, following from their definition within the system in question. To accord with the doctrine of analytic and synthetic *concepts,* Kant's doctrine of analytic and

1. *Analytic and Synthetic Judgments and Concepts*

JUDGMENT	CONCEPT
Analytic a priori	Pure Category ("substance")
Analytic a posteriori	Empirical concept ("water")
Synthetic a priori	Axiom (Russell's definition of "number")
Synthetic a posteriori	Formal concept ("4")

synthetic *judgments* should be developed as in Table 1. This makes clear the relation between category and axiom, and the difference between the analyticity of empirical and of formal concepts. The axiom is synthetic a priori precisely by virtue of its identification (Kant's "synthetic unity") of something logical (a priori) with something nonlogical (synthetic, in terms of judgment).[30] The analyticity of empirical concepts is based on intensional containment (analytic a posteriori judgment), that of formal concepts on axiomatic deduction (systemic "containment"). Only the first, analytic a posteriori judgments, may be what Kant calls analytic and synthetic judgments. In the light of the Kantian *Logic* this distinction is merely one between judgments whose empirical subjects are definitionally and those whose empirical subjects are descriptionally determined.

The Kantian *Logic* enables us to put analytic and synthetic concepts under a microscope to find out how they are related. We have already seen how they differ; now we must see how they resemble each other. Since we want to make the transition from a philosophy to a new science, we have to see how analytic concepts can become synthetic concepts. How do the latter arise out of the former, or how do the former lead to the latter? How, specifically, is material predication related to formal relation? How does science arise *out of philosophy itself?* [31]

Science, we have noted, uses concepts of generality *with* precision while philosophy uses concepts of generality *without* precision. The intension of the concept "motion" in Galileo is *precise* while it is imprecise in Aristotle. Precision itself is an intensional concept, and we must go back to Kant to find a logical account of it. Actually, the problem is the old Pythagorean-Platonic-Aristotelian one of limitation, of the *horismos* — the demarcation of territory, be it a piece of land or a realm of thought.

Kant was one of the last to investigate it within a text of logic. Precision hinges on the difference between analytic and synthetic concepts — the former those of everyday discourse and of philosophy, the latter those of science. The analytic concepts are *given*, either a posteriori, as, for example, the concept "water," or a priori, as, for example, the concept "substance." The synthetic concepts, on the other hand, are constructed by the human mind, as, for example, the concept "circle." The precision of any concept consists in the complete determination of its meaning in a minimum of terms, "*conceptus rei adaequatus in minimis terminis, complete determinatus.*" Given concepts, whether a priori or a posteriori, cannot have *complete* determination because with them it can never be ascertained whether the concept has been exhausted. Since the concept is one thing and what it refers to — the concrete or transcendental reality from which it is derived — is another, in the latter there may be elements not included in the former (the accidental and the noumenal, respectively). Only with constructive concepts can one be absolutely certain that they contain all that their object does, for these constructs come about together with, and actually are, their object. They possess *complete* precision, for they are creations of the human mind itself rather than abstractions. Thus philosophical concepts cannot possess the same precision as scientific concepts.

Kant does not explain in what exactly consists the major precision of synthetic concepts nor what is the exact logical relation between the two classes of concepts. But three fundamental results follow from Kant's *Logic*, together with observations he makes in the Transcendental Doctrine of Method in the *Critique of Pure Reason:* first, that there *are* these two kinds of concepts; second, that there is between them this difference of precision; third, that there is within the class of *analytic* concepts a hierarchy of precisions. This hierarchy begins with mere *Description*, which lacks all precision and "has no rules and only contains material for the definition"; continues in *Exposition*, which is "the successive representation of the properties of the concept found through analysis"; and ends in *Definition*, which is "the concept completely determined in a minimum of terms." Since the concept here in question is the analytic one, it can never be *completely* distinct and precise; for "one can never know by any proof that all the properties of a given concept have been exhausted by analysis. Therefore all analytic definition must be considered uncertain." [32] Hence the lack of precision in philosophical concepts; they are but a class of *given* concepts, the same as the empirical ones of everyday life. They are, as philosophers of modern science have expressed it, nothing more or less than abstractions from common sense.

Synthetic concepts cannot be arranged in a hierarchy of clarities because they are transparently clear by their very origin. They are "*made* precise" while analytic or philosophical concepts are "*to be made* precise" through analysis. How one arrives at synthetic concepts, as for example, the axiom of mathematics, Kant does not tell us, though already Galileo had told it, and we find it repeated by contemporary scientists and philosophers.[33] The logic of Kant gives us the foundation, but not the details of an intensional logic. It gives us the hierarchy of analytic clarities which provides us with an analysis of intensional rather than extensional precision: with clarification of intensional *Teilung* (analysis) rather than classificatory *Einteilung* (division).[34] But it does not help us to understand the *process* which transforms the "uncertain" precision of analytico-philosophical concepts into the "certain" precision of synthetico-scientific ones: the *logic* of the process which led from Aristotle to Galileo, from natural philosophy to natural science. The process which leads from moral philosophy to moral science must be logically and methodologically the same.

To understand it we must turn to Kant's predecessors, such as Descartes, or successors, such as Cassirer. There we find that *synthetic concepts follow the hierarchy of precision of the analytic concepts*. The differentiation of the given material, which leads from Description to Exposition to Definition, does not stop at the last. Rather, at this point — analytic definition — it rests, to take the leap into synthetic construction. The creative scientist must have exhausted the analytic concept to the point at which he is able to condense even the "minimi termini" of the analytic definition to one single term which, as an axiom, serves as point of departure for a system — illuminating the original darkness as the beacon of a lighthouse the waters of the sea. Description, so to speak, cuts out from the darkness of the unknown, the bottom of the sea, certain materials which it drags to the surface and builds with them a base for knowledge, the base of the lighthouse. Exposition selects from this base those materials which are relevant to the problem and builds a higher and more exact structure. Definition further narrows the problem and, with a minimum of materials, puts the roof on the building. Construction then grasps the unique essence of the whole work, tops it with the searchlight, the axiom, which it expands into a System, the beacon. This, in turn, illuminates vast new regions of the original darkness. Or, to use a different simile, the procedure is like an hourglass. As the sand filters down from the broad upper rim, the analytic, by description and exposition to definition, one corn, the axiom — the absolute simple in Descartes' language — enters the narrow waist, followed by all the other corns, which now replenish the lower half, the synthetic, and present the analyzed reality "in a different order" (*Regulae,* XII).

This procedure has phenomenological and psychological implications which have been widely discussed and do not need to concern us; but its *logical* implications must be stated.

The difference between the uncertain precision of the philosophical concepts and the certain precision of the scientific ones is, as we have seen, the different structure of the intensions of these respective concepts. The intension of the analytic concept contains within itself other concepts equally abstracted each representing a scale of abstraction of potentially tremendous dimensions. Thus, to mention again the Aristotelian concept of movement, defined as "the transition from potentiality to actuality," if we want to understand it we must explain the three concepts contained in it and subsequently the concepts contained in these explanations, and the concepts contained in the explanations *of* the explanations, and thus successively until we arrive at that of which the whole chain is predicated, some individual whose analytic intension is the totality of all the explanations given. The exact structure of this process is not known — the classical attempts at "division" are unsatisfactory. We have said that this procedure is similar to moving from one definition of the dictionary to the next. In the present case, the dictionary should be a philosophic one. By working his way from definition to definition, any intelligent layman could write a treatise about movement — or any other philosophical subject — once he is given but one definition. Logically, this process represents a specification or analytic deduction, the process opposed to abstraction: that of extracting from a given concept all the concepts contained in it, and of the contents of these the contents contained in them, and so on — a process practically without end, fundamentally different from synthetic deduction which is strictly ruled by the system in question.

The analytic definition, seen in analytic purity, is like an iceberg whose larger portion, the process of generalization, is submerged. Kant well says that the philosophical argument ought to lead *to* definition rather than start *from* it — as does the lexical method.[35] To develop a philosophical concept one must develop all the implications of the concepts contained in it. Through such development — like the gradual peeling, skin by skin, of an onion or the finding of a box within a box of a nest of Chinese boxes — we develop what is called a philosophy. A philosophical "system" is no more than the development of an analytic concept.

A scientific system is very different. The "content" of a synthetic concept does not consist of concepts which contain concepts containing concepts. It consists of *terms which are related to terms*. The model of a synthetic intension is a network rather than an onion or a nest of Chinese boxes. The concepts of an analytic intension do not, as we have seen, have any intrinsic relation one with another; and this, precisely, is what makes them logically equal to the concepts of everyday life. The concepts contained in the definition of "man" — "animal" and "rational" — have no intrinsic relationship to one another, that is, a relationship which arises from their own meaning — the concept "animal," as such, does not necessitate the concept "rational" nor inversely the concept "rational" that of "animal." Rather, the relation between the two is based on the referent of the two concepts, which is also the referent of the concept "man." Thus, as Aristotle observed, the unity of the (analytic) definition is based on the entity defined.[36] This means that the concepts contained in an analytic definition are interrelated *vertically*, by the totality of abstractions which reach from the entity defined up to these concepts or, inversely, by the totality of specifications which reach downward from these concepts to that entity. The last Chinese box, so to speak, is the entity. It is, as Aristotle rightly held, inaccessible to (analytic) cognition, which begins with the welter of properties Kant calls description. The entity to which it belongs is, as such, the unity of these properties; it can be grasped only by an intuition which is nondiscursive. The last Chinese box is never reached; it is continuously approached.[37] In our simile of the lighthouse, it is at the bottom of the sea. When the creative imagination of the scientist leaps from the analytic to the synthetic, when it puts the light on the tower, it must first have dived by intuition to the bottom. Otherwise the light will have no power, and instead of illuminating the depths will idly play on the surface. Our (analytic) cognition only reaches down to the level of Description — the base of the lighthouse, the simple properties of things. Once there, it can only take soundings of the depths from which the materials of the base have been dragged up; and it can never, as cognition, dive into it. We need a different power for that. For our cognition, the iceberg of which we spoke has infinite depth; the onion is continuously being peeled, its core is infinitely small.

Thus, analytic thinking is deeply frustrating, its only salvation being synthetic thinking. Here the depths seem to be skirted; but their contents are brought up and examined by minute methods of intellectual microscopy and spectroscopy. Synthetic thinking is, so to speak, the microchemistry of analytic thinking. What seems inaccessible to analytic thought, e.g. the cognitive nature of the individual, becomes a matter of course for synthetic thinking.

The terms of a synthetic definition are interrelated *horizontally* in a network of relations. They have no depth but float over the deep, as we have seen. For this reason creators of systems are often reproached for the facility with which they solve "profound" problems. Sagredo expresses this when he says to Salviati in Galileo's *Two New Sciences:* "You present these recondite matters with too much evidence and ease; this great facility makes them less appreciated than they would be had they been presented in a more abstruse manner." Salviati answers to the effect that "profundity" is only an alibi for fallacious thinking.[38]

The difference between *term* and *concept* is that the term has neither intension nor extension.[39] All its significance derives from its position within the system, its interrelationship with other terms. The term is so formal that it does not refer to anything — it is neither abstract nor abstracted, it is *constructed.* The only "reference" to reality which concerns it is the applicability of the network, of which it is a part, to a *totality* of phenomena which through this application become interrelated but do not derive any individuality from it. The term, in other words, is not a name referring to anything; it is a *variable* which can be applied to any individual whatsoever that fits into the network of interrelations in question. As we have seen, in the equation for "motion" of Galileo a *general name* — Aristotle's "movement" — was transformed into a relation between two terms, the arithmetical division $\frac{s}{t}$. In a logically higher system, this relation may be transformed into a mathematically higher expression, as is Galilean "motion" in subsequent physical systems. Also a *proper name* can be transformed into a relation between terms — when it is regarded as an infinity of properties and "infinity" is interpreted in the mathematical sense as a transfinite number.[40]

For example, let us transform a general name and a proper name (or a general concept and a singular concept) into terms of a system, the general concept "man" and the singular concept "Socrates." The concept "man" *refers to* all men, including Socrates. Every member of the class of men, including Socrates, fulfils the analytic definition of "man," say, "rational animal." But in saying all this we do not speak of either man or Socrates — we say nothing about the nature or evolution of man, or the life and activities of Socrates. We discuss the *logical interrelationships* of the *terms* "man" and "Socrates." We could express all this in a syllogism: "All men are rational animals, Socrates is a man, therefore Socrates is a rational animal." Here "man," "Socrates," and "rational animal" mean nothing more but knots in the network of logical relations which is the syllogism — they are terms, not concepts. The *term* "man" does not refer to any man — it has no extension — nor does it mean anything — it has no intension — as has the analytic concept "man." Even this, of course, is more abstract than the actual being to which it refers. "If I say, 'I met a man,' the proposition is not about *a man:* this is a concept which does not walk the streets, but lives in the shadowy limbo of the logic-books. What I met was a thing, not a concept, [this] actual man with a tailor and a bank-account or a public house and a drunken wife."[41] The only "meaning" of the term is the relation in which it stands to the other terms. The same is the case with the *terms* "Socrates" and "rational animal." Therefore all these terms can be replaced by any other terms, if only the relations are maintained in which they stand to each other. The only thing that counts are these relations. Thus, we may just as well write "All featherless bipeds are rational animals, Socrates is a featherless biped, therefore Socrates is a rational animal." Here we have replaced the term "man" by "featherless biped," but the relations have been maintained. In the same way all the other terms can be replaced. Actually the *form* of the syllogism does not need any terms at all. They can all be replaced by *symbols* as long as the form of the whole, the network of interrelations is maintained. Hence, "all *S* are *P*, X is an S, therefore X is a *P*" is as valid a syllogism as the first, indeed logically more valid. And if we also replace the words which indicate the relations — "all," "are," "is a," "therefore" — we arrive at the symbolic expression of the syllogism within certain systems, such as "AAA," "EAE," etc., "$(a,b,c) : a < b \,.\, b < c \,.\, \supset .\, a < c$," "$\vdash :. q \supset r \,.\, \supset : p \supset q \,.\, \supset .\, p \supset r$," etc. These symbols can in turn be manipulated according to their own laws and thus originate new forms, such as "$(a,b,c) :. a - b = 0 \,.\, b - c = 0 \,.\, \supset : a - c = 0$," "$\vdash : (x) \,.\, \phi x \supset \psi x : (x) \,.\, \psi x \supset \chi x : \supset .\, (x) \,.\, \phi x \supset \chi x$," which elaborate the original syllogism in new aspects without having to take into account this original: the symbolism itself takes over and pushes thought according to its own laws. The formal relations, by their very essence, by their *being relations,* advance thinking, for the form of thinking is, precisely, relating.

The predicative analytic concept is then nothing more than the first, still-sensible stage of thinking. It must be overcome by the second stage, that of synthetic relation — a transition which has been made in natural philosophy and is to be made in moral philosophy. Galileo declared over and over that his mathematical thinking enabled him to advance thought without referring to the original sense observation. "The knowledge of a single fact acquired through a discovery of its causes prepares the mind to understand and ascertain other facts without need of recourse to experiment, precisely as in the present case where by argumentation alone the Author proves with certainty that the maximum range [of the shot] occurs when the elevation is 45°. He thus demonstrates *what has perhaps never been observed in experience,* namely, that of other shots those which exceed or fall short of 45° by equal amounts have equal ranges."[42] In the same way we must free ethical thinking from "common sense" and give it the wings to soar up to its own realm. Its present analytic concepts drag it along in the welter of sense observation. Only its transformation into synthetic relations can give it the necessary uplift. The intellectual currents are around to lift it — the systematic relations of axiologic — and have been around for a long time. Plato, Leibniz,[43] and others have felt their breeze — but none has yet set out seriously to fashion the aerodynamic structures appropriate to them.

The decisive step is the transition from concept to term,

from analytic to synthetic relations. Even the term is only a make-shift, a place-holder, a lieutenant for the symbol. The logical "meaning" of the term is formal: the term is nothing but a point where relations cross.[44] Its entire meaning lies in the fact of this crossing, the relations in question. The point of intersection contains these relations; it is the differential of the relationship. Strictly speaking, as long as the syllogism contains terms it is no statement of logic. It becomes one only when the terms are replaced by symbols. "No proposition of logic can mention any particular object. The statement 'If Socrates is a man and all men are mortal, then Socrates is mortal' is not a proposition of logic; the logical proposition of which the above is a particular case is: 'If x has the property of ϕ, and whatever has the property ϕ has the property ψ, then x has the property ψ, whatever x, ϕ, ψ, may be.' The word 'property,' which occurs here, disappears from the correct symbolic statement of the proposition; but 'if-then,' or something serving the same purpose, remains."[45] Less strictly speaking, of course, propositions containing terms are propositions of logic, the terms taking the place of the symbols. Thus, "Australia is large" is a proposition of geography, but "Australia is large or Australia is not large" is one of logic.[46] In the first, "Australia" is a concept, in the second it is a term.

The difference between analytic and synthetic concept, then, is that the former is material while the latter is formal. The analytic concept is material in the sense that its intension consists of other analytic concepts each of which has meaning in its own content. The synthetic concept is formal in the sense that its intension consists of terms which have no meaning in themselves but only in the system of which they form part. The term is a point connected with other points, the concept is a volume containing other volumes which in turn contain other volumes, and so on. Yet, as we have seen, the "material" analytic concept becomes practically more and more irrelevant in the degree that it develops while the "formal" synthetic concept becomes in the same degree practically more and more relevant. We know the logical reason for this, the difference in the extensional-intensional relationship of the two concepts. Let us now try to discover the genetic reason: what is the connection with reality that the synthetic concept has and the analytic concept lacks?

From the analogy of a term as a point and an analytic concept as a set of volumes there derives a profound analysis of the relation between the analytic and the synthetic concept, or, as we may say, the analytic concept and the term.[47] If the concept is like a volume that contains volumes which in turn contain volumes and so on, *these volumes must become smaller and smaller and infinitely approach toward a point* — and this means: a *term.* Infinite sets of analytic implications, in other words, approach toward one term as their limit. Here we see that a "term" is the limit of the total meaning of a total set of concepts — and "limit" was the original meaning of "term." [48] In this analogy *the scientific concept is the ideal limit of the more and more intensive specifications of philosophical concepts.* Indeed, as anyone familiar with the notion of ideal limits will immediately conclude, the scientific concept *is* this series of more and more intensive specifications.

This explains at one stroke the power of the synthetic concept. The system of terms it signifies represents a whole realm of phenomena and their infinite analytic possibilities. At the same time it explains the procedure by which to arrive from analytic to synthetic concepts: infinite concentration of analytic content. Our result ties together the psychology of scientific creation, as witnessed by the testimonies of creative scientists, who "dwelled in" their theories,[49] with theoretical investigations in the relation between the world of sensible fact and the world of symbols that is science.

The analogy of an infinite approximation of volumes to a point has been elaborated in a particular logico-mathematical aspect by Whitehead. A generalization of his procedure gives us the result just mentioned, and thus the relation we are looking for, that between concept and term — between philosophical and scientific thinking, which is fundamental for our subject. Whitehead's method is that of "extensive abstraction," based on the principle of "convergence to simplicity with diminution of extent." It is, fundamentally, an application of the relation between whole and part to the relation between the empirical world of common sense and the constructed world of science. The former consists of "thought-objects of perception," the latter of "thought-objects of science." The latter arise out of the former as ideal limits of "convergent sets of enclosure objects," and indeed, *as* these sets. Simply expressed, Whitehead formulates the enclosure relation of a nest of Chinese boxes as a *logical* relation drawing all the consequences of this formulation. Groups of sets of *temporal* enclosure objects converge to, and define, a *moment,* groups of sets of *spatial* enclosure objects converge to, and define, a *point.* "Moment" and "point" are *defined as* the corresponding groups of convergent sets. Whitehead's procedure may be applied also to *concepts.* Groups of sets of *conceptual* enclosure objects converge to, and define, a *term.* Thus the three levels of science — in Whitehead the "first thought-object of perception," the "second thought-object of perception," and the "thought-object of science" — become connected by a logical relation. The result is the one mentioned, that *sets of conceptual enclosure objects converge to and define a term.* We may call this the method of "intensive abstraction" based on the principle of "convergence to simplicity with diminution of content (or intension)."

This method has far-reaching consequences for logic, phenomenology, and axiology. In logic, the classical treatment of the problem is that of division leading to essence. It can be shown that this Platonic-Aristotelian train of thought corresponds to the interpretation of Whitehead's argument given.[50] More obvious, and axiologically more important, is the connection with phenomenology. Our interpretation of Whitehead's procedure shows up this procedure as *identical with the method of phenomenological reduction.* The structure of the process of *Wesensschau* (as given, for example, by Nicolai Hartmann) presents the salient features of the method of "intensive abstraction": the limital nature of the ideal object, the "concentration toward a point" (*punktuelle Konzentration*) which leads up to it, and the leap from the totality of "categorial elements" which make up the *Hinleitung* to the immediate *Schau* itself, in an act *sui generis,* which abstracts from yet represents this totality.[51]

The term thus reached is, of course, not any term within a system but *the* term *of* the system, the term of terms from which the system itself and all its terms originate. It is the matrix of the system, the *axiom.* The process described is that of axiomatic identification. The infinite analytic aspects of the phenomenon all "come down to" the one limiting concept which contains the logical structure of the phenomenon and, as such structure, is equal to a logical element.

The Greek expression *axiom,* then, meaning that which is worth being thought, is the limit, in the precise mathematical sense, of an infinite totality of analytic contents. The worth of thought, that which gives thinking its intrinsic value, is this concentration, this com-prehension of an infinite multitude of categorial contents, empirically abstracted, in one single point infinitely remote from all abstraction. It is the con-struct, the intrinsic meaning of all abstraction. The axiom, thus, is of a different dimension than the category, in the exact mathematical sense that *one axiom is worth an infinity of categories.* The intension of an axiom, a synthetic system, is worth an infinity of intensions of a category, or sets of analytic concepts. If value is, as we define it to be, the fulfillment of an intension, then we have here two different kinds or dimensions of value: the fulfillment of an axiom and the fulfillment of a category. The former would be a value infinitely greater than the latter. Since, as we have seen, the world fulfilling a system is a world of science and the world pertaining to, and fulfilling categories is one of philosophy, the former world adds infinite value to the latter and makes it infinitely richer.[52]

Moreover, as we now see, there arises from the distinction between axiom and category, and their exact mathematical relation, a hierarchy of values. If we call the fulfillment of an axiom *intrinsic value* and that of a category or concept *extrinsic value* — a thing's goodness in its kind rather than in itself — *then each intrinsic value is the limital point of an infinite set of extrinsic values.*

Our interpretation, then, ties together phenomenological, epistemological, psychological, and axiological discussions, at present widely dispersed and intellectually separate, in a unified picture which may be called a *mathematical logic of axiomatic discovery:* A logic of the worth of thinking as well as the thinking of worth — *a logic of value.*[53]

Thus, Whitehead and Nicolai Hartmann give implicitly the logical details of the relation between analytic and synthetic concepts, which we find examined explicitly in Kant and Ernst Cassirer. Cassirer, in particular, has discussed the difference between conceptual analytic and symbolical synthetic thinking, and shown the significance of the transition from the one to the other. Symbolic thought reflects the conceptual content in an entirely new medium, that of formal relations, which have their own laws. Following them, thought can do without the original sensible perception of the phenomenon. The symbolism, the calculus, advances thought in a new dimension. The exact relation between the two dimensions is to Cassirer "the mystery of intellectual activity itself"; and although he speaks of the symbol's "power of condensation and concentration" and presents the process of intensive abstraction, so to speak, metaphorically,[54] it is only with Whitehead and Nicolai Hartmann that we learn its exact nature: the formal is the ideal, the infinite ultra-microscopic limit of the material — indeed it *is* the entire empirical process that leads up — and down — to it. Inversely, the material contains the formal as a germinal point[55] within itself.

Naturally, we do not have to take either Cassirer's or Whitehead's words. We can see with our own eyes that the core of the material world is found in some formulae — which are nothing but small black signs on white paper — of Newton, Einstein, and others, and that from these symbols arose the technological world of today and its nuclear power. The whole development began with Galileo, in whose mind the fundamental transformation of which we have spoken took place, and Aristotelian concepts became mathematical symbols. It is thus beyond doubt that the difference between philosophy and science is defined by the difference between abstractions *from* the commonsense world and constructions of ideal relations *applicable to* this world. It is equally beyond doubt that the world will not be morally efficient unless the same transition takes place in moral philosophy.

Value language must proceed from today's Aristotelian abstractions *from* commonsense value notions to the construction of a system of axiological relations *applicable to* the commonsense value world. The analytic determinations of value must all be brought down to the axiom of value science.

The process of "intensive abstraction" in the field of value was begun by G. E. Moore. He dealt with all possible analytic value-determinations and showed that *none of them could serve as a definition of value;* for each of them defined something other than value (the naturalistic fallacy). He showed, secondly, that the definition of good could not be analytic, and that anything analytic could not be a definition of good.[56] Thirdly, he gave a synthetic determination of value based on "*one* of the senses" of the notion of "description."[57] But this determination was logically not sufficiently clear for him to bring about the axiomatic identification of value with an element of logic. Thus, he was unable to found the Science of Ethics to which he wrote the Prolegomena. In the following Part we shall bring about this identification, developing the axiom of value from Moore's close approximation to it.

PART TWO

The Foundations of Value Science

3

THE CONCEPT OF AXIOLOGICAL SCIENCE

Ethics in all its forms comes down to the question, what is the principle of the "Good." . . . Positive morals cannot be appealed to, for each answers the question materially in a different manner. One sees happiness, another satisfaction, a third justice, a fourth love as the Good. . . . Philosophy has early recognized the complete onesidedness of positive morality and consequently searched for the Principle of the Good as something more general, superordinated to these fragmentary insights. It was looked for as the genus to the manifold of the species. The Platonic "Idea of the Good" was the most radical such attempt. But what is the content of such an "Idea of the Good"? It has been looked for in vain. Neither Plato nor any later philosopher has been able to determine it.

— NICOLAI HARTMANN

I have endeavoured to write "Prolegomena to any future Ethics that can possibly pretend to be scientific." In other words, I have endeavoured to discover what are the fundamental principles of ethical reasoning. — G. E. MOORE [1]

1. *Generic and Specific Value*

Any understanding of value, whether categorial or axiomatic, presupposes the answer to the question: What have all value phenomena in common? What, in other words, is the genus of which the various value phenomena are species? Before asking ourselves what is the relation between generic and specific value we might ask ourselves what is the relation between the generic and the specific itself. Before discussing the axiological relation between the general and the specific we might discuss the logical relation between the two, the axiological relation being a specific case of the logical. But in order to bring the discussion more directly to our subject, we shall begin with a specific case of the logical relation which is *not* the axiological but which throws light upon this relation. We mean the ontological relation between genus and species: the relation between Fact in general and the specific facts. Facts are better known than values, so this approach will at least clarify our method of inquiry.

There are three levels of fact: the generic (or conceptual) fact; the particular fact; and the singular fact.[2] Obviously, the three levels of fact correspond to a scale of richness in kinds of fact, the singular fact being richer in qualities than the particular, and the particular richer in qualities than the generic. The singular fact has the full concreteness of all its properties. The particular fact has only the properties of the class to which it belongs. The generic fact has only the property or properties contained in the definition of the concept "Fact." The concrete fact thus has a richer intension than the particular fact, and the particular a richer intension than the generic. On the other hand, and corresponding to the scale of intensional richness, there is an inverse scale of extensional richness: the generic fact is richer in extension than the particular, and the particular is richer in extension than the singular fact.

Both particular and singular are facts by virtue of possessing the property or properties which define the concept "Fact." What is this property? We shall define a fact as anything that is. The property, then, that fact in general possesses is "is-ness" or being. This definition converts the logical relation between the generic and the specific into the ontological relation between the two. A particular fact has the generic property of being, and in addition specific properties which the class in question gives it: there are physical, psychological, social, historical, economic, ethical, botanical, geological facts, etc. The class of classes of particular facts is infinite. The class to which a particular fact belongs is defined by certain material determinations abstracted from a particular field of phenomena.

The singular fact has the properties of fact in general, namely being, plus the properties of a particular class, plus an infinity of properties which the individual has uniquely. This threefold enrichment of fact is an enrichment of generic fact, that is, of Being. And Being is specified in the increasing scale of intensions and the decreasing scale of extensions. In this scale, it may be held, the particular is to the singular as the generic is to the particular. The scale is therefore a scale of increasing specificity.

This scale would be exact if the status of a particular or a singular fact within it could be read off by the degree of its intension or the extent of its extension, as, for example, the status of a particular condition of temperature within the whole range of temperatures, can be read off the thermometer. But, in spite of attempts by philosophers from Plato to Leibniz, Lambert, Kant, Lotze, Erdmann and others, such an exactness of the intensional-extensional scale of Being has not been achieved.

The relationship between fact in general and specific fact would be exact if there were a definition of fact in general such that application of the definition to a specific set of phenomena would not only establish this set as a specific class but would also establish the specificity of the class, that is, its own specific definition. In the definition of Fact would be contained in systematic detail the totality of all facts: the definition of fact would be capable of being expanded into a system isomorphic with the total realm of facts.

Obviously, the definition of Fact as Being is not such a definition, hence ontology is no science but a field of philosophy. A definition that would give rise to a system isomorphic with a phenomenal field would give rise to a science; and such a definition would be an axiom. (We defined an *axiom* as a formula that gives rise to a system isomorphic with a field of phenomena, and a *science* as the combination of the intension and the extension of an axiom.) If the intension of an axiom, the system, is isomorphic with the extension of the axiom, the realm of phenomena in question, then the extension and intension of an axiom do not vary in inverse proportion, as do those of a concept or a category of philosophy, but in direct proportion. Thus, a scientific ontology would be one where Being is defined not as a category but as an axiom. In the corresponding system, Fact in general would be a variable, and both particular and singular fact would be values of the variable.

What has been said of generic and specific fact can be applied *mutatis mutandis* to generic and specific value.

Obviously, Value in general is that kind which corresponds to the concept "Value," while a specific value is either a particular or a singular value. So there are three levels of value: the generic (or conceptual) value; the particular value; and the singular value.[3] These correspond to a scale of richness in kinds of value: the singular value is richer in qualities than the particular, the particular is richer than the generic. The singular value has the full concreteness of all its properties, the particular value has only the properties of the class to which it belongs, the generic value has only the property or properties contained in the definition of the concept "Value." The concrete value has a richer intension than the particular, the particular a richer intension than the generic value. On the other hand, and corresponding to the scale of intensional richness, there is the inverse scale of extensional richness.

The particular and the singular value are values because they possess the property, or properties, which define the concept "Value." What is this property? Let us call it, for the moment, X, and let us call the discipline which deals with this kind of value Axiology. A particular value has the generic property X and in addition the specific properties which the class in question gives it: there are physical, psychological, social, historical, economic, ethical, aesthetic, religious values, etc. The class of classes of particular values is infinite. The class to which a particular value belongs is defined by certain material value determinations abstracted from a particular field of value phenomena. The singular value has the property of Value in general, namely X, plus the properties of a particular class of values, plus an infinity of properties which the individual value in question has uniquely. This threefold enrichment of value is an enrichment of generic value, that is of X, and X is specified in the increasing scale of intensions and the decreasing scale of extensions, where particular is to singular value as generic is to particular value. The scale is therefore a scale of increasing specificity of value.

This scale, too, would be exact if the status of a particular or a singular value within it could be read off by the degree of its intension or the extent of its extension. But in spite of attempts by philosophers from Plato to Plotinus, Spinoza, Kant, Lotze, and others, such an exactness of the intensional-extensional scale of X has been achieved no better than that of the scale of Being.

The relationship between Value in general and specific value would be exact if there were a definition of Value in general such that application of the definition to a specific set of value phenomena would not only establish this set as a specific class of value but also establish the specificity of the class, that is, its own specific definition. This means that in the definition of Value would be contained in systematic detail the totality of all value: that the definition of Value was capable of being expanded into a system isomorphic with the total realm of values.

Obviously, the definition of Value as X is not such a definition; hence axiology is no science but a field of philosophy. A scientific axiology would be one where X were defined not as a category of philosophy but as an axiom of science. In the corresponding system, Value in general would be a variable, and both particular and singular value would be values — in the logical sense — of the variable.

We are interested not in the philosophical but in the scientific definition of value. We therefore want to determine X as an axiom, not as a category and develop the axiomatic definition of Value into a system, such that Value in general will appear as a variable of which the specific values are logical values. Value then becomes that variable the logical values of which are axiological values.

There have been several attempts in the history of philosophy to determine Value in this way, both implicit and explicit. "Good," as axiological writers from Aristotle to J. O. Urmson and Paul Edwards [4] have suggested, may be conceived as a homonym applicable in many different contexts, with different criteria for its employment in each. This is an exact, though implicit, description of the logical nature of a variable. Explicitly, philosophers from Plato to Descartes, Leibniz, Spinoza, and Kant, have tried to develop moral philosophy into moral science. Before the minds of the philosophical founders of natural science there stood with great clarity the vision of a science of Value or, as it was then called, a Moral Science, to be established by the side of natural science and based, like it, on the *mathesis universalis,* which today is called logic.

These attempts failed, mainly because generic value, particular value, and singular value had not been distinguished. Incredible though it may seem, generic value was constantly confused with particular value; it was defined in terms of a species of value, such as moral or psychological value. The reason is not a special opacity of thinking on value, but of philosophical thinking itself. As long as value was regarded as a category no exact relationship between generic and specific value could be established. The logical nature of the category and the inverse proportion of its intension and extension made the confusion almost inevitable. Only G. E. Moore, in 1903, made clear the range of this confusion throughout the whole of ethics, indeed, the whole of value theory. Any attempt to define value in general by specific kinds of value — ontological, teleological, epistemological, etc., as perfection, purpose, function, knowledge, God, pleasure, self-realization, preference — is such a confusion. Moore calls this the naturalistic fallacy. As we have seen, this is a fallacy not only of ethical but of all prescientific reasoning. Moreover, as we shall see, the naturalistic fallacy is but one of a cluster of fallacies found in any philosophy when it is contrasted to the corresponding science. Once there is such a science such confusions are impossible, for axiom and system make clear both the difference between the system itself and the phenomena to which it applies, and the method of application. The relation between the generic and the specific is itself specified.

It is thus impossible to deny that what Moore calls a fallacy really is a fallacy; for this would be to deny the logical or axiological possibility of a genus Value. This would mean that there is no value realm at all but a chaos of empirical determinations which could not validly be called by the generic name "Value," and the species of values could then not properly be called by this generic name. It would then be nonsense to speak, for example, of *economic value;* for the genus would be unknown and the differentia would be senseless. It would be like speaking of green flying saucers. If we wish to specify the relation between genus and species of value we must leave the realm of the categorial and enter the realm of the axiomatic. There is no other way, but to follow G. E. Moore.[5]

Moore did not produce a science of value, not because he confused the levels of generic and of specific value but because he did not know how to define generic value. This definition to him was an unknown, an X. In his first book he even held it *ought to be* an X, for pragmatic reasons, so to speak:

> If . . . we once recognise that we must start our Ethics without a definition, we shall be much more apt to look about us, before we adopt any ethical principle whatever; and the more we look about us, the less likely are we to adopt a false one. It may be replied to this: Yes, but we shall look about us just as much, before we settle on our definition, and are therefore just as likely to be right. But I

> will try to shew that this is not the case. If we start with the conviction that a definition of good can be found, we start with the conviction that good *can mean* nothing else than some one property of things; and our only business will then be to discover what that property is. But if we recognise that, so far as the meaning of good goes, anything whatever may be good, we start with a much more open mind.[6]

It is true, Moore also gave a kind of logical principle for the indefinability of good, but he recognized later that most of what he said on that score was false. He never gave the right reason, though, for this falsity. He said "good" was a simple, unanalyzable concept. Actually, it is a philosophical category, and as such it is unanalyzable because of the poverty of its intension and the infinity of its extension. Goodness is unanalyzable for the same reason that Being is. There is no property that can be attached to it except on the same logical level, as the scholastic transcendentals, and such attribution makes no logical sense. The transcendentals are attributes of Being; hence, according to the theory of types, they are of a higher type than Being. But if Being is the *summum genus*, then either the transcendentals are equal to Being and there are several highest genera, or they are higher than Being, then Being is not the highest genus but the transcendentals are, and again there are several higher genera.

As Being is to Fact, so X is to Value. X is the highest value genus. If it could be defined it would have to be the species of a higher genus, and this by definition is impossible. "Good" is indefinable *as a category*; it is *not* indefinable as an axiom.

Moore must have felt this; for he was insistent throughout that he was writing the prolegomena to a *science* of Ethics. Just as Newton wrote *Philosophiae Naturalis Principia Mathematica* so Moore hoped to write *Philosophiae Moralis Principia Ethica.*[7] Only Moore lacked what Newton had: an axiom, a formula giving rise to a system applicable to the reality in question. Yet, Moore did produce a formula. Although it was not an axiom it can be reformulated to serve as the axiom of value science.

2. *The Axiom of Formal Axiology*

An axiom, we said, is a formula giving rise to a system isomorphic with a field of phenomena. A system we found must always be a *logical* system; there is no other systematization but the logical one. Hence, an axiom must always connect a phenomenal field with a form of logic, either with logic itself, as did Frege and Russell who combined the field of number with logic, or with an application of logic, e.g. mathematics, as is done in natural science. If value theory is to be a science, Value must be determined by an axiom which identifies it with some notion of either logic or an application of logic. X, in other words, must be an axiom. It must be a formula which gives rise to a system applicable to the field of value phenomena.

After 1903 Moore tried to develop such a formula. In 1922 he enunciated one and in 1942 determined it more closely. The formula, to him, was a paradox: "Two different propositions are both true of *goodness,* namely 1] that it does depend *only* on the intrinsic nature of what possesses it . . . and 2] that, *though* that is so, it is yet not itself an intrinsic property." [8] This paradox, says Moore, could be resolved, and thus the nature of good determined, if we knew in which sense the natural properties do and the nonnatural — or value — properties do not describe an object. The key term of the problem, therefore, is the term *description.*[9]

To understand this we must think of the *logical* nature of the naturalistic fallacy. There is a difference between genus and species of values. What is true of the one is not true of the others. Goodness in general is what all specific goodnesses have in common — just as the number three is what all triads have in common. What all specific goodnesses have in common is, precisely, what Moore says is "true of *goodness,*" namely that it does, at the same time, depend *only* on the descriptive properties of the object and *yet* not describe the object at all.

This then is the formula for X, the determination of Value, in its form of Goodness.

The specific goodness of a thing must then be the specific way in which goodness depends only on the descriptive properties of the thing and yet does not describe the thing. Thus, if the thing is a car and I say "The car is good," then I must be saying something which depends only on the descriptive properties of the car yet does not describe the car.

This is obviously true. If I tell someone that I have my car outside and ask him to go there and wait for me, and he asks: "Which one is your car?" and I say: "It is a good car," then he will never find it, because he has received no information about it. I have given him no descriptive or natural properties: white or black, whether two-doored, convertible or sedan, etc. Yet he does know a great deal about my car; he knows it is *a good car,* that is to say, it has a motor, doors, and tires, and an accelerator which accelerates and brakes which brake.

There are, thus, two modes of knowledge, the natural and the valuational, and the task now is to define logically the distinction between them. The valuational, in this interpretation of Moore, does not refer to the individual thing — to my car, which my friend does not know — but to the concept "car" which he does know, or rather to *my car as possessing the properties of this concept. The value predicate "good," thus, is a property of concepts rather than of objects.* When a person understands that a thing "is good," it is not necessary that he know anything of the thing in question; but he must know something of the *concept* of which the thing is an *instance.* He must know *what is an* automobile but he does not have to know what is *my* automobile. The word "good" applies *not* to the knowledge of the *particular* automobile but to the knowledge of the concept "automobile."

Thus, whenever the word "good" is used, there a logical operation is performed. We combine the properties of the *concept* of the thing with the idea of the particular thing that is called good. When we hear of *a good automobile,* we combine the properties of the concept "automobile," which we have in our minds, with the idea of the particular automobile in question. We give to the particular automobile, of which we may know nothing, the properties of automobiles in general, of which we must know something. And thus we do whenever we hear that a thing is "good": we combine the properties of the *concept* of the thing with the idea of the *thing itself.* This logical operation is the *meaning* of the word "good." A *thing is good insofar as it exemplifies its concept.* This means that 1] the thing has a certain name, 2] this name has a meaning defined by a set of properties, and 3] the thing possesses all the properties contained in the meaning of the name. This is expressed in the *definition* of Good, that is, of that which all good things have in common. A *thing is good if it fulfills the definition of its concept.*

This is the Axiom, or fundamental principle, of Formal Axiology, developed from Moore's formula for X.[10]

From it follows the theorem that a thing is not good or

bad, when it does not fulfill the definition of its concept. The car that accelerates when I brake or brakes when I accelerate is "not a good car." The "natural properties," thus, on which the goodness of a thing "depends" are all those properties of it which correspond to the properties contained in the thing's concept.

The "depends on," in Moore's formula, therefore, has as its counterpart a "corresponds to." The value property of a thing depends on the natural properties: but the latter, in a good thing, correspond to the properties contained in the thing's concept. This correspondence is what makes the thing good. Moore saw that the value property *depends* on the natural properties of the thing; but he did not see the condition of this dependence: the logical reference of these properties as giving rise to the value predicate, their *correspondence to* the conceptual properties of the thing. Value, as Moore rightly saw, is *not* a natural property, yet it depends *only* on these natural properties, namely — and this is our addition to Moore's determination of the nature of goodness — *insofar as the natural properties correspond to the properties contained in the concept of the thing.*

Moore's "paradox" arose because the two propositions that "are both true of *goodness*" — that goodness is *not* a natural property *yet* depends *only* on the natural properties of the thing — refer to two different logical levels: the negative proposition — concerning what goodness is *not* — to the thing itself; and the positive proposition — concerning what goodness *is* — to the concept of the thing. The paradox thus finds its solution in the same way as other paradoxes of types. And in the solution is contained the Axiom of value.

Moore's paradox suggests that good is not a property of the thing itself but rather a property of the concept of the thing, namely the concept's being exemplified by the thing. Conversely, a thing is good insofar as it exemplifies its concept. Moore implicitly states this converse of the axiological Axiom. The Axiom is the converse of what Moore states.

This axiom defines Value in general. It allows us to develop a system of axiology isomorphic with the phenomenal realm of value, and thus to specify values scientifically. The axiom is scientific in four respects.

First, it opens up a new field, that of value, by defining the central term of the field, value, in terms of a formal system, namely logic itself. For the relation between a concept, its definition, and its referents is the fundamental relation of logic. Because logic is an exact, elaborate system, it can now be used to explicate value. This procedure is analogous to that of the natural sciences, where mathematics is used to explain nature. The explication of *value* in terms of the system of logic is what we call *scientific axiology* (*axiologic*): the framework for the explication of value phenomena.

Second, the axiom of axiology parallels that of mathematics. The recognition that value predicates apply to concepts rather than to objects corresponds to the recognition that arithmetical predicates — such as "four" — apply to concepts rather than to objects. This was the fundamental insight of Frege, and Whitehead and Russell.

Connected with this is the third scientific characteristic of the axiom: its formal nature. It consists of variables: not of specific values but of a form which determines the specifications of all possible value. It is, as we said, a formula. The term "good" in the axiom is a purely formal term, applicable to anything whatever. This universal nature of "good," we remember, has been observed throughout philosophical history. Aristotle called it the homonymity of the term, the scholastics its transcendental nature, modern value theoreticians its "polyguity" or its "punning meaning," and a modern semanticist calls it the syncategorematic meaning of "good."[11] Formal axiology recognizes in it the logical nature of "good" as a variable.

The fourth scientific characteristic of formal axiology is the most interesting of all, for it shows that to find a standard for measuring value axiology follows the method Galileo used to measure motion. He disregarded the secondary qualities of the phenomenon and concentrated on the primary qualities — that is, qualities amenable to measurement — so that what was measured was not the sense phenomenon of ordinary life with its secondary properties but a construct consisting of primary properties. In value measurement, what is to be measured is precisely what Galileo had to disregard, the ordinary sense object and this object not only *as possessing* its secondary properties, but rather *this very possession* is what measures its value. Hence for value measurement *the secondary properties must be used as primary ones*. The question, thus, is to find the standard which is related to the secondary properties as primary standards — of length, weight, etc. — are to primary properties. What, in other words, is it that contains the secondary properties as, say, the meter contains the centimeter? The answer is: *the intension of a concept.* The concept serves as the standard for value measurement; a thing has value in the degree that it fulfills the intension of its concept — the same result as the deduction from Moore. Value measurement, in a word, is measurement of, and by, conceptual qualities.

3. *Axiological Specificity*

Axiomatic or scientific specificity is richer than categorial or philosophical specificity. The latter is the progression from the general to the particular to the singular, and this progression is not very specific. Axiomatic specificity is specific on all three levels of science: the system itself, the applied science produced by the application of the system and the actual situation for which the applied science accounts.

Let us take as example the field of mathematics. Number appears on three levels, the logistic (*Number*), the mathematical (*number,* such as "3," "4") and the phenomenal (*numbering* or enumeration). The mathematical is the application of the logistic, and the phenomenal is the application of the mathematical. The logistic level is not directly applied to the phenomenal but through the medium of the mathematical. The logistic level consists of the axiom(s) and theorems deduced from them within the system. These theorems may be called specifics of the axiom; the various elements of the system are specific Numbers, or specifications of Number. We may call these *formal specifics.* They specify or differentiate Number in general. They develop in detail the axiomatic definition of Number as the class of all classes similar to given classes: the notion of class, the notion of similarity, which means the notions of relations, domains, fields, of products and sums of classes, and classes of classes, of identity and diversity, unit classes, selection, the ancestral relation, inductive relations, reflexive relations, cardinals, indeed the whole gamut of subjects covered in *Principia Mathematica.* All these are specifics within Number.

In the field of *number* there are other specifics: the definitions of the various numbers as given by the logistic, beginning with one, and the various operations with numbers. Here we have the whole field of pure mathematics as applied logistic: arithmetic, algebra, calculus, geometry, topology, etc., all consisting of specific numbers. These specifics differ from the logistic specifics, for they are mathematical or *theoretical specifics*, specifics of number, not of Number. Number is elaborated in mathematical logic, number in mathematics.

On the phenomenal level we have the application of mathematics to actual situations, applied mathematics or natural sciences. They use whatever parts of mathematics they need. In an engineering preview we find a chapter on mathematics dealing with the various kinds of numbers, the mathematical operations, equations, logarithms; and analytic geometry, trigonometry, calculus, which are then applied to phenomena. In a book on theoretical physics we have a first part, "Mathematical Introduction," with chapters on vector analysis, mathematical representation of periodic phenomena, complex variables, the calculus of variations, etc., and only then follow the physical parts, mechanics, field theory, theory of electricity, theory of heat, structure of atoms and molecules, etc. They are *material specifics.* Their problems can be solved in detail only because of the detailed mathematical framework isomorphic with the phenomenal field in question.

Let us apply this to Value. Here we have Value, value, and valuation. The first is elaborated in the logic of value or formal axiology (axiologistic or axiologic), the second in applied formal axiology or pure theoretical axiology, the third in applied pure axiology or applied axiology, which comprises the various value sciences. We have specific Values, specific values, and specific valuations (value phenomena or value situations): formal specifics, theoretical specifics, phenomenal (material) specifics.

On the highest level, that of Value (or "value"), we have formal axiological specifics: the definition of value in general and in particular; the definition of axiological terms such as "perfect," "good," "unique," "bad," "fair," "no good" and their interrelationships; the explanation of the logical use of these terms; the definition of value relations, such as "better," "worse," "good for," "bad for," "better for," "worse for," "it is good that," "ought"; the structure of axiological propositions and judgments; truth values of axiological propositions; dimensions of value; differences of value languages; the hierarchy of values; the calculus of value; rules of valuation as applied axiology; the universality or particularity of value; its absoluteness or relativity; its rationality or irrationality; its objectivity or subjectivity; the nature of the value world, its goodness or badness; the nature of the moral law; nature of axiological agreement and disagreement, and so on. Here would be found answers to such questions as these: Why ought I to be good? Is it better to be good than bad? Is it conceivable that it would be better to be bad than good? Ought the good to be? Ought what is to be good? Is the best better than the good? If the good ought to be, what about the best? Is the perfect better than the good? Is the perfect better than the best? Ought what is to be perfect? If the perfect ought to be, ought the good not to be? Is the best good enough? Is the best perfect? Is the perfect worse rather than better than the good? Is there any good at all? Is all good relative? What is the value of value? What is the value of fact? Is value? etc.

On the second level, that of pure or theoretical axiology, that is, of the definitions and interrelationships of the specific *values* (aesthetic and economic, ethical and psychological, metaphysical and technological), these must be exactly distinguished — thereby showing that the naturalistic fallacy belongs to value philosophy only and not to value science — and systematically interrelated within the consistent pattern of value logic, isomorphic with pure axiology. To this level belong the interrelationships between the various value sciences, and the value aspects of the natural sciences, e.g. the relation between music and astronomy in Pythagoras, between astronomy and theology in Plato, between theology and chemistry in alchemy, the confusion between ethics and other sciences, the confusion between sciences and their subject matter, statements such as "To be good is to do God's will," "To be good is to be preferred," "To be good is to feel satisfied," "To be good is to be a proletarian"; the value nature of Leibniz's or Spinoza's metaphysics, the value nature of Horney's, Maslow's or Fromm's psychology, the value nature of a value theory, etc. All these are specifics of pure axiology, pure specifics of value or specifics of pure value: theoretical specifics.

On the phenomenal level, *valuation*, are found applications of various specific values to corresponding phenomenal fields. Here we have phenomena of economics and aesthetics, ethics and theology, etc. These are material or phenomenal specifics of valuation, representing applications of pure axiology to its subject matter. The logic of value itself cannot be applied directly to valuation, any more than the logic of mathematics can be applied directly to counting: the *formal* nature of mathematical and axiological logistic signifies that what the logistic says can be applied to propositions of pure mathematics or pure axiology but not to referents of such propositions. Only the propositions indicate referent(s).[12] Thus between a formal system and a phenomenal field must be interposed propositions *constituted by* the formal system and *applicable to* the phenomenal field.

The logistic of Value, then, can be applied to the phenomena of value only through the medium of pure axiology. The application of pure axiology is the value world. It consists of value arguments (value statements, value judgments) and of value situations, all material specifics of value.

In the following, examples will be given for specifics on all three levels of value.

a) FORMAL SPECIFICS OF VALUE

As examples of formal specifics of value we shall examine some of the corollaries of the value Axiom. We shall mention seven, as follows:

Universal and Particular Value.

It is clear that the value defined by the Axiom is neither ethical, aesthetic, metaphysical, nor economic, nor indeed any other *kind* of value but that it is simply Value, formal or axiological value. Thus we avoid the naturalistic fallacy, that of defining value in general by confusing it with value in particular, or of explaining value in terms of natural predicates. We have created our own autonomous axiom for axiological science.

The Absoluteness or Relativity of Value.

The question is whether there is an absolute norm of value, that is, a universal measure, in comparison with which every other value is determined. The answer is affirmative: the universal norm of value for each thing is the thing's own name. *Norm equals name.* Whenever I judge a thing for its value, I compare the meaning of its name with the properties of the thing itself. The absolute norm of *each* value is the Axiom of value.

The Rationality or Irrationality of Value.

This implies that *value is rational*. I can value a thing only if I know it, that is, if I know its name and its properties. That this is true is confirmed by the fact that when we want to value something precisely we call in an expert. The difference between him and us is that he *knows* more about the thing than we do. Thus knowledge and valuation go hand in hand. It follows that the world as a whole, if it is to be judged valuationally, must be understood, and this in turn means that if value is possible the world *can* be understood. In other words, the world itself is rational insofar as it is valuable. This, of course, is the Platonic thesis put in a new demonstrative fashion and based upon a consistent theory. It also means that there is no difference in the *rationality* of the world of fact as it is accounted for in natural science, and the world of value as it is accounted for in moral science. If we are ignorant of a thing factually we cannot value it correctly, and if we are ignorant of the world factually we also misjudge it valuationally.

The Objectivity or Subjectivity of Value.

The axiom of value is *objective*. It is valid for every rational being whatever. Whenever a being thinks rationally, that is, combines concepts with objects, then he will have a term in his language connoting that a concept corresponds to an object and vice versa, and this is the term that in our languages is called "good," "bon," "gut," "jo," "khoroshii," etc. Formal or axiological value thus is objective. But its *application is subjective.* It may be possible that what I call good you call bad, and what I call bad you call good. But this is a matter of the application of axiology and not of axiology itself. If a drunk sees four apples where I see two he does not, for that matter, invalidate mathematics, he only uses it wrongly: he is really confirming that science, just as I do. What he saw did correspond to the number four, and what I saw corresponds to the number two. His mistake was in seeing, not in adding. In the same way, whenever anyone thinks that a thing fulfills its definition he will call it good, otherwise he will call it bad, and thus he confirms axiology. Whether he rightly or wrongly thinks that a thing fulfills its definition, is a different question — not one of axiology but one of its application. Axiological *interpretation* is subjective, axiological *formalization* is objective. Out of the differences in application arise axiological agreements and disagreements.

Axiological Agreement and Disagreement.

Being matters of the application of axiology, axiological agreements and disagreements are subject to axiology and can be classified. These disagreements may be either perceptual or conceptual: one can see the thing wrongly; one can believe that it has another name from what it has; one can misunderstand the definition of the concept; one can wrongly apply the concept to the thing; one can misunderstand the dependence of the value predicate on the natural predicates of the thing, and so on. Each of these cases has again subcases, all of which can be systematically studied. Thus, in every case of disagreement, axiology can be applied and the disagreement defined by it.[13]

Optimism and Pessimism.

One particular kind of disagreement is especially important in moral life, that between optimist and pessimist. The Axiom explains it simply. Anything which under one concept is good because it fulfills the concept may under another concept be bad because it does not fulfill that concept. Thus, as Spinoza observed, a good ruin is a bad house, and a good house is a bad ruin. It is the art of the optimist always to find that concept in terms of which the thing appears good, and that of the pessimist always to find that concept in terms of which the thing appears bad. The thing is always the same; optimism and pessimism appear in the art of naming and hence understanding it. Correct thinking, of course, is applying the proper concept to the thing, and the proper concept is the one which fits the thing and contains all the properties of the thing and no others. Such a concept makes the thing named *good.* Right thinking is finding things *good,* or optimistic thinking. The pessimist, on the other hand, suffers from incorrectness of thought and is, as Peirce has said, not "thoroughly sane." [14] He is out of tune with the world — a world, as we shall see now, which is good.

Goodness or Badness of the World.

A thing is good if it has all the properties of its concept. The proper concept of the world must contain all the natural properties there are, have been, or will be. The world is that which has all these properties and thus always fulfills its concept. Therefore it is good. If a concept of the world is posited that does not contain all the properties there are, then it is not the concept of the world, and wrong thinking results, in the light of which the world is bad because it does not fulfill the concept posited. The *goodness of the world* is, of course, not ethical but axiological goodness. Although the world as such is good, the things in it may, indeed must be, both good and bad; for as we have seen anything that is good under one concept may be bad under another. Badness thus is the transposition of concepts or the incompatibility of things which in themselves are good. The world, thus, axiologically good as it is, contains the maximum variety of good and bad things (cf. Leibniz's pre-established harmony).

b) THEORETICAL SPECIFICS OF VALUE

Theoretical specifics of value are the value sciences and their interrelationships.

Intensions are sets of predicates. The application of the combinatorial calculus to these sets makes possible the exact measurement of value. There are three possible kinds of sets, finite, denumerably infinite, and nondenumerably infinite (with cardinality, n, $\aleph_0$ and $\aleph_1$ respectively). The first kind of sets are called *definitions*, the second *expositions*, the third *descriptions* (or *depictions*). Each of these kinds of sets defines a specific kind of concept; and the fulfillment of each such concept defines a specific kind of value.

Finite intensional sets (definitions) define *formal concepts* (synthetic concepts in the sense of Leibniz and Newton and of Kant's *Logic*). The things corresponding to them are constructions of the human mind, such as geometrical circles. Such things either fulfill their concept or else they are not such things; they either are or they are not what they are said to be. There are no good or bad geometrical circles. A circle lacking a single one of the properties of the concept "circle" is not a circle. Constructions of the human mind thus have only two values, which we shall call *systemic values:* either perfection or nonvalue. This is the model of the black and white valuation of things, the simplest kind of valuation there is. Since it belongs to constructions of the mind it is obvious that when applied to actual things it "prejudges" them — it is the model of prejudice. This kind of thinking is based on the logical category of limitation; the variety of the world is limited to only two distinctions: A and non-A (e.g. white and non-white, Communist and non-Communist). Such persons are limited, *bornées*, value-blind, in the same sense that a person is unmusical who only knows two tunes, the one that is the Star-Spangled Banner and the other that is not. Systemic valuation is the model of schematic and dogmatic thinking.

Denumerably infinite sets of intensional predicates (expositions) define *abstract concepts* and *categories* (analytic concepts in the sense of Leibniz and Newton, and of Kant's *Logic*). Abstraction "draws off" properties common to at least two things. These properties are denumerable, for they must be abstracted one by one (in the process of learning to speak; a striking example is found in the autobiography of Helen Keller), but there is an infinity of such possible properties. Referents of such concepts are the things of the everyday world. Each such thing has potentially an infinite number of properties in common with other such things — depending on the extension of the class — but in practice valuation will turn upon only a few of these properties (a thing with only 10 properties considered for valuation has $2^{10} - 1 = 1023$ possible values; a glass of Burgundy, which, according to an expert, has 158 properties, has 3.6×10^{46} possible values.[15]) What is valued is not the thing in itself but its possession of the class predicates. Fulfillment by a thing of an abstract concept constitutes *extrinsic value*. Extrinsic valuation is the model of everyday pragmatic thinking.

Nondenumerably infinite sets of intensional predicates (descriptions or depictions) define *singular concepts* (also, as we have seen, axioms). Things corresponding to such concepts are unique. Uniqueness is the intensional counterpart to extensional singularity. The predicates of such intensions may be words of infinite meaning, that is, words which may mean any other word in the language. Such words are metaphors. A metaphor is a set of predicates used as a variable. Hence it can, in principle, replace every other word of the language — and even itself as an ordinary word rather than that of a metaphor, as in "a peach of a peach." Since the totality of all possible ordinary languages has $\aleph_0$ words, *each one* of which may signify as metaphor $\aleph_0$ senses or meanings, the total meaning of the metaphorical language is of $2^{\aleph_0} = \aleph_1$ meanings. The metaphorical language therefore is a nondenumerable infinity, a continuum. Since an element of a continuum may itself be a continuum, a metaphor may by itself be a continuum. A conjunction of a finite number of metaphors is a poem or a poetic novel.[16] The fulfillment by a thing of a singular concept, understood in this sense, constitutes *intrinsic value*. Intrinsic value is the valuation of poets and artists, lovers and mystics, magicians and advertisers, chefs de cuisine and politicians, theologians and creative scientists.[17] It is emphatic — and empathic — valuation.

Systemic value, extrinsic value, and intrinsic value are the three value dimensions. They constitute a hierarchy of richness, intrinsic being richer in qualities than extrinsic value, extrinsic richer in qualities than systemic value. "Richer in qualities" is the definition of "better," "poorer in qualities" is the definition of "worse." The contexual definition of "ought" is: "The worse ought to be better." Hence, intrinsic is better than extrinsic value, extrinsic better than systemic value; and systemic value ought to be extrinsic value, extrinsic ought to be intrinsic value. The hierarchy of value is a valuation of value; our science specifies and elaborates systematically what the old philosophy of value only suggested, the extensional-intensional scale of richness.

The value dimensions are *formal* specifics, they belong to value logistics, but their applications to fields of value phenomena constitute the specific values of pure axiology. These specific values are *theoretical* value specifics. Let us mention but four of them, two applications of extrinsic value, two of intrinsic. The first two are the value sciences of sociology and economy, the latter two the value sciences of ethics and aesthetics. Sociology may be defined as the application of extrinsic value to groups of persons, economics as the application of extrinsic value to individual things; ethics as the application of intrinsic value to individual persons, and aesthetics as the application of intrinsic value to individual things. Theoretical axiology may in this way precisely relate aesthetics and economics, ethics and sociology, ethics and aesthetics, ethics and economics, etc.

The relation between *aesthetics and economics* is, as indicated, that between intrinsic and extrinsic value applied to individual things. Because intrinsic value is infinitely more valuable than extrinsic value — in the precise sense of "infinitely more" that is defined by the difference between $\aleph_0$ and $\aleph_1$ — aesthetic value is infinitely more valuable than economic value. It is, therefore, infinitely more valuable to consider a thing aesthetically than to consider it economically; and therefore my enjoyment of viewing my Orozco is of an infinitely greater value than any sum of money that might be offered me for it. And the buying and selling of works of art has nothing to do with aesthetics.

The difference between *ethics and sociology* is the difference between intrinsic valuation applied to individual persons and extrinsic valuation applied to groups. As can be demonstrated, the value of a person is infinitely higher than that of a group. It is infinitely more valuable, in the strictly-defined sense of infinity, to be a morally good person than to be a good member of society, say a good conductor, baker or professor. To be sincere, honest, or authentic in whatever one does is infinitely more important than what one does. The relation between ethics and sociology, or between ethical value and social value, is minutely described by Kierkegaard in his *Point of View*, and in part of his *Sickness Unto Death*; and especially clearly in Martin Buber's *I and Thou*.

The relation between *ethics and aesthetics* lies in intrinsic valuation as applied to things and as applied to persons. It will be demonstrated in the next section that a person is of infinitely greater value than a thing. The subject matter of ethics, therefore, is of an infinitely higher value than the subject matter of aesthetics. This is the message of Kierkegaard's classic on the relation between ethics and aesthetics, *Either/Or*. Kierkegaard's ethics is precisely the ethics that results from the system of axiology.

The difference between *ethics and economics* is one between the application of intrinsic value to persons and the application of extrinsic value to things, the ethical aspect being infinitely more valuable than the economic. It is, therefore, profoundly bad to confuse moral and economic values, for example, to sell one's child, sell a person into slavery, or to degrade a moral value such as love, by selling it in prostitution. Intellectually, it is profoundly bad to subordinate a moral value to an economic value. When this occurs within a political system then the system, as measured by axiological science, is a *bad* system in the strict definition of this term.

Any other values and value fields can be similarly related once they have been defined in terms of theoretical axiology.

c) MATERIAL SPECIFICS OF VALUE

Rather than giving examples from the infinity of value situations, we shall examine four value arguments, Anselm's Proof of the Existence of God; a Proof of the Infinite Value of the Human Person; the Relationship between Moore's and the Schoolmen's Determination of Good, that is, between Good as a Nonnatural Property and Good as a Transcendental; and Moore's Open Question Test.

Anselm's Proof of the Existence of God.

God, says St. Anselm, is defined as that than which nothing greater can be thought, *quo maius cogitari nequit*. A thing which is only thought of as thought is not a thing than which nothing greater can be thought. For a thing which is thought of as existing is thought of as greater than one which is thought of merely as thought. Hence, by definition, God exists.

The difficulty with this proof is the word "greater," *maius*. It cannot be shown logically that existence is "greater than" mere thought or essence. And if this could be shown it would still have to be demonstrated that existence itself is that than which nothing greater can be thought. But, thanks to the axiological hierarchy of values, these difficulties can be overcome. We have seen that extrinsic value is better than systemic value and that, while systemic value refers only to mental constructions, extrinsic value refers to existing things. Hence, we can reformulate the Anselmian proof, substituting the word *better* for the word *greater* — as was actually the case in the original definition of God by St. Augustine on which Anselm bases his proof (*quo esse aut cogitari nihil melius possit*). Hence, we define God as that than which nothing better can be thought. If, then, God is thought merely as thought, that is to say, as systemic value, there is a being which can be thought as better, namely one which is thought as extrinsic value, and hence as existing. This would contradict the definition of God. Hence God must be thought as existing. This proof is consistent within the system of formal axiology. Moreover, it can be continued: Also an existing God is not that than which nothing better can be thought. This is rather God as intrinsic value. Hence, God must be thought not only as existing but as supreme value. He is the value of values.[18]

A *Proof for the Infinite* Value *of the Human Person.*

There are, axiologically, at least four proofs for the infinite value of the human person, the epistemological, the logical, the ontological, and the teleological. We shall mention here the *epistemological proof*.[19]

Our definition of man will be the time-honored: "Man is a rational being." By "rational" we mean the capacity to combine concepts with objects, which is really the capacity to find one's way in this world by representing it to oneself, that is, by giving names to material objects and interrelating the names. The intension, then, which a man must fulfill to be a man, and which measures his value as a man is "rational," or "the combining of concepts with objects." Let us call each such concept a "thought." We can then say that man is a being which can think thoughts. The intension of "man" now is "thinking thoughts." In order for this intension to be a *measure,* we must spell out the set of which it consists. This is the set of "thoughts," meaning the thought-items a man must be able to think in order to be a man. Fully elaborated, the intension of man is: "thinking thought *s* and thinking thought *t* and thinking thought *u* and . . . etc." Since each thought is the name of a thing, the number of thoughts a man must be able to think in order to be a man must correspond to the number of things. According to a theorem of transfinite mathematics, any collection of material objects is at most denumerably infinite, that is to say, can be put, at most, into one-to-one correspondence with the series of rational numbers. The cardinality of this series is $\aleph_0$.[20] The set of thought terms contained in the intension of "man," according to our first definition, has the cardinality $\aleph_0$.

So far, then, the intension of man is a denumerably infinite set of predicates; which means that man, according to this first definition, is essentially, that is by definition, a denumerable infinity. The cardinality of his *value,* as fulfillment of this definition, so far, is $\aleph_0$. Let us call the number of items in the intension of a concept the *characteristic number* of the corresponding thing. This term — "characteristic number" — was coined by Leibniz precisely for the purpose of signifying the cardinality of an intension.[21] Man's characteristic number then, so far, is $\aleph_0$.

However, the thought items contained in our definition of "man" do not exhaust this definition. These thought-items, *s, t, u,* etc., must also be *thought.* This may be understood extensionally or intensionally. Extensionally understood, the set in question becomes what Dedekind called *meine Gedankenwelt,* my world of thoughts, the set of possible objects of thought of a thinker (the existence of which Dedekind used as proof for the existence of infinite systems).[22] We shall not reproduce Dedekind's argument, which presupposes a definition of infinity too subtle for our present purposes, but give instead a simpler version.

Each of the $\aleph_0$ thought items may be thought, that is, it may be thought *that* they are being thought; and it may be thought *that* the latter thoughts are being thought, and so on ad infinitum. Each thought can be thought as thought $\aleph_0$ times: the item *s* can be thought as thought, and the thought *that s* is thought, s', may be thought as being thought, and the thought *that* s' is thought, s'', may be thought as being thought and so on theoretically $\aleph_0$ "times." Similarly t, t', t'', t''', t'''' . . . u', u'', u''', u''''. . . . Since there are $\aleph_0$ thought items thus capable of being thought of in $\aleph_0$ levels of being thought, each such thought combinable with every other, the total of my *Gedankenwelt* is $2^{\aleph_0} = \aleph_1$. The characteristic number of "man" then is $\aleph_1$. Man, as a rational being, is an infinity of cardinality $\aleph_1$, a thought continuum, a spiritual Gestalt. The same result can be reached extensionally in an even simpler manner. The $\aleph_0$ thought items can be thought not only individually but also in classes. The number of these classes is $2^{\aleph_0} = \aleph_1$.

Intensionally understood, the thinking of each of the $\aleph_0$ thought items of the original set gives rise to the same cardinality. "Thinking" here means not only thinking *that,* but thinking *what*. Each of the items is not only a name denoting a material object but also a concept connoting it, and each concept has its own intension consisting of the predicates signifying the properties of the thing, or things, in question. If these things are at least two, these properties must be common to, and abstracted from, the things as members of the class in question. The larger the number of class members, the smaller the number of predicates; the smaller the number of class members, the larger the number of predicates. With a class of two members, the number of common properties that may be abstracted is infinite. The cardinality of this infinity is $\aleph_0$, since, as we have seen, each of the common properties must be abstracted one by one. While the *process* of abstraction is potentially infinite, the *totality* of common properties abstractable is actually infinite. Understood intensionally, my *Gedankenwelt,* consists of $\aleph_0$ concepts each of which may contain $\aleph_0$ intensional predicates. Hence again, the intension of man, in this definition, consists of $2^{\aleph_0} = \aleph_1$ items. Man is essentially infinite — a spiritual *Gestalt* whose cardinality is that of the continuum.

This cardinality, however, is that of the entire space-time universe itself. The result of this axiological proof of the value of man is that every individual person is as infinite as the whole space-time universe. This point, made by philosophers from St.

Augustine through Pascal to Bergson, is here axiologically demonstrated. The demonstration in question is one segment of Dedekind's proof for the existence of infinite systems; Josiah Royce, in 1899, used it to demonstrate the infinity of the Self.[23]

Good as a Nonnatural Property and Good as a Transcendental.

Good as a nonnatural property means (in our interpretation of Moore's "paradox") that "good" is a second-order property of any intension. Transcendentals, according to scholastic doctrine, are attributes of Being. Logically, Being as the *summum genus* can have no attributes, for this would violate the theory of types. Hence in order to make sense the scholastic doctrine must be reinterpreted much as Moore's "paradox" had to be.

Being is extensionally the totality of all beings. Intensionally, it is the totality of all consistently thinkable properties: it is that than which nothing richer in properties can be thought.[24] But if Being is this totality, then by the definition of good given by the Axiom, Being is good. For if Being is the totality of all consistently thinkable properties, its goodness is the secondary property defined by this totality — good is a property of the set of properties that define Being.

Good then as a transcendental is the *second-order property of the maximum intension possible.* Good as a nonnatural property is the *second-order property of any intension.*

Good as a transcendental then stands at the top of a hierarchical order consisting of goods which are nonnatural properties. Any nonnatural good is a subdivision of transcendental Good, being the secondary property of *any* intension rather than of the *maximum* intension. As there is a hierarchy of intensions depending on the richness of the intensional contents, so there is a hierarchy of the corresponding secondary predicates, that is, of Goodness.[25]

Moore's Open Question Test.

According to Moore's open question test, whenever we define something as good, we may always ask whether it is good that what is defined as good is good. For example, if we define: "A is good means A is desirable," we may ask whether it is good that A is desirable; and thus we will not have advanced our knowledge. "Whatever definition be offered, it may be always asked, with significance, of the complex so defined, whether it is itself good. . . . It is apparent, on a little reflection, that this question is itself as intelligible, as the original question 'Is A good?' "[26] Let us now see what this "open question test" does to our axiom, and our axiom to it. We defined "*x* is a good *C*" as "*x* has the properties of the intension of C." We shall define "It is good that *xRy*" as meaning that *R* is part of the intension of the concept of either *x* or *y*. Then, "It is good that *x* is a good C" means that in this particular case *R* is the class membership relation, *y* is C, and the intension of C is the intension of the concept of *x*. The question then is whether the class-membership relation is "part of the intension of C." And this it is, according to the axiom, for "*x* is a good *C*" means that the intension of C "corresponds to" the set of properties of *x*; which in turn means that the class membership relation has become the relation of one-to-one correspondence between the intension of C and the set of properties of *x*. Instead of being, so to speak, a single ray between C and *x*, it is a cylindrical bundle of rays between the two domains. And this means, in turn, that the intension of C has become part of the class membership relation: the latter, so to speak, carries the intension of C into the set of properties of *x*, maps the former into the latter. Thus, to ask "Is it good that *x* is a good *C*?" does not leave the question open, but closes it. It is indeed good that *x* is a good C, according to both the definition of "*x* is a good C" and "it is good that." The open question test of Moore is passed by our axiom with flying colors. On the other hand, the "test" itself is invalidated by our definition of good: we do not define "a complex" *as* good, but goodness itself. That is to say, we define Good not materially with respect to something that *is* good, but formally as the logical procedure that makes anything good.

We have chosen these value arguments in part to illustrate material axiological specificity and thus give a preliminary idea of the power of the axiom; and in part in order to illuminate certain famous value arguments, among them Moore's. We now propose to test the axiom more extensively, and, using its full power, to elucidate the core of Moore's ethical theory, the naturalistic fallacy and the concept of intrinsic value.

4

THE AXIOLOGICAL REINTERPRETATION OF MOORE'S ETHICAL THEORY

When an idea which has grown familiar as an unanalyzed whole is first resolved accurately into its component parts – which is what we do when we define it – there is almost always a feeling of unfamiliarity produced by the analysis, which tends to cause a protest against the definition.

— BERTRAND RUSSELL

Until a person is able to separate the idea of good from all other things and define it with precision, and unless he can run the gauntlet of all objections, ready to disprove them not by appeals to opinion but to essential truth, never faltering at any step of the argument – unless he can do all this, you would say that he knows neither the good itself nor any particular good. He apprehends only a shadow, if anything at all, something given by opinion but not by science. — PLATO [1]

The account in the previous chapter had to be, by the very nature of the subject, a survey of formal axiology, both in its theoretical and its empirical import. The various subjects, examined on the three levels of formal, theoretical and material specificity, were all discussed in terms of the axiological axiom to give an idea of its range and power. Old difficulties of philosophy yielded to the axiom which, so to speak, smoothes out the rough places of problems arising not so much because of "difficulties" of the problem as because of inadequacies of the tools used to deal with them. This power of the axiom almost playfully to dispose of old difficulties is a feature that it has in common with any axiom. It has the same power that disturbed Galileo's contemporaries. Any new science seems to make things too easy and to lack profundity. But this is precisely due to the function of a new science;

it must change profundity into method, complexity into simplicity. "Profundity," says Husserl, "is the symptom of a chaos which true science must resolve into cosmos. . . . To reshape and transform the dark gropings of profundity into unequivocal rational propositions: that is the essential act in methodically constituting a new science." [2]

The axiom of the new science of axiology was developed from Moore's paradoxical formula, by explicating the *logical* meaning of his naturalistic fallacy; the difference between various levels of discourse, and hence solving his "paradox" in terms of these levels. As was said before, Moore's naturalistic fallacy is really one of philosophical method. It was also uncovered by Frege and Russell in their establishment of the science of mathematics. The history of science shows that whenever a philosophical definition of a subject matter is replaced by a scientific one, the philosophical one appears as a methodological fallacy, such as the naturalistic fallacy. This fallacy, a confusion of logical types, is a true logical fallacy. But it is only one of a cluster of such fallacies. Moore saw but one fallacy where in truth there are at least four, and he failed to see its methodological origin in a transition from the analytic to the synthetic, the categorical to the axiomatic — in short, the very transition from philosophy to science which he wanted to bring about in ethics.

Let us now both generalize and differentiate Moore's naturalistic fallacy and set it within the total field to which it belongs, the cluster of fallacies which appear from the vantage point of any new science, in this particular case, the new science of axiology. We shall therefore call this cluster the *axiological fallacies.*

1. *The Axiological Fallacies*

From the conception of a science as the combination of a formal frame of reference with a set of objects it follows that each science has its own frame of reference, hence its own set of objects. That is to say, a datum becomes an object of natural science if the frame of reference of natural science is applied to it, and it becomes an object of value science if the frame of reference of value science is applied to it. The same datum, thus, can appear either as an object of natural science or of axiological science, depending on the frame of reference applied. Thus, the world, although ontologically one, may appear in as many aspects as there are frames of reference applied to it, in the same way that, say, a curve may appear as either convex or concave, straight or curved, depending on the perspective or frame of reference (in the view of the differential calculus — if it is applicable — the curve appears as a straight line). Thus, it is always the same world, but it is viewed differently by each science. Each science, by its own frame of reference, cuts out its own subject matter.

From this follow the axiological fallacies, which are confusions, between different sciences,

A] of general frames of reference,
B] of specific frames of reference,
C] of general with specific frames of reference,
D] of either (α) general or (β) specific frames of reference with their subject matter.

The first of these fallacies we call the *metaphysical fallacy*, because it confuses different world views; the second we call the *naturalistic fallacy*, because its most usual form is the confusion of specific natural sciences with specific moral sciences, e.g. of biology with ethics; but it also confuses different natural sciences with each other or different moral sciences with each other, e.g. ethics with religion. The third is the *moral fallacy*, for its most usual form is the confusion of axiology with ethics; the fourth is the *fallacy of method*, the confusion of a science whether (*a*) natural or (*b*) moral, whether (α) general (mathematics, axiology) or (β) specific (physics, ethics, etc.) with its subject matter.

a) THE METAPHYSICAL FALLACY

The metaphysical fallacy, or the *fallacy of genera*, is the confusion of two general scientific frames of reference, or of two genera — rather than species — of science. When, for example, the mathematical frame of reference of the natural sciences is confused with the axiological frame of reference of the moral sciences, the naturalistic view is confused with the moral and we have as a result a fallacious metaphysical vision of the world. This fallacy, widely found, is committed among other ways in trying for conclusions valid for the world of value from observations or arguments valid only for the world of nature, and vice versa. Here we have, for example, the arguments of the church against Galileo and those of naturalists against the church; the writings of modern scientists like Jeans or Eddington about values, and those of ethicists about nature, as for example the argument that the principle of indeterminacy in quantum physics shows the free will of men, that relativity theory demonstrates the ideality of the world, or that the "survival of the fittest" exhibits its beastliness.[3] This fallacy also appears in the writings of philosophers such as Descartes, Leibniz, Spinoza, and others who have tried to apply (extensional) mathematics to ethics;[4] it appears in the writings of Freud, when he says that "religion has not stood the test of science" and in the half-serious jokes of materialistic dialecticians, such as the Soviet cosmonaut who said that in spite of his many revolutions around the Earth he had not encountered God.

The knowledge of this fallacy already enables us to bring order into many of the confused arguments of traditional — and contemporary — ethics, namely all those based on the matter, rather than the form, of science.

b) THE NATURALISTIC FALLACY

The second fallacy has similar scope. It is the confusion of specific frames of reference of particular natural or axiological sciences, such as ethics and psychology, chemistry and religion, and so on. Confusions concerning *natural* sciences are not frequent any more, since there exists the mathematical frame of reference which orders these sciences; but they were frequent before the modern age. Thus, music and astronomy were confused by Pythagoras and even by Kepler; astronomy and theology by Plato, Aristotle, and even Newton; theology and chemistry in alchemy.

In our day we still lack general knowledge of the frames of reference of the specific axiological sciences; hence the confusion between particular sciences is frequent, indeed usual. This confusion primarily concerns ethics, which is more often confused with other sciences than not. When it is said that the good is pleasure or preference, or God's will, or the interest of the proletariat, or a matter of human conduct, then we have respectively, confusions of ethics with psychology, theology, economics, and sociology. This fallacy we call the *fallacy of species,* or *naturalistic fallacy,* because it consists usually of confusing the nonnaturalistic sense of the good with its naturalistic sense.

c) THE MORAL FALLACY

This fallacy, as well as the preceding ones, is included in what Moore called the naturalistic fallacy. The moral fallacy, or *fallacy of types,* is the confusion of different types of frames of reference, such as a general with a specific one, for example the mathematical with the physical, or the axiological with the moral. This fallacy is infrequent nowadays in the natural sciences. Confusions between mathematics and physics or chemistry occur only occasionally, for example, in some popular writings, where mathematical formulae are "translated" into everyday language. But it was frequent before the time of modern science; thus, Pythagoras, Plato, and even Kepler and Newton[5] and pre-Fregean logicians, confused numbers with what numbers refer to. In moral thinking, on the other hand, the fallacy is extremely frequent, in particular when value in general or axiological value is confused with a specific value, such as goodness in general with moral goodness. To say, for example, that a murderer cannot be good is to commit this fallacy. A murderer cannot be morally good, but he can be good as a murderer, that is, he can murder well. Here also belong the examples given in (b) above, if "good" is interpreted as good in general rather than as moral good. Owing to the frequency of this particular confusion between good in general and moral good, we call this the moral fallacy; but it is committed whenever two different types of frames of reference are confused. Thus, to say that goodness in general is the goodness of God, is the beautiful or the true, or the classless society, is also committing the moral fallacy, although here axiological value is confused with theological, aesthetic, epistemological, and sociological value, respectively, rather than with moral value.

The two philosophers who most clearly warned against this fallacy are Plato and G. E. Moore. Plato's early writings are attempts to separate the Good from specific goods, thus avoiding the moral fallacy, his later writings are attempts to separate it from Being, thereby avoiding the metaphysical fallacy. The whole work of G. E. Moore in moral philosophy consists in establishing what we have called the metaphysical, the naturalistic, and the moral fallacies, lumped together by him under one name. Since neither Plato nor Moore make these fallacies systematically clear, they themselves commit them occasionally. When, for example, Moore in *Principia Ethica* calls his subject matter ethics rather than axiology, he commits the moral fallacy. And in doing so, he commits the fallacy of not distinguishing between the naturalistic and the moral fallacy. For in failing to distinguish between moral good and good in general he fails to distinguish between the confusions to which either of these are subject. To confuse moral good with pleasure is the naturalistic fallacy; but to confuse good in general with pleasure is the moral fallacy. In general, however, Moore is careful not to commit either fallacy; by saying that "good is good and not another thing" he makes sure to keep good strictly to the realm to which it belongs, namely the axiological.

d) THE FALLACY OF METHOD

Finally, there is the confusion between the frame of reference — either general or specific — and the subject matter of a science. This fallacy we call *the fallacy of method.* This again is one of types and thus a genuine logical fallacy. It is particularly frequent in cases where the scientist confuses himself or his acts or attitudes with the subject matter of his science. In the natural sciences this fallacy is both rare and common. It is rare in the sense that a natural scientist rarely confuses himself with his subject matter, for example, an electronic scientist with an electron. However, it was necessary to define this relation, as has been done with mathematical precision in Heisenberg's principle of indeterminacy. Also, this relation between the scientist and his subject matter is important in the operational philosophy of Bridgman and in the construction of the observer in relativity theory. Thus, this relation is clearly defined in natural science. It was not so in the Middle Ages, when the scientist was confused with his subject matter, for example, in alchemy, where certain personal attitudes and acts were considered necessary to bring about a chemical effect.

In another sense, however, this fallacy is very common in natural science today. Frequently natural science is confused with its subject matter in that some properties of the subject matter are thought to belong to the science itself. Thus it is said that "science" is "empirical" and therefore no axiological science is possible because it is not "empirical" — that is, it does not experiment, observe, or predict. We shall call this species of the fallacy of method the *empirical fallacy*. It may also be said to be a species of the moral fallacy, because it confuses two types of frames of reference, science in general with natural science.

As we said at the beginning, science in general is a method and has nothing to do with any specific content; if there is a formal frame of reference applicable to a set of objects, then there is a science, no matter whether the subject matter of the science is spatio-temporal — and hence empirically observable and predictable — or not. Thus, mathematics, music, and axiology are sciences, and they *do* include experimentation,[6] observation, and prediction even though they are not "empirical." For these operations are nothing but attempts to apply a frame of reference to a set of objects; whether these objects are spatio-temporal or not is irrelevant. Prediction, in particular, is nothing but the statement that the object conforms to the frame of reference. Thus, in music one can "predict" which note Wilhelm Backhaus is going to play next when he is playing a certain concerto; and in mathematics one can "predict" what will be the result of a certain operation. The same occurs in axiology. Given a certain situation and applying the axiological frame of reference, one can "predict" the axiological result. It is a peculiarity of natural science that its formal prediction coincides with a temporal process of its object; and this material process is confused with the formal predictability of any science as such. This confusion is the worse since the temporal process which is the subject matter of natural science is not temporal in the sense of having past, present, and future but is itself formal, being nothing but a linear geometrical relation. The empirical fallacy is, thus, a confusion of thought at the very core of natural science.

In the moral field the fallacy of method is extraordinarily frequent and with better reason for up to now it has lacked a formal frame of reference. Thus, it has been said — for example by certain British moralists and some positivists — that ethics is inaccessible to rational treatment because the moral sentiment is not rational, and cannot therefore be subject to rational analysis. But if this were true the psychiatrist ought to be mad and the cancer specialist the one with the most advanced cancer. It *is* difficult for the psychiatrist to keep himself apart from his subject — and practically almost impossible — but theoretically, and ultimately practically, he must make this distinction.[7]

That variant of the fallacy of method which holds that the irrational is not rationally accessible is the *positivistic* fallacy, and is part of a whole set of fallacies committed by the positivists. There are other variants of the fallacy of method; its *normative* species says that the moral sciences are normative while the natural sciences are not, though really there is little difference between the two in this respect;[8] its *existentialist* species says that the philosopher can explain human life only by his own being human, and his being human not in general but in his own particular selfhood.

There is a difference of type between the saying of something and what is said, hence, a type difference between a science and its subject matter. The neglect to clarify this and even more the powerful effort to obscure it — as is done not only in existentialism but also the so-called *Geisteswissenschaften,* for example in Dilthey — is one of the strongest reasons for the backwardness of axiological science. To *be* is one thing and to *analyze being* is another; the axiological scientist as such must see himself in distinction from his ontological being if he is to know — although knowing he is and being he knows. But he *knows* epistemologically, and he *is* ontologically. At bottom, therefore, the existentialist species of the fallacy of method is the naturalistic fallacy which confuses ontology and epistemology. Because in existentialist knowing ontology and epistemology fuse it is imperative, in order to understand this knowledge, to keep the two sciences apart.

Many objections against axiological science arise from the fallacy of method, that is the confusion of form and content of a science. This confusion, which arises from the overwhelming emphasis on the subject matter rather than the form of science (the former being so much more easily discernible than the latter), is a main obstacle to the creation of a science of ethics, and leads to misconceptions which invalidate much of traditional moral theory. They boil down to three objections against our enterprise, namely, 1] that knowing about value destroys the value experience, 2] that valuation is a matter of feeling, hence intellectually not accessible, and 3] that concrete phenomena can be known only by concrete observation.

The first objection is based upon the sound instinct that there is a difference between a subject matter and the knowledge thereof, and hence between valuation and the analysis of it, but also upon an incorrect view of the relation between the two elements of value knowledge. An axiological science, it is said, would make the moral life into a machine. But the moral life is the subject matter of axiology, not axiology itself. The moral life remains as it is, even though axiology tries to understand it. Understanding does not destroy what it understands. Things have been thrown before and after Galileo, as we said. But after him experts knew that the form of throwing was a parabola, and thus were instrumental in bringing about ever refined throwings — up to the sputniks of today. People will value after axiological science as they did before, but experts will know the *form* of valuing and thus be instrumental in bringing about ever refined valuings. Actually, knowledge deepens that which it knows. Knowledge of the musical score deepens the experiencing of music. Axiology is the score of the moral experience and deepens this experience.

At the basis of this objection is the suspicion of morally sensitive people against the kind of rationality that brought about the atomic bomb, and hence their escape into irrationality rather than into a different kind of rationality. This is the very attitude of the positivists toward morality, and is found among scientists and reformed positivists, who rebel against the content of positivism — its implied overvaluation of natural science — without giving up the method of positivism, to see the moral as irrational. We may call this the fallacy of the insufficiently converted positivist. He is, actually, a dilettante mystic — sharing with mysticism its superficial irrationalism rather than its profound rationalism.

The second objection, equally common, rests on the confusion of valuation with feeling. Valuation is no more nor less a matter of feeling than is music. It is a matter of feeling structured by laws — feeling following definite laws. The feeling of value is nothing arbitrary. In the words of Nicolai Hartmann: "The feeling of value is not free: once it has grasped the meaning of value it cannot feel differently. It cannot regard good faith as wicked, or cheating and deceit as honorable. It can be value-blind, but that is an entirely different matter: in this case it is not responsive to values at all and does not comprehend them." [9] Value thus is no more, and no less, arbitrary or subjective than is fact. Just as fact is a combination of perception with the theoretical framework of natural science, so value is a combination of perception with the theoretical framework of value science. The latter is as definite as the former.

The third objection, that concrete phenomena can only be known by concrete observation, is what we have called the empirical bias, which is empirically, methodologically, and logically fallacious. As we have seen from observing both the method of

Galileo and the systematic of science, it is the most abstract ideas that solve the most concrete problems.[10]

The axiological fallacies arise from the mere conception of an axiological science and are powerful instruments of intellectual analysis. They are like scalpels which enable us to cut with precision into the tangled arguments of traditional ethics and order their strands in the organic framework, the anatomy of axiological science.[11]

The fallacies are committed so frequently today because we have natural *science* and moral *philosophy*, but not natural *philosophy* or moral *science.* From the juxtaposition of natural science and moral philosophy follow the fallacies in a fivefold confusion, partly concerning natural science, partly concerning moral philosophy. From the fact that "science" today is natural science follow the fallacies when it is concluded 1] that no other kind of science is possible; the species "natural science" is taken for the genus "science." Consequently 2] the characteristics of natural science are taken to be the characteristics of science in general. These characteristics, however, arise from the specific subject matter of natural science; hence the *object* of a specific science is confused with science itself. The characteristics of "science" are said to be empirical observation and prediction. Actually, these are characteristics only of *natural* science, arising from the *subject matter* of this science, the natural phenomena, which can be "observed" because they are in space and time, and "predicted" for the same reason. A science which does not deal with spatio-temporal phenomena "observes" and "predicts" in a different sense and is science as well. The claim of natural science to be science pure and simple is therefore illegitimate. A similar illegitimacy arises on the side of moral philosophy. From the fact that there *is* no moral science it is concluded 3] that there *can be* no such science, and this conclusion 4] is applied to the *objects* of moral philosophy, namely values. It is then held to be impossible to account for values by a science, where by "science" is meant natural rather than moral science. Thus there is added 5] the fundamental confusion between natural science and moral science.

It is this confusion to which Moore addressed himself and for which he coined the term "naturalistic fallacy." This "fallacy" contains three of the four fallacies we have discussed. They arise as fallacies only when the concept of an axiological science is posited. For this reason, the majority of ethicists do not take

Moore's fallacy seriously enough to avoid committing it; on the contrary, they commit it freely and announce that they do,[12] but hold that it is no fallacy. They are right from their point of view; for their ethics are philosophies based on analytic concepts which *force* them to commit it. Only from a scientific point of view, Moore's and our own, does their procedure appear as a fallacy. Only when the method of axiological science is opposed to that of axiological philosophy does the fallacy show up clearly and branch out into the four axiological fallacies described.

These fallacies, we said, arose from the mere conception of an axiological science. We shall now use the axiom of this science and apply it to Moore's conception of intrinsic value.

2. *Moore's Paradox of Intrinsic Value*

For Moore, good was good and not another thing; but he could not define *what* good was. "Good is good and that is the end of the matter." We, on the other hand, wish to say that good is good and not another thing, and in addition define what it is. Good is good, and that is the beginning of the matter — the beginning of formal axiology.[13] Good, as Moore rightly saw, is neither an empirical nor a metaphysical term, it is, as we add, a *logical* term. In elaborating its logical nature we obtain a formal structure which can be shown to be isomorphous with value reality. The *formal structure* gives insight into the inner nature of goodness, the variety and plenitude of its intrinsic relations; the *isomorphism* with value reality enables us to analyze not only this reality but also theories concerned with it. While thus the intension of formal axiology is the elaborate, precisely-defined and structured inner nature of goodness, its extension is value reality and the value theories that deal with it. More specifically, if we limit ourselves to *moral* reality and the *ethical* theories dealing with it — rather than speaking of value in general and value theories concerned with it — formal axiology is a frame of reference giving the formal requirements of a theory of ethics and the critical standards for judging ethical theories.

This becomes especially clear if we apply the axiological axiom to Moore's own ethical theory. In his "Reply to My Critics," Moore formulates the problem of value as follows: "the question: What makes this experience good? *is* equivalent to the question:

From what intrinsic characteristics of this experience does it *follow* that it is good?, and the proposition that experiences with those intrinsic properties *are* good is not an empirical but a necessary one." [14] This question of Moore's has never been answered, except by our axiom: "That a thing is good follows from its possession of *all* its intrinsic properties." Let us now analyze this answer, not only as an exercise in formal axiology but also as a way to the deeper understanding of Moore's prolegomena.

By "follows from" we shall mean the relation of *entailment*, defined as any sequence of propositions with the same subject whose predicates stand in the analytic relationship. The analytic relationship is the containing of predicate by subject. What is meant by "containing" will become clear in the sequel. Thus, if of two propositions of the subject-predicate form *S-P* and *S-Q*, the predicates stand in the analytic relationship, then either the concept *P* contains the concept *Q*, or the reverse. Let us call the preceding proposition the *antecedent* and the succeeding proposition the *consequent*, and let us stipulate that *P* contains *Q* and *Q* is contained in *P*: the antecedent *entails* the consequent and the consequent *is entailed in* or *follows from*, the antecedent. Thus, the sequence of the expressions "*x* is rational" and "*x* is a man" may be regarded as an entailment.

By *intrinsic properties of a thing* we shall understand 1] all the properties which make up the description of the thing, and 2] all the properties which characterize this and no other thing, and 3] all the properties without which the thing would not be what it is. Let us see whether Moore's intrinsic properties have the same three characteristics. This question lands us in the middle of a few more of Moore's perplexing paradoxes. At the same time it will demonstrate the power of our axiological axiom.

Moore is vague about the relation between the *nature* of the thing and the *value* of the thing — which we have established through our definition of "good" — although, or rather just because, he has more deeply penetrated this relation than any thinker before him. In doing this he has so closely approximated the nature and the value of a thing to one another that — not knowing the exact, fine, logical line dividing the two — he is unable, in speaking of the thing, to keep apart its nature and its value. Out of this inability arise the paradoxes in question. Once this relation is known, however, Moore's statements of the paradoxes lose their paradoxical character.

Moore does see that a thing's nature — the totality of its intrinsic properties — has a very close connection with its value properties. He also realizes that he does not know the exact nature of this connection, so he gives the most startling expression to his ignorance and to its possible source.

In our theory the precise relation between the nature and the value of a thing is clear. The expression "*x* being good has all its intrinsic properties" is an entailment and, indeed, a mutual one, with antecedent and consequent interchangeable. Of the two propositions, "*x* is good" and "*x* has all its intrinsic properties," the first contains as predicate the definiendum and the second contains as predicate the definiens of the definition of 'good'. Hence it follows from our definition of entailment in conjunction with our definition of value that "*x* is good" *follows from* "*x* has all its intrinsic properties."

Thus Moore's question is answered in precisely the way that he thought it ought to be answered, so that, namely, "the proposition that experiences with those intrinsic properties *are* good is not an empirical but a necessary one." [15] Not only is the relation between the two propositions entailment, but also (a fact which Moore seems not to have suspected and, indeed, has denied [16]) this entailment is mutual. *Propositions about the intrinsic nature and about the intrinsic value of a thing are equivalent.*

This we shall now see in detail by examining Moore's discussion of the three characteristics of intrinsic properties which we find 1] in Moore's treatment of the relation between the natural intrinsic properties and the value properties of a thing, 2] in his treatment of the definition of a thing's intrinsic natural properties, and 3] in his treatment of the definition of a thing's intrinsic value.

1] Moore, in *Philosophical Studies*, twenty years after *Principia Ethica*, where he first approached the problem,[17] saw that the value properties of a thing have something to do with its natural properties and the latter something to do with the description of the thing. But his writing is so unclear about what the connection — or difference — might be that at that point he does not even distinguish between the two kinds of properties, the natural and the value properties, but calls them both "intrinsic," with the curious distinction of calling the former "intrinsic properties" while calling the latter "properties which are intrinsic" but not "intrinsic properties." As he explains still another twenty years later, in the "Reply":

> I had made a distinction between two kinds of properties which depend only on the intrinsic nature of a thing; but I had not called the one kind "intrinsic" and the other "non-intrinsic." I had called both "intrinsic," but had made a distinction between those which, besides being "intrinsic," were also "intrinsic properties," and those which, though "intrinsic," were not "intrinsic properties" — a manner of speaking which I admit to be very awkward, but which nevertheless did not express a self-contradictory view, since I was so using (very awkwardly) the expression "intrinsic property" that in order to be an "intrinsic property" it was not sufficient that a property should be "intrinsic" — it must also have some other feature which distinguishes it from other properties which are "intrinsic" but are not "intrinsic properties." [18]

Moore's device was ingenious. For value properties are not intrinsic properties in the same sense that the descriptional natural properties are, yet they are properties which are intrinsic — intrinsic to the thing. They are in one way identical with the natural intrinsic properties although in another way they are not. They are these properties in a new and different view — in their totality. The value properties are and are not identical with the natural properties, just as a whole is and is not identical with its parts. The value properties, in our analysis, depend on the quantity of the natural intrinsic properties, not on these properties themselves and as such — and it was this purely logical distinction which was hidden to Moore and which he expressed at the time in this absurd manner. Only gradually did he come to the conviction that the value properties were entirely different from the natural properties, and in the "Reply" he makes the difference clear by calling the value properties "non-natural intrinsic properties" and the original "intrinsic properties" "natural intrinsic properties." [19]

But he sees already in *Philosophical Studies* that the "other feature" by which the two kinds of intrinsic properties are distinguished has to do with their respective descriptive power. Here again Moore feels his way toward a distinction between the two kinds of properties which our analysis has brought out. Again he poses the problem in most suggestive terms without offering a solution.

> There must be some characteristic belonging to intrinsic [natural] properties which predicates of value never possess. And it seems to me quite obvious that there is; only I can't see *what* it is. . . . When you assert of a patch of colour that it is "yellow," the predicate you assert is not only *different* from "beautiful" but of quite a different kind. . . . But *what* the difference is, if we suppose, as I suppose, that goodness and beauty are *not* subjective and that they do share with "yellowness" the property of depending *solely* on the intrinsic nature of what possesses them, I confess I cannot say. I can only vaguely express the kind of difference I feel there to be by saying that intrinsic [natural] properties seem to *describe* the intrinsic nature of what possesses them in a sense in which predicates of value never do. If you could enumerate *all* the intrinsic [natural] properties a given thing possessed, you would have given a complete description of it, and would not need to mention any predicate of value it possessed; whereas no description of a given thing would be *complete* which omitted any intrinsic [natural] property.[20]

This is indeed puzzling: two kinds of intrinsic properties, the natural and the nonnatural (in his later terminology), are said to depend solely and entirely on the intrinsic nature of the thing possessing them, yet, the former describe the thing completely and in every respect and the latter describe it not at all and in no respect whatsoever.

Our analysis makes it clear why this is so. The natural intrinsic properties, namely the descriptive properties of the thing, describe its intrinsic nature as completely as possible. At the same time, their totality entails logically the predicate of value which, for the very reason that it is entailed by the totality of the natural intrinsic properties of the thing, is not and cannot possibly be included in the description. It does not describe at all, but is of a logical order higher than the description. Its exclusion from the description is required by Russell's vicious-circle principle according to which "whatever involves *all* of a collection must not be one of the collection."[21]

Since the value predicates are functions of the quantity of

the natural intrinsic predicates, and since these latter describe the thing *completely*, it follows that the value predicates cannot *describe* the thing — in this same sense of "describe" — no matter what else may be their relation to the thing. Again, Moore's intuition is vindicated by logical proof.

2] The second problem is that of the definition of the intrinsic properties. Moore's problem is, how can the natural intrinsic properties be defined without at the same time defining the non-natural intrinsic properties *in the same way*. In other words, what is the differentia that distinguishes the two kinds of properties? Not knowing the exact relation between the two, Moore is evidently unable to provide this differentia. All he can do is in some way distinguish the one set from the other, but he cannot *define* the first so that the definition will not also cover the second or vice versa. His only distinction between the two is their respective references to description. In the "Reply" he puts the matter succinctly:

> Properties which are intrinsic properties, but *not* natural ones, are distinguished from natural intrinsic properties, by the fact that, in ascribing a property of the former kind to a thing, you are not describing it *at all*, whereas, in ascribing a property of the latter kind to a thing, you are always describing it *to some extent*.
>
> . . . .
>
> an intrinsic property is "natural" if and only if, in ascribing it to a natural object, you are *to some extent* "describing" that object (where "describe" is used in one particular sense); and that hence an intrinsic property, e.g. the sense of "good" with which we are concerned, is not "natural" if, in ascribing it to a natural object you are not (in the same sense of "describe") describing that object *to any extent at all*. It is certainly the case that this account is vague and not clear. To make it clear it would be necessary to specify the sense of "describe" in question; and I am no more able to do this now than I was then [22] [twenty years earlier in *Philosophical Studies*].

The reference to description, then, is *a* distinction between the two kinds of properties, but not *the* distinction. It cannot serve as differentia. Moore sees, however, that the differentia may be found by clarity about the meaning of "description":

> I do think it important to emphasize that, *if* . . . "good," when used in the sense we are concerned with, is "the name of a characteristic at all" — *if*, that is to say, some view like Mr. Stevenson's . . . or my modification of it, is *not* the right account of this sense of "good" — then, there is a problem to be faced, since this sense of "good" certainly seems to be of an importantly different *kind* from those intrinsic properties which in *Principia* I called "natural" and in *Philosophical Studies* I spoke of, so awkwardly, as if they alone were "intrinsic properties," although there were other properties which were "intrinsic"! And I think it is possible that *a* solution of the problem is to be found in the different way in which they are related to one particular sense in which we use the word "description." [23]

Again, this becomes clear in the light of our analysis. On the one hand, it becomes clear that value predicates are excluded from description, on the other hand, it becomes clear that they depend on it, that is, on the totality of the natural intrinsic properties. Description, then, does become the differentia between the two kinds of properties. For a term "to describe" *means* "to be a natural intrinsic property" and "not to describe" *means* "to be a nonnatural intrinsic property." As long, therefore, as Moore was unclear about the value meaning of description he could not be clear about that of intrinsic properties.

Moore has made another attempt to define natural intrinsic properties which seems at first sight more obvious and promising, namely through reference not to the description of the thing but to the thing itself. But here again he encountered the old difficulty, with a new twist: as long as the exact relation between the nature and value of a thing is unknown it is impossible even to know what a thing is or how to distinguish between two things. Should we distinguish between the two natures of the things or between their two values? And how then is one to distinguish between the two differences, that of the natures and that of the values?

The solution of this problem presupposes knowledge of the precise relation between a thing's nature and its value. Not knowing this, Moore loses himself in a labyrinth of difficulties out of which no Ariadne is able to provide him a clue. Let us, however, follow him, secure in our possession of Ariadne's thread, the definition of "good."

Assuming a difference between the two things, one may employ this *either* for demonstrating that the things have different natures or that they have different values, but not for both. For the difference between the two things is only *one* relation whereas — as long as we do not know the relation between the things' natures and values — the difference between the natures of the two things, on the one hand, and the difference between the values of the two things, on the other, seem at least *two* relations. How then can the difference between the things, which is one relation, account for both of the other relations? Again, how can either of the two relations be used to account for the difference of the things?

Moore has progressed far enough, in *Philosophical Studies*, in the exposition of the relation between the thing's nature and the thing's value to see that the distinction between the two is so minute — considering the tremendous similarity between them, of depending solely on the intrinsic nature of the thing that possesses them — that if both sets were defined with reference to the thing, or by referring to the difference between two things, there would be danger that the distinction between nature and value, of the thing or things, would be lost altogether. His problem is that on the one hand he cannot account by one feature — the difference between two things — for two other features *each* of which, in this account, would consume *all* of the feature which is supposed to account for *both*, and on the other hand, if he *would* account for the two features by the one, the two would become like one and the distinction so laboriously made between them would disappear.

Moore struggles with this problem at the end of "The Conception of Intrinsic Value" in *Philosophical Studies* and finds it insoluble. Having used the difference between the two things to account for their respective *values*, he is unable to use the same difference again to account for their respective *natures:*

> Owing to the fact that predicates of intrinsic value are not themselves intrinsic properties, you cannot define "intrinsic property," in the way which at first sight seems obviously the right one. You cannot say that an intrinsic property is a property such that, if one thing possesses it and another

> does not, the intrinsic *nature* of the two things *must* be different. For this is the very thing which we are maintaining to be true of predicates of intrinsic *value*, while at the same time we say that they are *not* intrinsic properties.[24]

In other words, since he defined the nonnatural intrinsic properties by reference to the difference between two things, and at the same time maintained that these nonnatural intrinsic properties, the "predicates of intrinsic value," were *not* "intrinsic properties," that is, intrinsic natural properties, he cannot define the latter by what "at first sight seems obviously the right" way of defining them, namely by the difference of two things. This differentia has already been used and consumed by the definition of intrinsic value. On the other hand, since Moore insisted that intrinsic value and intrinsic nature are two different aspects of things, there ought to be two kinds of difference between two things, one based on the intrinsic values and the other based on the intrinsic natures of the two things. But what could these be? It would mean that *if* the intrinsic natural properties would be defined in the way that at first sight seems obviously the right one, namely, by referring to the difference between two things, then there must be *some other kind* of difference accounting for the two *values* of the things.

> Such a definition of "intrinsic property" would therefore only be possible if we could say that the necessity there is that, if x and y possess different intrinsic *properties*, their *nature* must be different, is a necessity of a *different kind* from the necessity there is that, if x and y are of different intrinsic *values*, their nature must be different, although both necessities are unconditional. And it seems to me possible that this is the true explanation. But, if so, it obviously adds to the difficulty of explaining the meaning of the unconditional "must," since, in this case, there would be two different meanings of "must," both unconditional, and yet neither, apparently, identical with logical "must."[25]

This problem which, if possible, seems even more perplexing than the previous ones, nevertheless finds an easy solution. We have no difficulty at all in defining an intrinsic natural property as one which if one thing x has it and another thing y does not have it then x and y are different things. We may use this definition because we have not used its differentia in any other definition, in particular, we have not used it in the definition of intrinsic value, which was defined in an entirely different, independent way. On the other hand, the difference between two things is precisely the differentia which we have used, as characteristic (2), for the definition of intrinsic property. The set of a thing's intrinsic properties has as its very essence the fact that if one thing *x* has it and another thing *y* does not have it, then *x* and *y* *must* be two different things — and the "must" is unconditional. At the same time, however, this same set determines the value pattern of the thing. Thus the difference in the values of two things is based on that very difference which makes the natures of the two things different, namely, the difference of their natural intrinsic properties, so that the value properties are just as different as are the things themselves. Or, in other words, the difference between the two things may just as well be based upon the difference between the two values as upon the difference between the two natures. And the former difference rests on the same "must" as the latter.

Thus, knowing the relation between the thing's nature and value, we can relate things by their natures as well as by their values. Moore could not do this, so was unable to distinguish one thing from another with reference to its value; and he had to reduce the difference of values to one of intrinsic natures, losing thereby all criteria for the difference of natures. Again, he clearly saw the problem but, lacking the fundamental relation between intrinsic nature and value, was unable to solve it.

3] The reason is that he so closely combined intrinsic nature and intrinsic value that he could not separate them again. The welding of the two constitutes his attempt at defining the intrinsicness of value: Moore defines the in-itself-ness of "good," which comes as close as possible to defining "good in itself" or intrinsic value. But he comes even closer to defining goodness itself by bringing the exposition of "good" down to two propositions, separated only by their paradoxical and seemingly incompatible content. They *could* become one — and thereby presumably a definition — if only the nature of description would be clear.

> What is meant by saying with regard to a kind of value that it is "intrinsic?" To express roughly what is

> meant is, I think, simple enough; and everybody will recognize it at once, as a notion which is constantly in people's heads; but I want to dwell upon it at some length, because I know of no place where it is expressly explained and defined, and because, though it seems very simple and fundamental, the task of defining it precisely is by no means easy and involves some difficulties which I must confess that I do not know how to solve.[26]

He consequently sets up "the following definition. *To say that a kind of value is 'intrinsic' means merely that the question whether a thing possesses it, depends solely on the intrinsic nature of the thing in question.*" This, then, is Moore's *definition* of intrinsicness of value. But he feels that it expresses more a feeling than a certainty:

> Though this definition does, I think, convey exactly what I mean, I want to dwell upon its meaning, partly because the conception "differing in intrinsic nature" which I believe to be of fundamental importance, is liable to be confused with other conceptions, and partly because the definition involves notions, which I do not know how to define exactly.[27]

Thus, the "definition" is at best an exposition, leading very close to the definition, but never to its heart. The dependence of intrinsic value on intrinsic nature, which Moore expounds more or less intuitively, gives exactly the relation between intrinsic nature and value we have deduced from our definition. Thus Moore shows a unique insight into a relation whose very nature is to him unknown. It may even be held that while his intellect works toward the definition, his original intuition is dimming. Thus, in the "Reply" he comes very close to the definition, yet he almost gives up his lifelong quest in his concession to Stevenson (but in the same passage he takes back with one hand what he has given Stevenson with the other).[28] This wavering shows that the concept of value set up in "The Conception of Intrinsic Value" is still in the fluid stages of development, oscillating between description and definition — an exposition. In this respect, the "intuitionism" of Moore's value theory refers to his mind rather than to the nature of value. Though he has seen value more clearly than any man before him he yet sees it only dimly, tantalizingly.

Moore labors to explain what he means by a difference in intrinsic nature and a difference in quality, finally summarizing this in a question:

> what is meant by the words "impossible" and "necessary" in the statement: A kind of value is intrinsic if and only if, it is *impossible* that x and y should have different values of the kind, unless they differ in intrinsic nature; and in the equivalent statement: A kind of value is intrinsic if and only if, when anything possess it, that same thing or anything exactly like it would *necessarily* or *must* always, under all circumstances, possess it in exactly the same degree.[29]

Moore cannot answer because he is unable to define the fundamental relation between a thing's nature and value. He does not know what, apart from intrinsic nature, intrinsic value is; and although Moore knows what an intrinsic natural property is — a property such that if one thing possesses it and another does not, the intrinsic nature of the two things *must* be different — this definition is inadmissible because its differentia has already been used for the definition of intrinsicness of value. For this reason Moore refrains from using this definition of intrinsic natural property, although it seems obviously the right one.

The solution is to render the definition of intrinsic natural property admissible by defining intrinsicness of value so it will not preclude use of the "obvious" differentia. For us the prohibition does not exist: we have defined intrinsic value, and intrinsicness of value, not in terms of the difference between the intrinsic nature of two things but in terms of an exactly defined relation to the quantity of intrinsic properties. We are therefore free to use the "obvious" definition for natural intrinsic property[30] as well as our definition of value; and the way to the solution is clear.

Both the "impossible" and the "necessary" in Moore's question *are* of the logical nature which Moore doubted them to be. First the "impossible." In our analysis, x and y differ in intrinsic nature if and only if they have different sets of intrinsic natural properties. But if they have such different sets they must also have different kinds of value, for their values are functions of their sets of intrinsic natural properties and vice versa. Therefore it is *logically impossible* that x and y should have different intrinsic values *unless* they differ in their intrinsic nature, that is, have different sets of natural intrinsic properties. Here "value" is meant phenomenally; *x* and *y* *could* have different *formal* values even though the same intrinsic natures, depending on whether these natures were fulfilled or not. Thus, *x* could be a good C and *y* a bad C. But the values *must* be *phenomenally* different — even though they may be formally the same, e.g., both good C's — since the one depends on the C-ness of *x* and the other on that of *y*, when *x* and *y* are different.

The "necessary" is explained in the same way. The set of intrinsic natural properties determines the intrinsic nature of the thing, the latter changing with every change within this set. But what is true of the thing and set is also true of its value in relation to the set. This value of the thing depends on the set of its intrinsic properties, so must follow the thing's nature throughout all its changes. Therefore the thing *must necessarily* and under all circumstances possess this value in exactly the same degree — namely, as depending on its nature, no matter what this nature may be, and in always that degree which corresponds to the nature. Here "value" is meant formally.

These simple answers, closed to Moore through his vagueness about the natural and the nonnatural intrinsic properties, are open to us since we defined their exact relation. Moore, not coming to these answers, arrives instead at the ingenious result from which we started, which again testifies to the depth of his insight without possessing the key to the problem: "Two different propositions are both true of goodness, namely: 1] that it does depend *only* on the intrinsic nature of what possesses it . . . and 2] that, *though* this is so, it is yet not itself an intrinsic property."[31] To these two propositions the exposition of goodness itself comes down, and they contain the principal paradox of Moore's theory. The paradox says that although propositions with the predicate good are all of them synthetic and not analytic [Proposition (2)] there is yet a necessary connection between the subject and that predicate [Proposition (1)]. This paradox is deepened by the consideration that, as we have seen, only a posteriori analytic judgments with empirical analytic concepts can be either analytic or synthetic, so that "good," if it makes any sense at all to call propositions with it synthetic, ought to be such a concept; but this contradicts what Moore says about it. On the other hand, his emphasis on the self-evidence of "good" as the basic notion of a *science* of ethics, that is, its axiomatic nature for such a science, and his classifying "good" with number seems to indicate that "good" is a synthetic concept. Thus, Moore stands between the analytic and the synthetic view of "good." Methodologically, this means that he performs what Descartes calls the reduction to a "simple" — by cleaning out the Augean stable of ethical theories and clearing the way for the unique simple notion of goodness — but that he does not perform, at least not consciously, the leap to the synthetic formula. Logically, it means that propositions with "good" either are both analytic and synthetic or neither analytic nor synthetic; and this can only mean that they are synthetic a priori: There is *no* connection between subject and predicate *except* the one which an a priori supplies,[32] that is, the formal system of ethics that Moore has in mind. Since, however, it is the intrinsic nature of the thing on which goodness depends and which is the link between subject and predicate, it follows that the *a priori, the formal system connecting both, is the intrinsic nature of the thing itself, the set of its secondary properties!* It is this set whose structure is the structure of value. Hence our statement, in one proposition, of "what is true of goodness" (in Ch. 3, sec. 2) is not only a definition of Goodness but the axiom of a new science. This is the methodological solution of Moore's paradox.

Moore, although he divined this science, and stated emphatically in *Principia Ethica* that it must be based on the notion of goodness, never reaches the axiom: "There must be some characteristic belonging to intrinsic properties which predicates of value never possess . . . only I can't see *what* it is."[33] All he would have had to do at this point was to follow up his account of "complete description" and translate it into logical terms by showing that the completeness of the description excludes by the vicious-circle principle the value predicate which involves reference to this completeness. Thus the value predicate can, indeed must, never describe. In characterizing the set of descriptive predicates the value predicate is of a higher logical order than is a descriptive predicate. At the same time, in characterizing the set it characterizes the thing and thus *seems* to be of the same logical order as a descriptive predicate but actually it is not. This is the logical solution of Moore's paradox of intrinsic value.

The science corresponding to these solutions is not, as Moore believed, ethics but axiology, the latter being the formal science of which the former is but one application. Moore sees that the advance of ethics must pass through an understanding of the word "good" in its most general meaning; but he does not know how to open up this meaning, thinking it unanalyzable. Some of his British colleagues have thought it could be analyzed, even though, as Wittgenstein has said, it is "a terrible business — just terrible! You can at best stammer when you talk of it." [34] Some of Wittgenstein's followers and some semanticists have thought they could get at the meaning of "good" by the classification of its uses. Their tool is not logic but ad hoc linguistic constructions.[35] Occasionally, the *logical* nature of "good" is recognized, even its dependence on "the logical relation between member and class," [36] but it is not taken seriously.

Of the three recent efforts to penetrate the nature of "good," the intuitive approach of Moore, the semantic of the Oxford ethicists, and the syntactic of the formally logical method, the latter, it seems, not only promises to open up the full meaning of the word, but also to give a fuller meaning to the other two approaches. The reason for this power of the formal method is its higher abstraction. If natural science has developed through ever higher and higher abstraction, there is no reason why moral science should not. "Nothing is more impressive," writes Whitehead, "than the fact that as mathematics withdrew increasingly into the upper regions of ever greater extremes of abstract thought, it returned back to earth with a corresponding growth of importance for the analysis of concrete fact. The history of the seventeenth century science reads as though it were some vivid dream of Plato or Pythagoras." [37] Knowledge of moral nature, we hold, lies in the same direction. Hence, we regard as mistaken those schools of moral philosophy which believe that knowledge of the concrete lies in the concrete, whether of linguistic analysis or situational observation. Since the good is not a descriptive property but a logical one it cannot be grasped by the senses. It is in this respect an "irreal" quality, as the phenomenologists hold. Trying to catch it by naturalistic methods is like catching an electron by observation. The unreality or nondescriptiveness of good is like an uncertainty relation of moral science. The only way to penetrate to goodness is by the eye of the mind, a *Gedankenexperiment* in the sense of Planck, an intuition in the sense of Moore.[38] One has to capture its *formal structure.*

Moore was not capable of doing this. Yet, he came to the threshold of the axiom of a moral science. He did, indeed, write its Prolegomena. He did so, as we said, no more and no less obscurely, than Kepler wrote those of the science of nature. Moore's role in today's moral theory — as the bridge between value philosophy and value science — illustrates well the formal analogy between natural science and moral science discussed earlier.

There is in contemporary value theory the same division of labor and talents that existed in natural science in the seventeenth century. Some like Tycho Brahe excel in observation and accumulation of materials, others, like Kepler, Galileo and Newton, in analysis and synthesis. Some eventually reach the goal, as did Moore, others wind around and around it, but all make their contribution, either positively or negatively, just as out of the fantasies of the alchemists arose chemistry, and in the labyrinth of Kepler's obscure and labored works on astronomy were buried the planetary laws — still, in the deductive context of Newton's *Principia,* the cornerstones of all natural science. In the same way, in the labyrinth of Moore's obscure and labored works on axiology there are buried the laws of value which, in the deductive treatment of formal axiology, will be the cornerstones of moral science.

The analogy between natural science and moral science, thus, is very close indeed. The present opposition in ethics between naturalists and nonnaturalists had its counterpart in the opposition between the Aristotelians, on the one hand, and Kepler and Galileo on the other. The Aristotelians continued seeing the world within the medieval frame of reference, analytic commonsense concepts which, in alchemy, astrology, and the like, became more and more bizarre, and out of focus with reality. But, to their minds, it was Galileo and Kepler whose fantasies were out of touch with reality.[39] So today the naturalists, as Kecskemeti has pointed out, "argue from faith in simple analogy, not supported by factual evidence," [40] while to them it is the nonnaturalists who disregard valuational reality and substitute for it an illusion. The differences between Tycho Brahe and Kepler, and Galileo and Kepler correspond to those in the nonnaturalist camp. Tycho Brahe saw well the difference between Ptolemaic and Copernican cosmology and was a tireless collector of observational material. But he never quite knew which system it confirmed; and, lacking the vision of Kepler and being excessively cautious in not going beyond his observations, he produced his own fanciful system, with the planets revolving around the sun, but *in addition* the sun revolving around the earth — a mixture of the old and the new cosmology. So we have in today's value theory writers who see the difference between fact and value but not enough of it to construct a new, autonomous, purely valuational nonnaturalistic system. Rather, they remain stuck in a refined form of naturalism, especially what we called the normative fallacy, which seduces them to operate with fanciful imperative and similar "logics," corresponding to the epicyclical fantasies of the half-hearted Copernicans — epiaxiological logics of half-hearted Mooreans.[41]

The difference between Kepler and Galileo is found within Moore himself. Moore has, in today's axiology, the position that all three, Copernicus, Kepler and Galileo, had in the beginnings of astronomy; and this in itself is a measure of his significance. His Copernican thesis is the unobservable, nonnaturalistic uniqueness of Good, his Keplerian refinements are his closer and closer *logical* determinations of the thesis, and his Galilean confirmations are his critical analyses of traditional ethical theory.[42] Neither Copernicus nor Kepler nor Galileo knew the true system of the heavens — which was to be the Newtonian — but all three knew that they were laying the foundations of some such new science. Galileo did so empirically, by confirming the Copernican hypothesis, Kepler did so theoretically, by refining it through his three laws. Moore does not know the true system of axiology, but he knows that he is laying the foundations of a new science. He does so both empirically, by the critical observation of traditional ethical theories and their analysis in terms of his Copernican thesis, and his instrument, the naturalistic fallacy; and theoretically, by refining his thesis in three steps, first, the statement of the nonnaturalistic nature of good; second, the connection of this nature with the logic of description; and, third, the clarification of the meaning of "natural" and "nonnatural" intrinsic properties, the statement of the logical connection between the two, and — the outlook to the future — the statement that if the logical nature of description were known the nature of value would be known. As Kepler's writings were stretched over many years, the three laws following in three works, the "prolegomena" in 1596, the first two definitions in 1609, the third and crowning definition in 1619,[43] so Moore's statements stretched over many years, the "prolegomena" in 1903, the first two determinations in 1922, the third in 1942.[44] As Kepler's three laws were buried like pearls among miscellaneous oddments he collected with tireless zeal among the refuse of the past — Pythagorean and Platonic mysticism, disdained by good Aristotelians — and as he insists on showing the reader through all the by-ways and alleys where he found them, all the clues he followed and lost, the trails and triumphs of his soul, so Moore's exact logical determinations of value are hidden pearls among oddities gleaned from traditional ethics, appeals to intuition, and an insistence to let the reader partake of the trials and tribulations — there were no triumphs — of his thought. As the layers and layers of complexities Kepler puts in the way of his readers prevented them from penetrating to the priceless core of his work and from recognizing the true significance of it — including the otherwise so astute Galileo — so Moore's axiological contemporaries were unable in the obscurities of his work to detect the crystal-clear core of it and never bothered *logically* to investigate what he logically — and yet so enigmatically — proposed.[45] And, to make the analogy complete, since Moore is his own Galileo, he himself never really understood his own theoretical thought, which fought with his empirical ideas; his Keplerian nature with his Galilean nature, to the extent that at one point in his career he betrayed himself with himself. The Galileo in him betrayed the Kepler, his empirical betrayed his theoretical thought, his naturalistic betrayed his nonnaturalistic view of value. In the "Reply to My Critics" he professed to be drawn equally toward ethical objectivism and to emotivism, to nonnaturalism and to naturalism — a temporary confusion which he never bothered to correct because he completely forgot it; so completely did he go back to his original nonnaturalistic, Keplerian, position.[46]

That the true logical significance of Moore's theory is not understood is analogous not only to Kepler but also to Galileo. In spite of Galileo's savage and empirically irrefutable arguments against Aristotelian physics, his observations, since they did not fit into the Aristotelian framework, were never understood by his Aristotelian contemporaries who even refused to look through his telescope, or refused to see anything if they looked. So Moore, in spite of his acute and logically irrefutable arguments against naturalism is not understood by his naturalistic contemporaries who even refuse to make the thought-experiment of accepting his thesis as an hypothesis and see where it would lead. Rather, they simply refuse to look and, shrugging their shoulders, deny that they see what he pretends to be showing them.[47] Finally, just as Copernicus, Kepler, and Galileo were predecessors of Newton, so Moore is the predecessor of a future Newton of axiology who will combine all the strands found in Moore into one all-encompassing theory of value.

As is seen, these analogies between science and ethics are formal, not material. Materially, of course, no connection can be seen between, say, Moore's entailment of nonnatural intrinsic by natural intrinsic properties and Kepler's third law, or between Galileo's mistaken disregard of Kepler, and Moore's mistaken disregard of his own original thesis. In general, materially, no connection between factual and valuational knowledge can be detected. But formally this connection is not only obvious but compelling. It is, indeed, logically necessary; for all knowledge is methodologically one. It is built on the process of axiomatic identification.

We have developed the axiom of value science from Moore's formula, which ties goodness to description, by identifying Value with one-to-one correspondence of intensions;[48] and Goodness with an intensional interpretation of the class-membership relation. We have seen how this axiom not only solves old puzzles of value philosophy but also Moore's own paradoxes. In the next Part we shall develop the axiom into the system of formal axiology.

PART THREE

The Structure of Value

5

THE ELEMENTS OF THE AXIOLOGICAL SYSTEM

> Formal logic may be extended to include a *formal axiology and method*. This means the rise of a formal logic of goods or values. Each propositional sphere has its "syntactic" categories, its particular fundamental modalities, its deductive forms; and hence each has its own "formal logic" or "analytic."
>
> — EDMUND HUSSERL [1]

> *Good*. . . . Of things: Having in adequate degree those properties which a thing of the kind ought to have.
>
> – *Oxford English Dictionary*

Before examining the systematic structure of formal axiology let us summarize our argument. Moore's two different propositions, both true of *goodness*, that it does depend *only* on the intrinsic nature of what possesses it, and that, *though* this is so, it is yet not itself an intrinsic property, are taken by us not as Keplerian guesses but as Galilean foundations of a new science and, we suggest, precisely that science which Moore himself had in mind. We eliminate the paradox of the conjunction of these two propositions by regarding them as pertaining to two different logical levels. We thus obtain the *axiom of value* which defines "good" as a logical term, namely as *the predicate of any subject said to fulfill the intension of its concept*. This axiom has the fundamental properties that make it scientific, and thus has both systematic and empirical import. It is the basis of a *formal system applicable to its corresponding reality*, that of value phenomena; a system, that is, in the strict sense of consisting of synthetic constructs, or terms deduced from the axiom by formal relations. Thus, it is not the basis of a philosophy whose propositions arise by implication from the intension of "good" as an analytic concept.

Both, then, the axiom and the system following from it have systematic and empirical import. The systematic import of the axiom is its internal logical structure, its resolving the Moorean paradoxes, and its capacity of bringing forth the system. The empirical import of the axiom are those practical consequences for the value world which follow directly from it. The systematic import of the system is its logical structure which, being isomorphous with the value realm, represents at the same time the structure of value. The empirical import of the system is its capacity of accounting for the value realm, its applicability. We shall, in this and the following chapter, examine the systematic import of the system and later some of the aspects of its empirical import. We will begin with the primary elements of extrinsic value, the Value Terms.

1. *The Value Terms*

"Value" is a term that has no intension in the analytic sense but only in the synthetic sense: it is defined as a formal relation, namely, the correspondence between the properties possessed by a subject and the predicates contained in the intension of the subject's concept. Thus, the proposition "The chairs in this room are good" means that there are some things in this room which are called "chairs" and have *all* the properties connected with the concept "chair." Or, more precisely, some things in this room are members of the class of chairs; the class of chairs is defined by the concept, or name (depending on the kind of definition one prefers), "chair," and is characterized by certain properties; and the things in this room called "chairs" have the properties in question. More generally — and this is the logical elaboration of the axiom with respect to the value "good" —

> a thing x is a good C if and only if (I) x is a C, (II) C has the intension Φ consisting of properties ϕ, (III) x has all the properties ϕ.

Other values than "good" arise from the modification of "all" in (III).

The axiom thus contains not just one formal relation but a set of three of them, interrelated in a small network of relations.

All these relations are *logical* so that the axiom, to paraphrase Hall,[2] is a miniature logic devised to account for value. Relation (I) is that of class membership, (II) is that of conceptual analysis, (III) that of predication. The system combines the extensional and the intensional views of a class. In (I) we have class extension, signified by "$x \varepsilon C$"; in (II) we have class intension, which we shall discuss presently; in (III) we have the combination of both, "ϕx," where "ϕ" is the intensional and "x" the extensional constituent.

The relationship between a concept and its analytic parts has not yet been systematically investigated in modern logic. The concept as consisting of such parts is said to be "complex" [3] and to "contain" these parts. Let us symbolize the relation "is contained in," in the Kantian-Moorean sense, by some sign, say "ω." "$\phi \omega C$" then means that any or all of the values of ϕ are contained in C. "$\phi \omega C$" in other words, corresponds on the intensional side to "$x \varepsilon C$" on the extensional side. Just as "$x \varepsilon C$," through either quantification or specification of "x," can become an extensional proposition, so "$\phi \omega C$," through quantification or specification of "ϕ," can become an intensional proposition. Thus, "$(\phi)\ \phi \omega C$" signifies that all values of ϕ — α, β, γ, δ, ζ and so on — are contained in C, "$\gamma \omega C$" means that the specific value γ, say "rational," is contained in C, which may stand for "man," and so on. Through quantification and specification of "ϕ" the concept C can thus be used in whole or in part intensionally, just as through quantification and specification of "x" it can so be used extensionally. If by "C" in "$x \varepsilon C$" is meant the extension of the class concept rather than this concept itself — strictly speaking it should — then ϕ is contained in the intensional part of the class concept, which we designate by "Φ." Rather than "$\phi \omega C$" we should then write "$\phi \omega \Phi$." Then "C" signifies Extension, "Φ" Intension, "ε" the relation between C and the extensional instances, x, y, etc., and "ω" the relation between Φ and the intensional properties, ϕ, ψ, etc. Just as $C = \hat{x}\ (x \varepsilon C)$, so $\Phi = \hat{\phi}\ (\phi \omega \Phi)$. Just as "man" is Peter, Paul, etc., so "human" is "rational," "animal," etc.

The axiological pattern, then, consists of the sequence of the functions (I): $x \varepsilon C$; (II): $\phi \omega \Phi$; and (III): ϕx, which latter is entailed by the conjunction of (I) and (II): if x is a member of C and ϕ is contained in Φ, then x is ϕ. If this entailment is quantified with respect to "x" we call it *logically quantified,* and if with respect to "ϕ" we call it *axiologically quantified.* To be *universally axiologically quantified* means that x, being a member of C, has all the properties ϕ contained in C (or Φ). In this case x is a good member of C. This is our formal definition of "good."

The expression "x is a good C" is axiologically a proposition because it is axiologically quantified ("good"), but logically a propositional function because it is not logically quantified. "Some x are good C's" would be a proposition both logically and axiologically. An *axiological propositional function* would be the expression: "x is a C of some (unspecified) value" or "x is a C," for any x has some value as a C. Thus any logical proposition or propositional function is *as such* an axiological propositional function. Since a proposition is a logical value of a propositional function we may say that an axiological proposition is an axiological value (a quantification or specification with respect to value) of a logical proposition or a logical propositional function. On the other hand, we have seen above that value is that variable the logical values of which are axiological values. *Thus axiological values are either logical values of axiological propositional functions, or axiological values of logical propositions or propositional functions.* If logical expressions are regarded as factual and axiological expressions as valuational, this defines with precision the relation between fact and value.[4]

Thus, the simple form of the axiological proposition "x is a good C" disguises a certain complexity. It means the following things:

I. x is a member of C with intension Φ
II. Φ contains $\alpha, \beta, \gamma, \delta, \zeta, \ldots$
III. 1] x is α
2] x is β
3] x is γ
4] x is δ
5] x is ζ
. . . . . .

This total pattern is what we call the *value pattern* of the proposition x is a good C. If and only if this pattern is meant does the proposition mean that x is a good C.

There is, then, a difference between "x is a C" and "x is a good C." "x is a C" (I) does not entail $(\phi)\phi \ \omega \ \Phi$ (III). A propositional function does not entail any of its propositions (or logical values). "x is a C" does not necessarily mean that x must have *all* the properties contained in C (or Φ). If, however, x does have all the properties, then x is a C as well as a good C. The reason that these two expressions have either been distinguished too much or too little is that the logical distinction between them is almost imperceptible. We have here the razorsharp, razor-thin distinction between fact and value of which we spoke in the Introduction. In "x is a C," every predicate contained in C (or Φ) is one of x. But in "x is a good C," the predicate "good" is *not* a predicate of x, even though it may look as if it were, but a predicate of the predicate(s) of x, a predicate of C (or Φ) — if not a predicate *of* a predicate, or predicates, of C (or Φ). Thus, "x is a good C" is at least a second-order function, that is, a function *of* a first-order function. In other words, when a thing is what it is, it *is* what it is, and that is the end of the matter. But when a thing *fully* is what it is, it *not only is* what it is but is *well* what it is: value has been added to its factuality. A thing that has all the chair qualities is not only a chair but also a good chair. To be fully a chair and to be a good chair is the same thing; but to be fully a chair and to be a chair are different things. To be a chair is a function of the form "$\phi\, x$," to be fully a chair is a function of the form "$(\phi)\phi\, x$." [5] "Fully a chair," "a real chair," "a good chair," "ah, a chair!" etc., are not the same as merely "a chair." That they seem to be the same is due to a fundamental characteristic of language, not always recognized in its universality, although Bertrand Russell has touched upon it in his Axiom of Reducibility (introduced for purely mathematical reasons): that language contains an infinity of words which appear as first-order functions, or as predicative functions of any order, which are equivalent to functions of a higher order. They are higher-order functions "reduced" to predicative functions — single predicates equivalent to combinations of predicates.[6]

"Good" is such a reduced function; it hides within it a whole hierarchy of logical orders. *x is good* is a first-order function equivalent to a function, or functions, of higher orders. It is this higher-order content in the seemingly simple predicative function "good" that has been hidden so far, and which is being made explicit by formal axiology.

Moore's paradox of the two different propositions both true of goodness, that 1] good is not a natural intrinsic property but

2] depends only on the natural intrinsic properties of the thing that has it, may be stated logically as (1) *x is good* is a predicative function without a descriptive predicate which (2) is equivalent to a set of such predicates. And our addition to Moore's statement, and the solution of its paradox — that the value of a thing depends on the correspondence of its properties with those contained in its concept — may be rendered by saying that 3] value is a second-order function. The secret of value, then, is *not to be a simple predicative function but to appear as one* (Moore) or *to be (at least) a second-order function reduced to a first-order function* (formal axiology). In one statement: *the secret of value is to be a reduced higher-order function.*

"Good" is only one of an infinity of such functions throughout all realms of language. A whitewashed house, in the second-order sense, may or may not be actually whitewashed in the first-order sense; a nurse on duty, in the second-order sense, may actually be asleep; a ship at anchor may actually drag its anchor; an article in stock, in the second-order sense, may actually be out of stock. Such double-edged — indeed, multi-edged — expressions have recently been said to refer to a new irreducible "major category of human thought," called "continuous states." [7] Actually, they are multi-order logical statements capable of reduction to predicative statements. Their mercurial character, making them capable of now appearing as predicative functions, now as higher-order functions makes them highly disturbing elements of language, like comets swishing through its well-ordered circles. Like comets, they appear as harbingers of disorder, of linguistic confusion: second-order (and higher-order) statements appearing misleadingly in the form of first-order statements, statements about *concepts* appearing in the misleading form of statements about *instances*. Like comets, they were hard to discover, and it seems Rees — following glimpses and preliminary reports of Ryle on occurrences in one particular field — is the first to have discovered them. Like comets, their regularity and orderly nature was not immediately seen. Rees says that "these expressions fall under some new and independent category." However, their category seems simply to be logic. In all cases, the ambiguity of the expression is due not, as Rees says, to the fact "that they may signify either an event or a continuous state initiated by the event" and the like, but to the fact that they may appear as *either instantial or conceptual expressions.* Rees sees only specific cases but not the general law. Actually, "continuous states" are *classes of* events, *of* dispositional, relational or other characteristics "normally associated" with them. They are of a higher logical order *than* these events, characteristics, etc. Hence, Rees says the same of them that Moore says of the value property in relation to the descriptive property — and this in itself is a proof of our view: "If my furniture is in store no amount of *inspection* of my furniture will reveal that it is in store," "We cannot meaningfully describe any of these [states] in the language which is appropriate to the description of events," etc. But he does not see the logical nature of the "category." Indeed, Ryle seems to be closer to it, for his "frame of mind" is a subjective version of "frame of reference" and his "mock murder" theory actually calls these states "a class of higher order events, that is, ordinary events involving in some way the thought of other events, just as a stage murder is not a real murder but an ordinary event which involves in some way the thought of real murder." This doctrine, says Rees, "is now unnecessary." [8] It is, not because the events mentioned by Ryle are "continuous states," but because they represent *second-order functions reduced to predicative functions.* They are, in all cases, conceptual expressions capable of appearing in instantial form.

It may be that all instances when regarded in conceptual form, or all predicative functions when regarded nonpredicatively — e.g. "Her nose is red" (because of the cold) as against "She's got a red nose," [9] "Deaf-mute burglar refuses to talk," [10] "Meeting of Local Democrats Abroad," [11] "A rarely resident writer-in-residence," [12] etc. — are valuations of some sort; and some of Rees' examples seem to bear this out: "their doctors are no doctors at all," or "we live our lives from the day we are born until the day we die, but in another sense we do not really *live* unless we engage in life's normal activities and enjoy its normal enjoyments." With other examples this is not so clear; yet our axiom shows that once the "continuous state" is regarded as a concept with respect to its instantial "events," this state serves as *norm* for the event or events. These things then are members of the continuous state, having its intension, *but not fully.* Thus, a whitewashed house not actually white, a nurse on duty who is asleep, a ship at anchor dragging its anchor, an article in stock that happens to be out of stock, and a person in love who happens not to feel tender, or one "in prison" who happens to be at home — without *being* at home, in the second-order sense — all these do not fulfill their respective concepts and hence are not, in the instant in question, good whitewashed houses, nurses on duty, lifers, etc. And those who only vegetate and do not live "fully" do not fulfill the intension of "life" and are not good "livers, not living the good life."

Even though this seems true we do not insist on it; and rather than saying that all reducible higher-order functions — or reduced predicative functions — are valuations, we shall say that all valuations are such functions and hence that they form part of a large and fundamental department of language, the one glimpsed by Russell in his axiom of reducibility. But we shall not say, although it seems almost certain, that this is the department of valuation; and that any reduced function, that is, any predicative function equivalent to a higher-order function, is a valuation.[13]

The transition from (I) x is a member of C, with intension Φ, to (II) Φ contains properties ϕ, to (III) x has all the properties ϕ contained in Φ, then, is a transition from a lower to a higher logical type, and back again,[14] and to a higher logical order. The relation between (I) and (II), taken together, and any of the propositions in (III) — x is α, x is β, x is γ, etc. — is that of entailment. This entailment, as again Kant has shown (in *Die falsche Spitzfindigkeit der vier syllogistischen Figuren,* which may well be translated as The Synthetic Subtlety of the Four Syllogistic Figures in both the early Kantian and the modern sense of "synthetic" as "artificial") represents the original and natural function of the human mind, expressed by the rule *nota notae rei ipsius nota* — itself a nice formulation of the axiom of reducibility, also called by Russell "the axiom of classes." If x is a C and C through Φ is α, β, γ, δ, ζ, then the propositions in (III) are analytic propositions following from the premise that x is a C, and they are all the analytic propositions possible.[15]

The axiological proposition "x is a good C," then, stands for a totality of analytic entailments or a value pattern, which defines "good" as a logical term. The definitions of the other value terms follow logically. If x *has* some of the properties ϕ of C, then x is a *fair* or a *so-so* member of C. If x *lacks* some of the properties ϕ of C, then x is a *bad* member of C. And if x lacks most of the properties ϕ of C (or all of the expositional properties, that is, all the properties beyond the first two, which are definitional) then x is a *no good* member of C.

The question is, whether "bad" means that a subject partly fulfills or partly does not fulfill its definition. The answer must be the latter: the emphasis must be on nonfulfillment. On the one hand, partial fulfillment has a positive emphasis and therefore means "fair," and the like. On the other hand, "bad" is a negative term, "not bad" a positive one. The difference is well illustrated by a little dialogue. *Obstetrician:* "Mrs. Jones, I have very good news for you." *Patient:* "I am not Mrs. Jones, I am Miss Jones." *Obstetrician:* "Miss Jones, I have very bad news for you." A term x is "good for" another term y if x partly fulfills the definition (exposition) of y. A child is "good for" Mrs. Jones but "bad for" Miss Jones — it partly fulfills the exposition of "Mrs. Jones" but does not fulfill the exposition of "Miss Jones." This means that "bad" means partial nonfulfillment (not having some properties of C) rather than partial fulfillment (having some properties of C). To *have* a husband and child fulfills the exposition of "Mrs. Jones," to *lack* husband and child fulfills the exposition of "Miss Jones." Not to lack a child is partial nonfulfillment of the exposition of "Miss Jones," whereas to lack a husband is partial fulfillment of this exposition. The comparison between Miss Jones and Mrs. Jones shows that what actually is *bad* is the *transposition* of fulfillments and nonfulfillments. In having no husband but a child, Miss Jones partly fulfills her exposition and partly that of Mrs. Jones, or, she partly fulfills (having a child) and partly does not fulfill (having no husband) the exposition of Mrs. Jones. According to our definition of "good for," insofar as Miss Jones partly fulfills the exposition of Mrs. Jones, Miss Jones is good for Mrs. Jones. That is, having the baby is good for her future marriage, for the baby's father may marry her. But insofar as Miss Jones partly does not fulfill the exposition of Mrs. Jones, Miss Jones is not good for Mrs. Jones. Having no husband, yet a baby, is not good for her future marriage to another man. (Both men, for argument's sake, have the name "Jones.") The question could be simplified by defining "fair" as "having more properties than lacking them" and "bad" as "lacking more properties than having them."

Axiological and logical classification must be strictly distinguished. Any axiological proposition must be analyzed both logically and axiologically. There is, as we have seen, both logical and axiological quantification and qualification. *All x are good members of C* is logically as well as axiologically universal and positive; it means that all *x* have all the properties of C. *All x are bad members of* C is logically universal and positive, axiologically particular and negative; *No x are good* is logically universal and negative, and axiologically universal and positive; *Some x are bad* is logically particular and positive, axiologically particular and negative; and so on. Also, *x is* C is both logically and axiologically a propositional function; *All x have some kind of value* is logically a proposition and axiologically a propositional function; *x is a good* C is logically a propositional function and axiologically a proposition, etc.

While "good," "fair," "bad," and "no good" are the basic value terms, there are a great number of equivalents which are to them as the linguistic quantifiers — "few," "only," and so forth — are to the four basic quantifiers. Such linguistic axiological quantifiers (qualifiers) are "excellent," "fine," and so on for *good;* "so-so," "not bad," "o.k.," and so forth for *fair;* "poor," "not good," "inferior," "deficient," and so forth for *bad;* "rotten," "lousy," "miserable," and so on for *no good*. Note that "no good" is the contrary and "not good" or "bad" the contradictory of "good." Similarly, "not-bad" is the contrary rather than the contradictory of "bad." The axiological square of oppositions shows "good" and "no good" as contraries, "fair" and "bad" as subcontraries, "good" and "bad," and "no good" and "fair" as contradictories.[16] Sometimes expressions identical with, or similar to, logical quantifiers are used as axiological quantifiers, for instance, "some . . . ," as in "some boy!", "quite a . . . ," "quite the . . . ," "not much of a . . . ," and so on. Sometimes axiological quantifiers are used as logical quantifiers, for example, "lousy with . . . ," or the French "n'importe . . . ," for "some." If the qualifier is "more than good" it means that "a new concept" of the thing has been established, and the thing is "in a class by itself."

2. *The Value Relations*

All these expressions are axiological predicates. There are also axiological relations. *x is a better* C *than y* means that *x* has more expositional properties of C than *y* and is therefore "more of" a C than *y*. "Better than," in other words, relates two members of the same class, the first of which has more of the class properties than the second. The converse axiological relation, that between *y* and *x*, is the relation "worse than"; *y is worse than x* means that *y* has fewer expositional class properties than *x*.

x is good for y means, as we have seen, that *x* and *y* are in different classes but have overlapping intensions such that the intension of *x* is part of that of *y*. *Hay is good for horses* means that "hay" and "horse" are in different classes but that the intension of "hay" is part of that of "horse"; the digestive tract of horses has an affinity for something that is hay. *x is bad for y* means that *x* is contrary to some part of the intension of *y*. *Arsenic is not good for horses* means that arsenic is contrary to something which is good for horses. *y is better for x than z* means that *x* is in a different class from both *y* and *z*, but that the intensions of *x* and *y* overlap while those of *x* and *z* either do not overlap or do not overlap in the same degree. *Hay is better for horses than arsenic* may mean either that the intension of "arsensic" does not at all overlap with that of "horse" or that it does so very much less than that of "hay"; in a very specific case horses may be given very small doses of arsenic as medicine. The converse relation, that between *z*, *x*, and *y*, is *is worse for . . . than; Arsenic is worse for horses than hay.*

It is good that xRy means that the relation R is part of the intension of one or all of its terms. *It is good that John reads "Ivanhoe"* means that reading *"Ivanhoe"* is good for John or that such reading belongs to the nature or intension of John. *It is good that John loves Betty* means that love is good for either John or Betty or both. *It is good that members of society love each other* means that love is good for either members of society or society or both. Predicates may here be regarded as monadic relations.

In all cases of "*it is good that*," e.g. *it is good that x is C*, goodness is attributed to the referent of a proposition, a situation or state of affairs (a *Sachlage*), whereas in the case of "good," e.g. *x is a good C*, goodness is attributed to the referent of a concept, an object (a *Gegenstand*), e.g. a chair. Whereas it can be said that the goodness of the object is the fulfillment of the concept in question, it cannot be said that the goodness of the situation is the fulfillment of the proposition in question. The "fulfillment" of the proposition, as Kant, Husserl, and others have taught, is the fulfillment of its modal range, of the range of its modalities, from negatoriness to apodicticness. A completely fulfilled proposition has the value *truth* — in Husserl's sense of the term—and not the value *goodness.* Conversely, truth, in this same "full" sense, is the fulfillment of a proposition, goodness the fulfillment of a concept.[17] Thus, what is *truth-value to a proposition is good-value to a concept.* The scholastic saying that *bonum et verum convertuntur* has then an interesting axiological interpretation: the fulfillment of a proposition is *verum,* the fulfillment of a concept is *bonum.* While syntactically *bonum* and *verum* are parallel, semantically they are not. It is the proposition, not its referent, that is called *verum* while it is the referent of the concept, not the concept itself, that is called *bonum.* The *goodness* of a situation is the fulfillment of its *concept,* not of its corresponding proposition. On the other hand, just as there are synthetic, analytic, and singular intensions or contents of concepts, leading, respectively, to systemic, extrinsic, and intrinsic value, so there are systemic ("apophantic"), objective, and subjective intentions of judgments, with their corresponding modal truth values. The modes of fulfillment of *intention* correspond to the modes of fulfillment of *intension.*

It is better that . . . than that . . . has two forms. *It is better that* xRy *than that* xRz (or *it is better for x-to-Ry than for x-to-Rz*) means that the relation R is part of the intension of *x* and *y*, but either not at all or not in the same degree of the intension of *z. It is better that John read "Ivanhoe" than that he read "Lady Killer Comics"* means that reading "*Ivanhoe*" does but reading "*Lady Killer Comics*" does not, or does not in the same degree, belong to John's intension. *It is better for John to date Betty than Lou* means that John's and Betty's dispositions are more compatible than John's and Lou's. The converse relation is *it is worse that* xRz *than that* xRy. The second form of the relation in question is *it is better that* xRy *than that* xSy. This means that the relation R but not, or not in the same degree, the relation S is part of the intension of either or both *x* and *y. It is better that John love Betty than that he beat her* means that it is better for John and/or Betty and/or both that he love rather than beat her. Predicates may again be regarded as monadic relations. *It is better that John studies than that he loafs* means that studying does, or does to a higher degree, fulfill his intension than loafing.

Since for anything to be good means fulfilling its intension, *it is better for x to be good than to be bad* is always true. The converse relation is *it is worse that* xRy *than that* xSy. *It is worse for x to be bad than to be good* is always true. [*It is worse for x not to fulfill its intension* (xRy) *than to fulfill it* (xSy).]

In summary, "better" and "worse" are axiological relations *between members of the same class;* "good for" and "bad for" and "better for" and "worse for" are axiological relations *between members of different classes;* and "it is good that" and "it is bad that," and "it is better that" and "it is worse that" are axiological *relations between relations.* What is missing, then, is an axiological relation between a member and its class.

This relation, *the axiological copula,* is the relation "ought."

"Ought" does not relate two things, such as *x* and *y*. To say that *x* ought to be *y* is meaningless, since it says that one thing ought to be another. Rather, "ought" relates things to concepts or relations. *x ought to be a C* or *x ought to be* Φ means that either *x* is a member of *C* and is or is not deficient in its C-ness, or that *x* is not a member of *C* but, say, of *B*, and that it would be "better for *x* to be" a *C*. "Ought" thus is equivalent to the relation "it is better that." *x ought to be* Φ is equivalent to *it is better for x to be* ϕ and *x ought to be a C* is equivalent to *it is better for x to be a C.*[18] *x ought to . . .* thus is an elliptic expression for *it is better that* xRy *than that* xRz or *it is better that* xRy *than that* xSy. In both cases the "than that . . ." part is suppressed. Thus, *John ought to read "Ivanhoe"* is equivalent to either of the two forms of *it is better that,* namely, 1] *it is better for John to read "Ivanhoe" than to read something else, say, "Lady Killer Comics,"* and/or 2] *it is better for John to read "Ivanhoe" than not to read it* (or *to burn it,* or *eat it,* or the like). *John ought to love Betty* may mean 1] *it is better for John to love Betty than to love Lou,* and/or 2] *it is better for John to love Betty than to beat her* (or *not to love her,* and so on).

Again, *x ought to be good* is always true; for it is equivalent to *it is better for x to be good than to be bad,* which we found to be always true. Since this is a monadic relation, there is but one form of it.[19] Since "ought" is equivalent to "it is better that," *x ought to be . . .* means *it is better for x to be.* In other words, what *x* ought to be is what is better for *x* to be. What is better for *x* to be is what better fulfills *x*'s intension. Thus, if *x*'s intension is γ, δ, ζ, what *x* ought to be must add to this intension some properties which make *x* better. Hence, what *x* ought to be is better than what *x* is. Conversely, what *x* ought not to be is worse than what x is. "Ought" thus relates the worseness of a thing to its betterness, and "ought not" relates the betterness of a thing to its worseness: the worse ought to be better, the better ought not to be worse.

In terms of *choice* or *preference* this means that we ought to choose or prefer what is good; [20] that we ought not to choose what is bad; that we ought to choose what is better; and that we ought not to choose what is worse. It also means that 1] it is good for us to choose what is good for us rather than what is bad for us, 2] it is good for us to choose what is better for us rather than what is worse for us, 3] it is better for us to choose what is good for us rather than what is bad for us, and 4] it is better for us to choose what is better for us rather than what is worse for us. In terms of *existence* it means that, since an existing thing has more properties than a nonexisting thing, existence is better than nonexistence, it is better to exist than not to exist, things ought to exist rather than not exist.

Besides axiological predicates and relations there are axiological terms which are combinations of predicates and relations, such as, for example, "best" and "worst." *x is the best* C means that x is the one and only C that has the maximum of intensional properties of C, and *x is the worst* C means that x is the one and only C that has the minimum of intensional properties of C—which does not have to mean that x is no good. The worst C may be a fair one. The axiological relations are summarized in Table 2.

2. *Axiological Relations*

RELATIONS BETWEEN	AXIOLOGICAL RELATIONS
Members of same class	*better than; worse than*
Members of different classes	*good for; bad for; better for; worse for*
Relations	*it is good that; it is bad that; it is better that; it is worse that*
Members and class	*ought*

A great deal could be said about all these relations, but we shall confine ourselves to one of them, the axiological copula "ought." This copula, as do all axiological relations, has what we call a positive sense and a negative one. The *positive sense* is the direction toward fulfillment of the intension, the *negative sense* being the direction toward nonfulfillment of the intension. The intension may be of one of the terms of the proposition, either the predicate or, if the predicate refers to the intension of the subject, the latter. Thus, *x ought to be good* uses "ought" in the positive sense: the predicate refers to the subject and goodness means fulfillment of the subject's intension. On the other hand, *x ought to be bad* uses "ought" in the negative sense: the predicate refers again to the subject, but badness means nonfulfillment of the subject's intension. *These houses ought to be roofed, bachelors ought not to marry* are also positive uses of "ought," for "roofed" and "not to marry" refer to the intension of the subjects, and in a positive way, for houses that are roofed and bachelors that do not marry fulfill their respective intensions. On the other hand, *these houses ought not to be roofed, bachelors ought to marry* are negative uses of "ought." (This does not mean that bachelors ought not to marry; it only means that if they marry they are no longer bachelors). Examples of where not the intension of the subject but that of the predicate is in question are propositions of the type *x ought to be a C, x ought not to be a not-C,* which are both positive uses of "ought," and *x ought not to be a C, x ought to be a not-C,* which are negative uses of "ought." Here the predicate does not refer, positively or negatively, to the intension of the subject. All *axiological* predicates — "good," "bad," and so forth — refer to the intension of the subject, and therefore the positive or negative sense of "ought," in the case of all axiological propositions with an axiological predicate, depends on the intension of the subject; whereas the sense of "ought" in the case of axiological propositions with *nonaxiological* predicate — "C" — depends on the nature of the predicate. In the former case, the logical *and* the axiological quality of the proposition determines the sense of "ought"; in the latter case only the logical quality does, for there is no axiological quality.

Let us call any predicate other than an axiological predicate a *logical* predicate. Then *x ought to be a C* is an axiological proposition with a logical predicate whereas *x ought to be good* is an axiological proposition with an axiological predicate. Both are simple value propositions. Conjunctions of both or compound axiological propositions would be x *ought to be good and a* C, x *ought to be good as a* C, x *ought to be a good* C, and the like. In the following, we shall deal mainly with simple value propositions.

The *subjects* of all simple value propositions are logical, not axiological terms; x *is a good* C, x *ought to be a* C, etc., all start with "x." An axiological term can be the subject of a value proposition only if the proposition is compound, for example: "*the good* x . . . ." A subject such as *the Good* . . . is a highly complex term whose meaning can only be ascertained by axiological analysis.[21]

3. *The Value Propositions*

Value propositions now fall into four categories: 1] propositions with logical predicate, 2] propositions with axiological predicate, 3] propositions with logical copula, and 4] propositions with axiological copula — the logical copula being "is," and the axiological copula "ought." Value propositions with logical copula will be called *logical value propositions*, value propositions with axiological copula *axiological value propositions*, or *axiological propositions proper*. Logical value propositions with logical predicate will be called *pure logical* value propositions and logical value propositions with axiological predicate *mixed logical* value propositions. Axiological value propositions with axiological predicate will be called *pure axiological* value propositions and axiological value propositions with logical predicate *mixed axiological* value propositions. Thus, x *is a* C is a pure logical, x *is good* is a mixed logical, x *ought to be good* a pure axiological, and x *ought to be a* C a mixed axiological proposition (or propositional function). This is summarized in Table 3.

3. *Value Propositions*

Logical Propositions (Copula *Is*)		Axiological Propositions (Copula *Ought*)	
Pure Logical (Logical Predicates)	Mixed Logical (Axiological Predicates)	Pure Axiological (Axiological Predicates)	Mixed Axiological (Logical Predicates)

Value propositions are, as we have seen, subject to both logical and axiological quantification and qualification. Logical quantifiers and qualifiers are sometimes explicit words, sometimes implicit. Thus, the word "no" is both a quantifier and a qualifier, and the affirmative qualifier is always implicit in the copula. Axiological quantification and qualification are always implicit in the value predicate. For purposes of analysis, the implicit quantifications and qualifications must be made explicit, just as the quantifiers "all," "no," "some," "some . . . not," are made in school logics, by means of the letters A, E, I, O and their explanations. "Good," as we have seen, is the universal affirmative, "no good" the universal negative, "fair" the particular affirmative, and "bad" the particular negative quantifier and qualifier. Let us call axiologically universal and affirmative propositions G-propositions, axiologically universal and negative propositions T-propositions, axiologically particular and affirmative propositions B-propositions, and axiologically particular and negative propositions D-propositions.[22] The axiological square of oppositions follows:

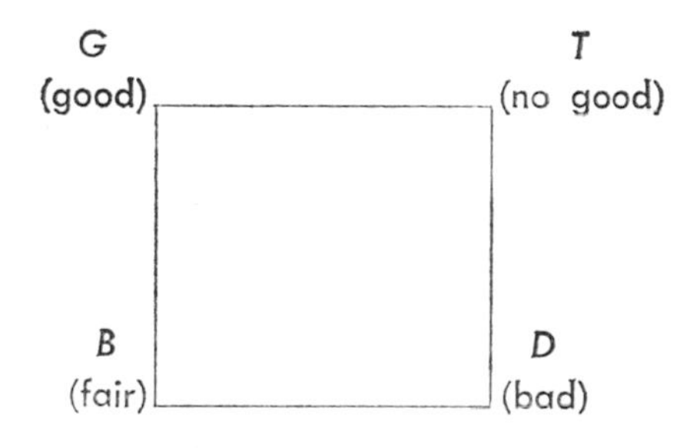

The combination of logical with axiological letters gives all the possible kinds of propositions with axiological predicates, that is, mixed logical and pure axiological propositions. Using "-" for the logical relation between subject and predicate and "→" for the axiological relation between them, we have A-G (*all x are good*), A-T (*all x are no good*), A-B (*all x are fair* or *all x are not-bad*), A-D (*all x are bad*), and correspondingly E-G, E-T, E-B, E-D; I-G, I-T, I-B, I-D; O-G, O-T, O-B, O-D. The forms of the pure axiological propositions are A→G (*all x ought to be good*), A→T (*all x ought to be no good*), A→B (*all x ought to be fair*), A→D (*all x ought to be bad*), and E→G,

$E \rightarrow T$, $E \rightarrow B$, $E \rightarrow D$; $I \rightarrow G$, $I \rightarrow T$, $I \rightarrow B$, $I \rightarrow D$; $O \rightarrow G$, $O \rightarrow T$, $O \rightarrow B$, $O \rightarrow D$.

Following this pattern we may symbolize the value propositions with logical predicates, that is, pure logical and mixed axiological propositions. The former are A-C (*all x are* C), E-C, I-C, O-C; the latter $A \rightarrow C$ (*all x ought to be* C), $E \rightarrow C$, $I \rightarrow C$, $O \rightarrow C$.

Axiological value propositions, that is, propositions with the copula "ought," are called *positive* if the copula is used in the positive sense and *negative* if it is used in the negative sense. Thus, positive axiological propositions are $A \rightarrow G$, $E \rightarrow T$, $A \rightarrow B$,[23] $E \rightarrow D$, $I \rightarrow G$, $O \rightarrow T$, $I \rightarrow B$, $O \rightarrow D$; $A \rightarrow C$, $E \rightarrow \overline{C}$, $I \rightarrow C$, and $O \rightarrow \overline{C}$, where "$\overline{C}$" means "non-C." Negative axiological propositions are $A \rightarrow T$, $E \rightarrow B$, $A \rightarrow D$, $E \rightarrow G$, $I \rightarrow T$, $O \rightarrow B$, $I \rightarrow D$, $O \rightarrow G$; $A \rightarrow \overline{C}$, $E \rightarrow C$, $I \rightarrow \overline{C}$, and $O \rightarrow C$.

It is obvious that some of these propositions are true and some are false. *All x ought to be no good* ($A \rightarrow T$), for example, seems obviously false, for if no *x* ought to fulfill its intension ($E \rightarrow G$) then there ought to be no class of *x*; and since valuation is based on classification, this means that there ought to be no valuation. This, obviously, cannot be axiologically true. Again, *no x are good* (E-G) would mean that there is no class of which *x* is a member since no *x* fulfills any intension. Thus, again, valuation would be impossible and the proposition must be axiologically false. *All x are good* (A-G) seems false for a different reason. In this case all *x* would fulfill their intension and if this is the case the class would be a systemic one and hence extrinsic valuation and the extrinsic values "good," "bad," etc. would not be applicable. Thus, the nature of valuation depends on the nature of the intension in question — and here our fundamental distinction between analytic and synthetic concepts comes in in a new role. This distinction is not only basic for the difference between philosophy and science, but fundamental in the differentiation of values. Other propositions seem obviously true, such as *some x ought to be good* ($I \rightarrow G$), *some x ought to be bad* ($I \rightarrow D$), *some x are good* (I-G), *some x are no good* (I-*T*). Thus emerges a pattern of axiological validity. It must, obviously, have a connection with the value pattern. Let us return to our "depth analysis" of axiological propositions.[24]

As we have seen, an axiological proposition of the form

"x is a good C" has a complex pattern which includes three types of what we now call pure logical propositions, namely, (I) x is a member of C with intension Φ; (II) Φ contains ϕ; (III) x *is* ϕ. Depending on the quantification of ϕ, x is a good, fair, bad, or no good C. The proposition *x is a good* C is what we now call a *compound axiological proposition,* namely, the conjunction of the pure logical propositions *x is a* C and the mixed logical proposition *x is good.*

We shall now extend our analysis of this compound axiological proposition to all value propositions and say that all value propositions follow the value pattern. That is to say, no value proposition is ever merely what it seems to be, namely, the mere proposition, but each such proposition is more than it seems to be: it always means the whole pattern. *The pattern consists exclusively of pure logical propositions.*

These propositions are not *stated* in the value proposition but merely *assumed.* A value proposition is like a woman or an iceberg; it hides more than it shows. In order to understand the validity of the value propositions we must examine the value pattern.

Since the pure logical propositions which form the value pattern underlying the value proposition are not expressed, we shall call them, again following the example of Kant, *judgments* rather than propositions. Judgment and proposition are, for Kant, "moments of thought": judgment is a thought not yet clearly understood, where the relation between subject and predicate is still problematic; proposition is the thought understood, and stated assertorily or apodictically. The difference between judgment and proposition thus lies in the *modality* of thought.[25]

Since the judgments underlying an axiological proposition are not expressed, more likely than not they are not clearly thought out. Their modality, therefore, plays a role and must also be of importance for the meaning of the axiological proposition itself. It must make a difference for that meaning whether the underlying judgment is thought assertorily, say, "x is a C," problematically, "x may (or may not) be a C," or apodictically, "x must be a C." We shall not make any distinction between assertory and apodictic modality, but instead add the modality of negatoriness which is to negation and limitation as assertoriness is to affirmation. It is, in other words, negative assertoriness. By problematicalness we shall mean both positive and negative possibility; for if x may be C then it is doubtful that x *is* C, and this means that x may not be C. The three modalities of the first of the three kinds of judgment underlying an axiological proposition (I) are then *a*] the assertory, in which it is assumed that x *is* a C, that is, that the copula expresses an existing relation between subject and predicate; *b*] the problematical, in which nothing specific is assumed concerning the relation between x and C, that is, it is assumed that x may or may not be a C; *c*] the negatory in which it is assumed that x is *not* a C, that is, that the copula expresses a nonexisting relation between x and C. In this case the judgment would be "x is not a C." Not only the first pure logical judgment of the value pattern, "x is a C," has a modality but every single one of them — "ϕ is contained in Φ" (II), "x is ϕ," "x is α," etc. (III) — may have different kinds of modality. This gives a great variety to the value pattern.

We shall now classify the value propositions according to this pattern, that is, the modalities of the underlying pure logical judgments. If the modality of such a judgment is assertory, we shall call the corresponding value proposition axiologically *analytic*; if the underlying modality is problematical, we shall call the corresponding value proposition axiologically *hypothetical;* and if the underlying modality is negatory, we shall call the corresponding value proposition axiologically *synthetic*. Axiological analyticity, hypotheticity, and syntheticity will be called the *modes* or *moods* of value propositions.

Axiological analyticity, hypotheticity, and syntheticity must, of course, be distinguished from *logical* analyticity, hypotheticity, and syntheticity, for the latter are relations within the logical proposition itself, whereas the former are relations not of the value proposition itself but of its underlying logical judgments. In the following, whenever we speak of analyticity, hypotheticity, and syntheticity we mean the axiological moods.

The importance of these moods will appear at once. To take a compound axiological proposition, say *John is a good student,* we find that the proposition is axiologically *analytic* if the speaker assumes that (I) John *is* a student (x *is* a C); and/or (II) what he says John is corresponds to the content of the concept "student" (ϕ *is* Φ); and/or (III) John has all the properties which the speaker takes to be the properties of a student (x *is* ϕ). But the same axiological proposition is *hypothetical* if the speaker assumes that (I) John may or may not be a student; and/or (II) what he says John is may or may not correspond to the content of the concept "student"; and/or (III) John may or may not have any or all of the properties of a student. [In these cases the proposition will be stated *questioningly;* with the stress on "student" in cases (I) and (II), and on "good" in case (III)]. The same axiological proposition is *synthetic* if the speaker assumes that (I) John is not a student; and/or (II) what he says John is does not correspond to the concept "student"; or (III) John does not have all of the properties of a student. (In these cases the proposition will be stated *ironically,* with the same emphasis as above). A synthetic axiological proposition is not a *false* one. The modality is an assumption, not a fact.

Let us now exchange the copula "is" in our example for the copula "ought" and discuss the proposition *John ought to be a good student.* This proposition is a compound axiological proposition of greater complexity. It may be composed of a variety of propositions, such as the pure logical proposition *John is a student* and the pure axiological proposition *John ought to be good;* or the mixed axiological proposition *John ought to be a student* and the mixed logical proposition *John is good;* or the mixed axiological proposition *John ought to be a student* and the pure axiological proposition *John ought to be good.* The exact composition makes, of course, a difference in the analysis of the compound proposition which here, however, does not need to concern us. Again, the proposition is *analytic* insofar as the underlying modalities are assertory. This refers, of course, to the modalities of the underlying pure logical judgments. Where the constituent parts of the compound axiological proposition are themselves axiological propositions, *their* underlying modalities must be analyzed. The iceberg of this kind of compound axiological proposition reaches down deeper than that of the previous kind. Insofar as the underlying modalities are problematical the proposition is *hypothetical,* and insofar as they are negatory the proposition is *synthetic.* (Here the different moods will not always be indicated by tone of voice, though they might be. But the shadings are lighter; "ought" itself becomes the indicator of the attitude.)

In the first case — if the proposition is analytic — the "ought" is an *analytic ought.* It does not say anything new, it only confirms what is assumed as true anyway: that John is a student, that a student studies, that therefore John is studious —

hence that he ought to be a good student. The proposition therefore is true; it is not different in this respect from the proposition *circles ought to be round,*[26] which is equally true, for it simply means that *circles ought to be circles,* which is an axiological truism, based on the logical truism *circles are circles.* In other words, the analytic ought-proposition assumes the corresponding logical (factual) judgment. A→C presupposes *A-C,* *E*→C presupposes *E-C,* etc.

Circles ought to be round is an analytic mixed axiological proposition; *John ought to be a good student,* as an analytic compound value proposition, is composed of value propositions at least one of which is analytic. Other analytic mixed axiological propositions are *kings ought to be regal, justices ought to be just, students ought to be studious, teachers ought to be instructive, you ought to be yourself,* and so forth.

But *John ought to be a good student* can also be hypothetical. In this case nothing is assumed about John's being a student or his study habits. He may be a student or he may not be one, he may be studious or he may not be. But if he is a student, or (depending on the speaker's imagination concerning John's being a student) if he is to graduate, or if he is what his parents think he is, or if the college is a bad one, and so on — then *John ought to be a good student.* The truth-value of such a proposition is not so easy to ascertain as that of the corresponding analytic proposition. The hypothetical ought-proposition assumes the corresponding logical (factual) judgments to be both true and false, or neither true nor false. Thus, if we denote hypothetical ought by "↔", A↔C presupposes both *A-C* and *E-C.*

Finally, the proposition may be a synthetic one, if I assume that John is not a student or not studious. In this case *John ought to be a good student* adds something new to my assumption: John will have to change. Whether he will change or not nobody knows, hence the truth-value of synthetic value propositions is indeterminate. On the one hand, the present intension or nature of John, if fulfilled, would not make him a good student, on the other, what I assume John's intension or nature to be may not be so at all. I assume what John is *not,* I do not assume what John is. The negatory modality corresponds to both negation and limitation. The synthetic ought-proposition assumes the corresponding negative logical (factual) proposition. Thus, if we denote synthetic "ought" by "←", A←C presupposes *E-C* (or *O-C*).

4. *Value Propositions and Fact Propositions*

These determinations of the three kinds of axiological propositions define necessarily and sufficiently the famous relation between "is" and "ought," beyond the derivation of "ought" itself from "is good." The pattern of the "ought" proposition applies both to mixed and to pure axiological propositions, that is, to those whose predicate is logical (*C*) as to those whose predicate is axiological (*G, T, B, D*). In the latter case, the pattern of "good" (or "bad," etc.) must be added to the pattern of "ought." Thus, $A \rightarrow G$, the analytic "ought" proposition that all *x* ought to be good (*C*'s), presupposes that every *x* is a member of *C* and has all the properties of *C*, that, in other words, every *x* is α, every *x* is β, etc. and thus fulfills the total value pattern of the mixed logical proposition *A-G*. For analytic "ought" confirms what is the case, that is, the corresponding pure logical propositions. Similarly $A \rightarrow T$ presupposes *A-T*, $A \rightarrow B$ presupposes *A-B*, etc. "All *x* ought to be no good," analytically, presupposes "all *x* are no good;" "all *x* ought to be fair," analytically, presupposes "all *x* are fair," etc. Again, $A \leftrightarrow G$ presupposes both *A-G* and *E-G* and/or *A-T* and *E-T*; for "all *x* ought to be good," hypothetically, presupposes both "all *x* are good" and "no *x* are good" and/or both "all *x* are no good" and "no *x* are no good." Finally, $A \leftarrow G$ presupposes any form except *A-G* or *E-T*; for "all *x* ought to be good," synthetically, presupposes that no *x* *are* good, or that all, or at least some, *x* are *not* good, but not that all *x* are good or no *x* are no good. For in these latter cases, "ought" would be analytic rather than synthetic. Hence, $A \leftarrow G$ presupposes *E-G, O-G, A-T, I-T, A-B, I-B*, all in turn with their corresponding underlying patterns, which form *secondary* underlying judgments — underlying judgments *of* underlying judgments — in the case of *pure* axiological propositions. In all, the pattern of axiological propositions, of ought-propositions, with their primary underlying logical or "is"-judgments, presents for *each* axiological proposition 84 possibilities of modal interpretation: there are three judgments in the pattern of each proposition, each of which may appear in one out of three modalities. We thus have nine elements which may be combined in groups of three, or the formula ${}_9C_3 = 84$ different modal interpretations of the proposition *John ought to be a good student*, not counting the modalities of the judgments *x is* α, *x is* β, etc., but only those of *x is* ϕ.

4. *Axiological Propositions* (*Copula Ought*)

Mixed Axiological Propositions Logical (Factual) Predicate		Pure Axiological Propositions Axiological (Value) Predicate	
Axiological Propositions	*Underlying Logical Judgments*	*Axiological Propositions*	*Underlying Logical Judgments*
Analytic $A \rightarrow C$ $E \rightarrow C$ $I \rightarrow C$ $O \rightarrow C$	A-C E-C I-C O-C	$A \rightarrow G$ $E \rightarrow G$ $I \rightarrow G$ $O \rightarrow G$ $A \rightarrow T$ $E \rightarrow T$ etc.	A-G [Secondary E-G underlying I-G logical O-G judgments: A-T $(\phi)\phi\omega\Phi$; $\sim(\exists\phi)\phi\omega\Phi$; E-T $(\exists\phi)\phi\omega\Phi$; $\sim(\phi)\phi\omega\Phi$] etc.
Hypothetic $A \leftrightarrow C$, $E \leftrightarrow C$ $I \leftrightarrow C$, $O \leftrightarrow C$	A-C *and* E-C A-$\overline{C}$ *and* E-$\overline{C}$ I-C *and* O-C I-$\overline{C}$ *and* O-$\overline{C}$	$A \leftrightarrow G$, $E \leftrightarrow G$; $A \leftrightarrow T$, $E \leftrightarrow T$, $I \leftrightarrow G$, $O \leftrightarrow G$, $I \leftrightarrow T$, $O \leftrightarrow T$ $A \leftrightarrow B$, $E \leftrightarrow B$, $A \leftrightarrow D$, $E \leftrightarrow D$, $I \leftrightarrow B$, $O \leftrightarrow B$, $I \leftrightarrow D$, $O \leftrightarrow D$	A-G *and* E-G A-T *and* E-T I-G *and* O-G I-T *and* O-T A-B *and* E-B A-D *and* E-D I-B *and* O-B I-D *and* O-D
Synthetic $A \leftarrow C$ $E \leftarrow C$ $I \leftarrow C$ $O \leftarrow C$	E-C, O-C A-C, I-C O-C, E-C I-C, A-C	$A \leftarrow G$, $E \leftarrow T$ $E \leftarrow G$, $A \leftarrow T$ $I \leftarrow G$, $O \leftarrow T$ $O \leftarrow G$, $I \leftarrow T$ $A \leftarrow B$ $E \leftarrow B$ etc.	E-G,O-G,A-T,I-T A-B,I-B,A-D,I-D A-G,I-G,E-T,O-T E-B,O-B,E-D,O-D O-G,E-G,I-T,I-B I-D I-G,A-G,O-T,O-B O-D E-B,O-B,A-D,I-D A-B,I-B,E-D,O-D etc.

The total pattern of axiological propositions and their underlying logical judgments is shown in Table 4 which gives the pattern by which "ought" is related to "is," the axiological to the logical, the so-called normative to the so-called declarative, the valuational to the factual. There are no other relationships between these terms than the ones given in the Table. In particular, the terms "normative" and "declarative," as usually taken, namely as supposedly distinguishing between value and fact, make no sense axiologically. "John is good" is declarative but valuational while "circular discs ought to be round" is normative but factual. The relationship between fact and value appears in *pure* axiological propositions only in the *secondary* underlying judgments. As is seen, to every "normative" proposition there belong one or several "factual" judgments which underlie it, or are embedded in it like ore in a rock, and in several layers, different for *mixed* and for *pure* axiological propositions. These layers, though not obvious, give the rock its axiological character, just as layers of ore, equally hard to discern, give rocks their geological character.

The way in which *any one* axiological proposition can be modally interpreted may be seen in Table 5. A proposition such as

5. *Modal Interpretation of an Axiological Proposition*

PROPOSITION	Underlying Judgment		
	I ($x \varepsilon C$)	II ($\phi \omega \Phi$)	III (ϕx)
Analytic	1	1	1
Hypothetical	2	2	2
Synthetic	3	3	3

John ought to be a good student may be analytic in all three ways, that is, the speaker supposes that John actually is a student, that "student" does mean what he means to say, and that John is it. We may signify such a proposition by p_{111}. Again, John may not be a student, the speaker is uncertain as to what a student is, and John does study assiduously (though not formally). Here we have p_{321}. In this case the proposition means that John ought to register at a college and formalize his assiduous studies, whatever they may be. The indices of "p" thus give the modal structure of the axiological proposition. In all, as we have seen, there are 84 different structures such as these for any one proposition, each of which can be indicated by the three-figured index. As is obvious, a logical proposition (one with the copula "is") will always show the first index figure, "1", i.e. be of the form p_{111}. On the other hand, there are axiological modalities of logical propositions, such as doubt, questioning, and the like (p_2), irony, lying, and the like (p_3), but these belong in a special chapter of formal axiology.[27] We are here only concerned with the modalities of axiological, that is, ought-propositions. Each such proposition appears as a matrix in the mathematical sense, which means that combinations of such propositions can be analyzed in accordance with, and by analogy to, the calculus of matrices.

5. *The Truth Values of Value Propositions*

The combinations of the three axiological moods with the four kinds of value propositions determine the truth-values of these propositions in an exact pattern which we shall now develop.

Classification of the axiological *propositions* according to moods corresponds to the classification of axiological *terms*, defined by their reference to the expositional properties. "Good" meant the subject's possession of all these properties, "fair" and "bad" meant the partial possession or nonpossession of these properties, and "no good" meant their nonpossession. As we saw, a difference had to be made between definitional and expositional properties. The definitional properties must in all cases be fulfilled; the value differentiations adhere to the expositional properties exclusively. This means, however, that the definitional properties must always be asserted, whereas the expositional properties must not. In other words, any value judgment *asserts* that a thing is a member of the logical class of which it is said to be a member, for there is no qualification to the logical membership. But it does not in the same sense assert that the thing is a member of the expositional class, for the possession of the expositional properties, and hence the belonging to the expositional class, and the kind of this belonging are, precisely, what is in question. Therefore the belonging to the expositional class is *problematical*. The thing may or may not, fully or partially, belong to this class. The problematical character of this belonging is precisely what makes the judgment a value judgment. For if that belonging were not problematical, the class in question would be systemic and the proposition concerning membership in it a logical, not an axiological one. In this case, the value character of the proposition would be *negated*. The relation between definition and exposition is thus similar to that between proposition and judgment — as also becomes clear in Kant's own analysis: [28] it is a relation of modality.

Valuation, then, comes about through the "modalities" of exposition; that is, the unexpressed status of the relation between subject and predicate. When that modality is *assertory*, that is, it is assumed that x has all the expositional properties of C, then x is said to be a *good* C. If, however, the modality is *problematical*, that is, it is assumed that x may or may not have some or all the expositional properties of C, then x is said to be a *fair*, *so-so*, a *poor*, or a *bad* C. And if the modality is *negatory* that is, it is assumed that x has none of the expositional properties of C, then x is said to be a *no good* C. Valuation, in this sense, is the modalization of class-membership. Axiological quantification and qualification, then, represent a certain kind of modality.

We have, thus, a relationship between the modes of axiological propositions and the axiological terms, which themselves express modifications of the class-membership relation. Analytic propositions correspond in logical nature to the predicate "good"; just as "good" means fulfillment, by the subject, of the exposition of the predicate, so the analytic axiological proposition assumes a definite relation between subject and predicate of the underlying logical judgments. Hypothetical propositions correspond in logical nature to the predicates "fair" and "bad"; just as these predicates are indeterminate as to fulfillment, by the subject, of the exposition of the predicate, so the hypothetical axiological proposition assumes an indeterminate relationship between subject and predicate of some or all of the underlying judgments. Synthetic propositions correspond in logical nature to the predicate "no good"; just as "no good" means nonfulfillment, by the subject, of the exposition of the predicate, so the synthetic axiological proposition assumes no relation between the subject and predicate of some or all of the underlying logical judgments.

We can now assign the three propositional modes, the analytic, hypothetic, and synthetic, to the four kinds of value propositions.

1] It might seem that *pure logical propositions,* based on the form *x is C,* can have no axiological analyticity, hypotheticity, or syntheticity, but only logical analyticity and syntheticity. Actually, however, pure logical propositions *can* have the axiological modes, that is, they can have underlying logical judgments whose unexpressed modalities influence the meaning of the pure logical propositions. In this direction may lie answers to the questions about the logical nature of contrary-to-fact conditionals, interrogative, imperative, and nonsensical propositions. We have, here, axiological relationships as significant for logical propositions. Conversely, logical propositions may be significant for axiological relations and propositions. Just as an axiological proposition assumes unexpressed logical judgments, so expressed logical propositions may assume an unexpressed axiological judgment — when, namely, the logical proposition is part of a value pattern the axiological proposition of which is not expressed. Such pure logical propositions are, so to speak, *enthymematic axiological propositions.* Another axiological meaning of a logical proposition is given in those cases in which the logical predicate implies an axiological predicate, as in the case of "honor student," "prize bull," and the like. Here belong all terms with both fact and value meaning, such as epithets; perjorative, euphemistic, metaphorical terms; etc. Other such cases are the predicates which although not formally axiological are what we called *applied axiological terms,* that is, terms which do not belong to the formal science of value, as do "good," "fair," and so forth, but to one of the specific value sciences, such as ethics, aesthetics, economics, religion, and so on. Such applied (or interpreted) axiological terms are "honest" (ethics), "beautiful" (aesthetics), "expensive" (economics), "holy" (religion), and so on. These applied axiological terms are to the formal axiological terms as physical are to geometrical terms, for example, the path of a ray of light in a homogeneous medium is to a straight line. Thus, "beautiful" (aesthetic "good") is, in this sense, an aesthetic interpretation of axiological "good," and moral "good" is an ethical such interpretation. All these are applied axiological predicates. There are also applied axiological relations, such as *all men follow their consciences, John is true to himself,*[29] *Rita likes ice skating,* and the like. Such applied axiological relations are interpretations of formal axiological relations. Applied axiological propositions, whether with applied axiological predicates or with applied axiological relations, have the form of pure logical propositions but follow the rules of mixed logical or mixed axiological propositions. They have, therefore, again all three axiological modes. However, these axiological modes of pure logical propositions need not concern us in the following.

2] *Mixed logical propositions,* based on the form *x is good,* mean that *x* has all the properties *x* is supposed to have; thus, the (logical) value of *x* can never be mere things or events as such but only things and events having properties. They may be designated by proper names, by descriptions, or by expositional class names. Strictly speaking, it is never a thing *x* as such that is good but always a thing as having some properties; just as in the form *x is a good (member of) C* it was presupposed that C contains ϕ and therefore *x* is ϕ, so here it is presupposed that *x* has a set of properties and that by virtue of having them *x* is good. Thus "*x*" here is not a thing variable in the usual sense.[30] But neither is *x is good* a compound axiological proposition as was *x is a good C.* The latter presupposed the underlying judgment "*x* is a C" and all the entailments thereof. *x is good,* on the other hand, simply means *x is good as x, x is good being x, x is a good x,* or the like. The underlying assumption here is simply "*x* is *x.*" But this assumption is not the same as the identical-looking assumption — often called a law of thought — underlying the logical proposition *x is a C.* This proposition, too, may be said to presuppose the assumption "*x* is *x.*" But whereas in the latter case "*x* is *x*" is a tautology and "*x*" *is* a mere thing-variable, in the former, the axiological case, "*x* is *x*" is not a tautology for "*x*" is not a mere thing-variable. The assumption "*x* is *x*" in this latter case guarantees that *x,* whatever its properties, will always be *x*; that is, it guarantees the identity of *x* throughout the whole range of its definition and exposition: it is a kind of substantial guarantee of *x,* a guarantee of *x* as substance in the old sense of the word, but without the metaphysical connotation. *x* is a "continuant" in the sense of W. E. Johnson,[31] it remains the same though its set of occurrent properties may change. For *x is good* is only one of many mixed logical propositions about *x,* others being *x is fair, x is bad, x is no good,* and so forth. In all these cases the assumption is that *x* is *x,* no matter whether *x* is good, bad, or indifferent, and that the *x* which in one value proposition is said to be good is the same which in another value proposition is said to be bad.

Unless it is assumed that *x* is one and the same throughout the whole range of its expositional determinations, and that this *x* is the same as the definitional *x*, the *x* of each expositional value might be thought of as being different from that of every other expositional value, and all of these different from the *x* which has the *definitional* properties. The axiological assumption "*x* is *x*" thus guarantees that the expositional "modalities" — the value terms — can be applied. In other words, the modality of the value pattern — the assumption "*x* is *x*" — guarantees the expositional modality. The only modality of the value pattern, however, which is applicable for this purpose is assertoriness, namely that *x is x*; and therefore mixed logical propositions are *analytic*.[32]

3] *Pure axiological propositions,* based on the form *x ought to be good,* have two underlying assumptions. One is, as above, that *x* is *x*, in which case the pure axiological proposition, like the mixed logical, is *analytic*. The assumption "*x* is *x*," we said, meant more than merely a relation of identity: it meant the guarantee of *x*'s identity *throughout its exposition and definition.* But if this is so, then *x* cannot merely be *x*, that is, a thing or event, because things or events as such have no exposition and definition. Exposition and definition are matters of *concepts*. Hence the "*x*" which has exposition and definition must be a concept, namely, the concept of *x*, say, "X". There are thus two kinds of *x* — *x* as thing or event, that is, as "*x*", and *x* as concept of that thing or event, and of that thing or event exclusively, that is, as "X." In other words, there is no thing without its concept — again something of a Kantian tenet. Indeed, if this were not so, *x ought to be good* would be meaningless. For we remember, "ought" is a relation exclusively between thing and concept. Thus, if *x ought to be good* is to have any meaning, it must mean *x ought to be a good* X. The assumption underlying *x ought to be good* is not merely "*x* is *x*," as in the case of the mixed logical proposition *x is good,* but "*x* is X" — and this is not an identity but a class-membership relation, where *x* is the only member of X. It spells out the curious identity of "*x* is *x*" which we discussed above and did not regard as a tautology. We cannot go into the details of this matter, which defines intrinsic as against extrinsic value.[33] The assumption "*x* is X" is again assertory. For unless *x is* (a member of) X, *x ought to be good* is meaningless. Hence, pure axiological propositions are *analytic,* just as are mixed logical propositions.

But there is another side to the matter. In order for *x ought to be good* to be valid I must assume that *x* is X. But do I necessarily have to assume that which makes what I say valid? It seems much more natural to assume that *x* is *x* than to assume that *x* is X. It may well be assumed that everybody assumes that *x* is *x*. But only a sophisticated mind assumes assertorily that *x* is X. There is, after all, a difference between *x* and X. Indeed, just a sophisticated mind may well doubt that *x* is X; *x* may be X and again, *x* may not be X. That in the first case the proposition *x ought to be good* may be valid and in the second invalid is an entirely different matter. My assumptions are one thing and the validity of what I say another. Thus, we cannot exclude the assumption that *x* is either X or is not X, that is, the problematic modality.

While *x ought to be good* thus leaves room for the problematic modality, it does not seem to leave room for the negatory modality, that *x* is not *x*, or X not X, or *x* not X. For if *x* ought to be good, certainly there must be *x*, and that *x* must be *x*, and at least *may* be X; it cannot be assumed that it is not X; for this would mean that there is no definition and exposition of *x*. Doubt of *x*'s being X may be permitted but not the assumption of its not being X. Thus, we must say that pure axiological propositions may also be *hypothetical* but not that they are synthetic. Our reasoning has brought out a close connection between the axiological mode and axiological validity.

4] *Mixed axiological propositions,* based on the form *x ought to be* C, may have all the three axiological moods, for the underlying modalities may be either assertory, problematical, or negatory. When I say "*x* ought to be C," then, as we have seen before, I can assume that *x* is C, in which case the mixed axiological proposition is *analytic;* or I may assume that *x* may or may not be C, in which case the axiological proposition is hypothetical; or I may assume that *x* is not C, in which case the mixed axiological proposition is *synthetic.*

Let us now refine somewhat the notion of axiological modality. Since in all value propositions the underlying logical judgment contains the subject and, explicitly or implicitly, the predicate of the value proposition itself, axiological *analyticity* means that there is assumed a connection between the subject and the predicate of the value proposition; axiological *hypotheticity* means that nothing is assumed about such a connection; and axiological *syntheticity* means that there is assumed not to be such a connection. In the first case, the subject is assumed to be a member of the class of, or referred to by, the predicate; in the second case nothing about such a membership is assumed; and in the third case there is assumed not to be such a membership. In the first case the exposition of the predicate class is normative for the subject, in the second case it may or may not be normative, and in the third case it is not normative.

Depending on the mode of the value proposition, the "ought" in it is either analytic, hypothetical, or synthetic. In analytic propositions, *analytic* or *logical* "ought" merely confirms an existing logical relationship, namely the one assumed in the assertory modality. In *hypothetical* propositions "ought" has two functions. Since the assumed logical relation is in the problematical mood and is assumed as possibly existing and as possibly non-existing, "ought" may confirm one of these possibilities and "ought not" the other. Both senses of ought thus confirm *something.* Depending on which sense is used, we shall call hypothetic "ought" either *existential* or *statistical* (the reason for the choice of terms will appear later). In *synthetic* propositions, synthetic "ought" appears in three varieties. The underlying assumption is in the negatory modality, "*x* is not *C*." If the proposition itself is positive, *x ought to be* C, the relation between *x* and C, which is assumed not to exist, is to be constituted. This we shall call *constitutive* or *constructive* "ought." Constitutive or constructive propositions are, for example, *All governments ought to merge in a world government, All airplanes ought to be jet planes, Latvia ought to be free.* If the proposition is negative, *x ought not to be* C, two cases must be distinguished: the connection between subject and predicate is to be *abolished* or it is to be *prevented.* In both cases it is assumed that this connection does not exist, in the first case partially, in the second case wholly. If the connection between subject and predicate is to be abolished, that is, if *x* ought not to be C, then since "ought not" relates the betterness of a thing to its worseness, *x* as something else, say *B*, is better than *x* as *C: x* is not a good *C*; which means that *x* does not fulfill the exposition of C. This not being the case, *x* may "just as well" not be a C at all: it is "better for" *x* not to be a C. This kind of logically negative synthetic axiological proposition we shall call *destructive.* The second kind, where *x* is assumed to be no C at all we shall call *obstructive.* Thus the assumption that the subject does *not* belong to the class of the predicate is common to all synthetic axiological propositions. In the case of constructive and obstructive propositions it refers to total not-belonging: the subject does wholly not belong to the predicate. In the case of destructive propositions it refers to partial not-belonging; the subject partially does not belong to the predicate. Destructive and obstructive propositions can be regarded as obversions of one another when they agree in subject and predicate: *Betty ought-not to steal* is destructive "ought," meaning that Betty is not a natural-born thief, although she steals; *Betty ought not-to-steal* is obstructive "ought," meaning that Betty ought not to do what she is assumed not to be doing.

The classification of value propositions according to their modalities is shown in Table 6.

6. *Modalities of Value Propositions*

PURE LOGICAL	MIXED LOGICAL	PURE AXIOLOGICAL	MIXED AXIOLOGICAL
Logically Analytic or Synthetic and Axiologically Analytic, Hypothetical; and Synthetic	Axiologically Analytic	Axiologically Analytic or Hypothetical (Existential or Statistical)	Axiologically Analytic or Hypothetical (Existential or Statistical) or Synthetic (Constructive, Obstructive, or Destructive)

We can now pull the threads together and determine the truth-values of value propositions. In logical propositions we distinguish between truth and validity, meaning by truth in this sense truth in terms of reference to reality and by validity truth in terms of the systemic pattern of logic. Logical propositions, in the first sense, appear against a background of empirical reality. Logic, in the second sense, has so constructed its rules that in sound reasoning, that is, reasoning from sound premisses, empirical truth and falsity and systemic truth and falsity coincide. It is this co-incidence, this *sym-ballein,* by which symbols enable us to deal with the world.

The same reflection will now lead us to the truth-values of axiological propositions. These propositions, also, appear against a background of empirical reality, even though that reality is, as it were, not quite so empirical as that against which the logical propositions appear. But it is reality, nevertheless, and it is empirical, even though of a more subjective kind. The background against which value propositions are presented, that is, pro-posited, *vor-gesetzt,* is the value-pattern — the pattern of assumptions and modalities which we have discussed.

Thus, we are now led to the concepts of axiological truth and axiological validity. An axiological proposition is true if it corresponds to the underlying value pattern and it is false if it does not; just as a logical proposition is true if it corresponds to the "facts," that is, the pattern of reality to which it refers, and is false if it does not. The difference between logical and axiological propositions, in this respect, is only that logical propositions semantically *refer* to their background — or foreground? — and axiological propositions pragmatically *arise from* it.

But, just as logical propositions *as such* can have nothing to do with this kind of truth and falsity, neither can axiological propositions. Indeed, they can even less, for their correspondence with the underlying pattern is generally much more vague — and less testable — than the correspondence of logical propositions with their empirical reference. Thus, the truth and falsity of value propositions *as such* must be based on the system of axiology, just as that of the logical propositions *as such* on the system of logic; it must be based on rules of validity and invalidity. These rules must be so constructed that, as in the case of logical propositions, sound argument, that is, in the case of valuation, not merely sound reasoning but sound reasoning on the basis of sound assumptions, will lead to correct results both empirically and axiologically. In other words, in the case of valuation on sound assumptions, truth and validity of value propositions must coincide.

This being understood, the formulation of the rules of axiological validity is simple and follows naturally and logically from our analysis of value propositions.

Value propositions are either axiologically true, axiologically false, or axiologically indeterminate ("a-true," "a-false," "a-indeterminate"). Axiological truth (T), axiological falsity (F), axiological indeterminacy (I) are the *axiological truth-values* or *the truth-values* of *value propositions.* Unless otherwise specified, by "true," "false," "indeterminate" we shall in the following understand "a-true," "a-false," "a-indeterminate."

Depending on the nature of "ought," axiological value propositions having the truth-value Truth may be logically (analytically), existentially, or statistically a-true. If they have the truth-value Falsity, they may be logically (analytically) or existentially a-false.

Propositions are axiologically *true* if they conform to the rules of axiology and *false* if they do not so conform. They do not conform if they contradict these rules *a*] implicitly, *b*] explicitly. The latter is the case if they are excluded from the range of these rules by an axiological rule itself. They are *indeterminate* if *c*] they are not axiological in nature, that is, neither have an axiological copula nor an axiological predicate, or if *d*] there is assumed to be no relation between subject and predicate.

1] *Pure logical propositions are indeterminate.* Pure logical propositions are not axiological in nature (*c*). Their truth is not axiological truth but logical truth (insofar as they *are* axiological in nature they fall under rules 2–4).

2] *Mixed logical propositions are true if logically particular and false if logically universal.* Mixed logical propositions are axiologically analytic. This means that it is a-false that all members of a class C (all those that have existed, do exist, or will exist in time) are good. For in this case they would all fulfill both exposition and definition, we would have a systemic class, and value predication would be impossible. Hence it is false that all members of a class are good (*a* or *b*).[34] If all members of a class are less than good or no member of a class is good, then no member fulfills the exposition and there would be no empirical class; for such a class presupposes that at least two members fulfill the exposition. The reason is the following: the minimum properties of the expositional class are given by the *definition.* Any being fulfilling the definition belongs to the definitional class. In addition, every such being has other properties. Those which it has in common with other members form part of the *exposition.* This means that at least one nondefinitional property must be held in common by at least two definitional class members for an expositional class to come into being. If each definitional class member had a different nondefinitional property, or a different set of such properties, there would not be an expositional class. If at least two members had one such property in common, there would be an expositional class of two members and an exposition of one property. If two other members had two such properties in common, one of which would be the property just mentioned, the expositional class would have four members and two properties. The members having both properties would be "good" class members while those having only one property would not be so good. In general, the exposition is the maximum set of non-definitional properties held by at least two definitional class members. But if there is no expositional class, valuation is again impossible, since valuation depends on the existence of that class (*a* or *b*). Hence it is false that all members of a class are less than good or that no member of the class is good. *If it is both false that all members of a class are good and that all members of a class are less than good, then it must be true, if there is to be an empirical class at all, that some members of the class are good and some are less than good.* This means that logically universal mixed logical propositions are false and logically particular mixed logical propositions are true.

Mixed logical propositions are all logical propositions with axiological predicate, such as the true propositions *Some men are good* (men), *Some automobiles are good* (automobiles), *Some things are no good* (things), and the false propositions *All men are good* (men), *No automobiles are good* (automobiles), *All things are no good* (things). All these propositions demonstrate the rule that generalization in axiologically analytic propositions is a-false. Violations of this rule may be called the *fallacy of illicit generalization.* This rule, which in logic proper has no systematic status, is a theorem in axiology, following from the axiological distinction between systemic and empirical propositions and the consequent exemption of systemic propositions from extrinsic value. Any class all of whose members are good or fulfill both exposition and definition, is a systemic class.[35]

3] A. *Analytic pure axiological propositions are true if positive and false if negative.* B. *Hypothetical pure axiological propositions are true if positive; if negative they are true if logically particular and false if logically universal.* Analytic pure axiological propositions are logically a-true and a-false. Hypothetical pure axiological propositions, if positive, are existentially a-true; if negative, they are statistically a-true when logically particular, and existentially a-false when logically universal.

A] In analytic propositions it is assumed that the subject *is* a member of its class, hence the copula "ought," used in its positive sense, merely confirms what is assumed to be the case. Positive analytic pure axiological propositions are therefore true. "Ought" used in the negative sense, on the other hand, contradicts what is assumed, and the proposition is therefore false.

B] In hypothetical pure axiological propositions it is indeterminate whether the subject belongs to its class or fulfills the exposition. Hence the copula "ought," used in the positive sense, confirms what may be the case, and the proposition is existentially true. "Ought" used in the negative sense confirms as negative what may *not* be the case; the proposition is therefore statistically true (to the axiological *rule* that in an existential class there *must* be some members which do not fulfill the exposition there corresponds the statistical *fact* that there usually *are* such members). This, however, can apply only to logically particular propositions; for, while it may be assumed that some members of a class may be bad it cannot be assumed that all of them are bad, and hence do not fulfill the exposition. This assumption would contradict the possibility of the empirical class itself. It may be assumed, however, that all members of the class may be good, for this assumption does not contradict the *possibility* of the empirical class — even though, if all *were* in fact good, the class would *actually* be a systemic one. But this class would be more than, and include, the expositional one. Therefore, if positive, *all* hypothetical pure axiological propositions are true, whether logically universal or particular. But if negative, only logically particular propositions are true, while logically universal ones are false.

Examples of true propositions of this kind are *All teachers* (being teachers) *ought to be good* (teachers) (A→G; analytically a-true); *All teachers* (if they are teachers, that is, belong to the expositional class "teacher") *ought to be good* (teachers) (A↔G: existentially true); *Some teachers* (teachers being what they are, either fulfilling or not fulfilling the exposition of "teacher") *ought to be bad* (teachers) (*I*↔*D*: statistically true).

Examples of false propositions of this kind are *All teachers* (being teachers) *ought to be no good* (teachers) (A→*T*: analytically a-false); *All teachers* (if they are teachers) *ought to be no good* (teachers) (A↔*T*: existentially a-false); *Some teachers* (being teachers) *ought to be no good* (teachers) (*I*→*T*: analytically a-false).

4] A. *Analytic mixed axiological propositions are true if positive and false if negative.* B. *Hypothetical mixed axiological propositions are true if positive; if negative, they are true if logically particular and false if logically universal.* C. *Synthetic mixed axiological propositions are indeterminate.* Analytic and hypothetical mixed axiological propositions are a-true and a-false following the pattern of pure axiological propositions.

A] In analytic mixed axiological propositions it is assumed that the subject *is* a member of the class of the predicate. Such propositions use the copula "ought" logically, and the same argument applies as for analytic pure axiological propositions.

B] In hypothetical mixed axiological propositions it is indeterminate whether the subject belongs to the class, or fulfills the exposition, of the predicate. The copula "ought" is used hypothetically, and the same argument applies as for hypothetical pure axiological propositions.

C] In synthetic mixed axiological propositions it is assumed that no relationship exists between subject and predicate. Such propositions are, therefore, indeterminate (*d*), for there is no exposition which can serve as norm for the subject. These propositions use "ought" either constructively, obstructively, or destructively, depending on the positive or negative sense of "ought."

True propositions of this kind are *All promises ought to be kept* (A→C: analytically a-true if it is part of a promise to be kept); *All promises* (if they are promises) *ought to be kept* (A↔C: existentially true); *Some promises* (promises being what they are, for example, some of them foolish) *ought not to be kept* (O↔C: statistically true); *Your duty in this situation* (whether or not it can be done) *ought to be done* (A↔C: existentially true); *Your duty* (defined as what ought to be done) *always ought to be done* (A→C: analytically true); *Your duty* (duties being what they are, that is, at times conflicting with each other) *sometimes ought not to be done* (O↔C: statistically true); *No bachelors ought to marry* (E→$\overline{C}$: analytically true).

False propositions of this kind are *No promises ought to be kept* (E→C or E↔C: analytically or existentially a-false); *Some promises ought not to be kept* (O→C: analytically false if "to be kept" is part of the definition or exposition of promise); *Your duty in this situation* (whether or not it can be done) *never ought to be done* (E↔C: existentially false); *Your duty* (defined as what ought to be done) *sometimes ought not to be done* (O→C, I→$\overline{C}$: analytically false).

Indeterminate propositions of this kind are *Promises ought to be kept* (A←C, I←C — if there is supposed to be no connection between promises and keeping them); *Promises ought not to be kept* (E←C, O←C — under the same Machiavellian presupposition); *Duties ought to be done* (A←C, I←C — if there is supposed to be no relation between duties and doing them); *There ought to be craters on the other side of the moon* (I←C — I know nothing of the other side of the moon and assume there are no craters, but I infer from what I know of this side that there should be craters: constructive or constitutive "ought"[36]); *You ought not (never) to become a doctor* (A←$\overline{C}$: obstructive "ought"); *You ought not to be a doctor* (E←C: destructive "ought"); *You ought to close the door* (A←C: constructive "ought").[37] The truth values of value propositions are summarized in Table 7.

Having examined the elements of the axiological system — value terms, value relations, and value propositions — we shall now turn to the system itself.

7. *Truth-Values of Value Propositions*

	Pure Logical	Mixed Logical		Pure Axiological					Mixed Axiological						
		Logical Universal	*Logical Particular*	*Positive*		*Negative*			*Positive*			*Negative*			
	A – C E – C I – C O – C	A – G A – *T* A – B A – *D* *E* – *G* *E* – *T* *E* – *B* *E* – *D*	I – *G* I – *T* I – B I – *D* O – G O – *T* O – B O – *D*	A→G E→*T* A→B E→*D* *I*→G O→*T* *I*→*B* O→*D*	A⟷G E⟷*T* A⟷B E⟷*D* *I*⟷G O⟷*T* *I*⟷*B* O⟷*D*	A→*T* E→B A→*D* E→G *I*→*T* O→*B* *I*→*D* O→*G*	A⟷*T* E⟷B A⟷*D* E⟷G	 *I*⟷*T* O⟷*B* *I*⟷*D* O⟷*G*	A→C E→$\bar{C}$ I→C O→$\bar{C}$	A⟷C E⟷$\bar{C}$ *I*⟷C O⟷$\bar{C}$	A←C E←$\bar{C}$ *I*←C O←$\bar{C}$	A→$\bar{C}$ E→C I→$\bar{C}$ O→C	A⟷$\bar{C}$ E⟷C	 *I*⟷$\bar{C}$ O⟷C	A←$\bar{C}$ E←C *I*←$\bar{C}$ O←C
Is	I	F	T												
Analytic Ought				T		F			T			F			
Hypothetical Ought (Existential)					T		F			T			F		
Hypothetical Ought (Statistical)								T						T	
Synthetic Ought											I				I

6

THE SYSTEMATIC IMPORT OF FORMAL AXIOLOGY

Everything is what it is, and not another thing. I have tried to shew, and I think it is too evident to be disputed, that such appreciation is an organic unity, a complex whole; and that, in its most undoubted instances, part of what is included in this whole is *a cognition of material qualities*, and particularly of a vast variety of what are called *secondary* qualities. If, then, it is *this* whole, which we know to be good, and not another thing, then we know that material qualities, even though they be perfectly worthless in themselves, are yet essential constituents of what is far from worthless. — G. E. MOORE

Fact is one of the possibilities of varying the given in imagination. — EDMUND HUSSERL [1]

The system of formal axiology is based, as we have said, upon the logical structure of intension. The intension is a set of predicates; and by applying certain rules of set theory to it we have arrived at various kinds of intensional structure: finite, denumerably infinite, and nondenumerably infinite. These kinds of structure are norms determining the kinds of value things with corresponding sets of properties possess: systemic, extrinsic, or intrinsic value. The correspondence between the set of predicates of an intension and the set of properties of a thing, and the normative nature of the former set for the latter, makes the structure of intension the measure of value. The intension, in formal axiology, is an *axiometric structure.*

The central role of the intension in formal axiology makes it imperative for us to examine in detail the intensional structure. The total such examination would be the total system of formal axiology. In the present context, where we are only trying to lay the foundation of this new science, we shall limit ourselves mainly to the discussion of one such structure, denumerable intensional structure which rules extrinsic value. Although in theory denumerable intensional structures have infinite content, in practice we deal with finite structures. Extrinsic valuation, in practice, is based upon relatively few properties, up to about two hundred in expert valuation of an everyday thing and upon far fewer in everyday valuation and discourse. Thus, in this chapter, we shall examine the structure of the intensional norm of everyday extrinsic value. We shall thus continue the discussion of the preceding chapter, where we examined the *elements* of extrinsic value.[2]

The old problem of the structure of intension, which is so far unsolved, is illuminated by formal axiology in several ways, three of which should be mentioned: 1] the dimensional pattern of value terms, based on the differences between analytic, synthetic, and singular intensions, 2] the logical pattern of value terms, based on their definition in terms of truth-values, 3] the arithmetical pattern of value terms, based on the relation between value properties and descriptive properties. All these patterns elaborate and clarify the relation ω, which defines the mutual relationship of intensional predicates.

1. *The Dimensional Pattern of the Value Terms*

Because, as we have seen, the synthetic concept — the term — has strictly speaking neither extension nor intension, things defined by such concepts are *constructs*, as are these concepts themselves. These things cannot *be* unless they are what the concept — or rather the system of relations of which it is a term — defines. Hence the corresponding things cannot fail to fulfill their concept for unless they do they are no such things. Geometric circles, triangles, electrons, numbers, and the like — *systemic* things — cannot *as such* be either good or bad. They can only be, or not be, such things. The values connected with systemic concepts, therefore, can only be synthetic being or not being, complete fulfillment or complete nonfulfillment, perfection or nonperfection. Systemic value is the value of *Perfection*.[3] As Aristotle rightly says, a number cannot be "mutilated"; for the loss of a unit makes it another number. But a cup can be mutilated; for the loss of a handle still leaves it a cup.[4]

Cups and other empirical things are subject to analytic

concepts and their valuation, *Goodness, Badness,* etc. Analytic concepts are not singular, but general concepts. An empirical thing determined by an analytic concept is always a particular thing, a member of a class of at least two, and never a singular thing. The values connected with singular things are again different from those connected with particular things, due to the different axiometric structure of the singular intension. If systemic value is the value of *Perfection* and extrinsic value that of *Goodness,* singular or *intrinsic* value is the value of *Uniqueness.*[5]

Only particular empirical things, then, can have the values of axiological *Goodness, Badness,* etc., *extrinsic* value, that is, value of a thing as member of its class. But even here there is a certain complexity which has to do with the intermediate position of this value between *Perfection* and *Uniqueness,* or the intermediate *logical* position of the analytic concept between the synthetic and the singular concept. This position arises from the structure of the respective intensions. The synthetic intension contains, strictly speaking, no properties at all but only formal relations; the singular intension contains a nondenumerable infinity of properties and thus is not discursive but a continuum, a Gestalt; and the analytic intension has theoretically a denumerable infinity of properties but practically a certain number of them which may be larger or smaller; it may be (potentially and actually) infinite; or it may be as small as three, that is, one more than a definition consisting of genus and species. In general, the structures of intensions depend on the levels of abstraction of the intension in question. The *definition* as the highest abstraction has a *minimum* of terms, while the *exposition,* on a lower level of abstraction, has more terms than the minimum. Since the minimum properties *defining* a thing must be a genus and a differentia, it follows that the *exposition* of a thing theoretically may contain any number of properties larger than two. The exact line where an exposition becomes a definition is not ascertainable, precisely for the reason that the thing is an empirical and not a systemic one. Exposition and definition, however, are easily discernible in practice. Thus, to use Moore's example, the expression "hoofed quadruped of the genus equus" is the analytic definiens of the concept "horse." The exposition of "horse," on the other hand, as found in a scientific dictionary, is "a solid-hoofed and odd-toed quadruped (*Equus caballus*), varying much in size, color, speed, and so forth, having horny patches or chestnuts on the inner side of both pairs of legs (above the knee and below the hock), a mane and tail of long coarse hair (which distinguish it from other species of the *Equidae*, as the asses and zebras), and . . . ," in all fifty properties, as against the three of Moore's definition.

If x is an empirical thing and belongs to the class C, the question whether x is a C is then not so easily answered as in the case of a systemic thing. Suppose the *definition* of C to be α and β, and the exposition γ, δ, and ζ. Then if x has α, β, γ, δ, and ζ, x will be a C according both to exposition and to definition. But if x has only α and β, it will be a C only according to definition and not to exposition; and if x has only γ, δ, ζ, or any of these, it will not be a C according to definition but only to exposition. Thus, if x is hoofed, quadrupedal, and of genus *Equus* it is a horse, even if it lacks a mane, a tail, and all its teeth. It is, in this case, "not much of" a horse — it is a horse by definition but not by exposition. If, on the other hand, it is neither hoofed nor of genus *Equus* it is not a horse, even though it may have a mane, a tail, and all its teeth — it is, say, a lion.

There is, of course, a difference between a horse and a lion, and the difference between the *definitions* of "horse" and of "lion," respectively, takes care of it. But there is also a difference between a horse with a mane and a tail and teeth and one without. This difference is taken care of by the difference between *definition* and *exposition*, and the differences within the exposition. In everyday language these differences are obscured. By "horse" may be meant either the defined or the expounded horse. Language, therefore, had to include words whose function it is to refer to the difference between definition and exposition, and within the exposition. *These words, precisely, are the extrinsic value terms.*

Thus, *a good horse* is one which is a horse not only according to definition — this is given by the noun "horse" — but also according to exposition — this is given by the adjective "good." *x is a good horse* means that x is a *horse* in having the definitional properties of "horse" and that x is a *good* one in having the expositional properties of "horse." *x is a bad horse* means that x is still a horse, in having the definitional attributes α and β, but not a good one, in lacking some of the expositional attributes, γ, δ, or ζ. On the other hand, *y is a good lion* does not, of course, mean that y is a horse, but that y is a lion; yet, *y as a good* lion has in common with x, which is a good horse, the *goodness* even though y's goodness is leonine and x's goodness is equine. But both goodnesses are alike in being the possession of the respective expositional properties. In general, therefore, *x is a good C* means that *x has all the expositional properties of* C, or that *x fulfills the exposition of* C. Since we are usually not concerned with the C-ness of x but only with its goodness, hence not the definition of C but only its exposition, we may use "exposition" and "definition" interchangeably. A class of empirical things whose definition is an exposition is an *empirical* or *existential class.*

The *distinction* between definitional and expositional properties of an empirical class must only be made when the loss of an intensional property means loss not of the goodness but of the particular existence of the thing. For, if x does not have the *definitional* properties of C, then x is not a C. Whereas, when x has none of the *expositional* properties but *does* have the definitional properties, then x is a C but a no good one. Thus, whenever we speak of the values of ϕ, we mean the expositional properties in this narrow sense. For, loss of the definitional properties would mean not loss of goodness but of C-ness, and we could not speak any more of x as a C, whether a good or a bad one. The same distinction applies in the case of "fair" and in that of "bad." For here, too, none of the definitional properties of C must be missing. Thus, if x is to be a fair C, among the properties x has must be α and β. And if x is to be a bad C, among the properties it lacks must not be α or β. Only in the case of "good" does the distinction not *have* to be made, because in that case x does have α and β anyway. Thus, in *all* cases the definitional properties must be present, and *valuation is exclusively a matter of the expositional properties.* Therefore the class defined by the exposition, which we called the empirical or existential class, may also be called the *axiological class;* while the class defined by the definition is a logical class in the sense defined above. The logical class is always a systemic class, even if the thing in question in other respects is empirical. The logical or systemic part of an empirical thing is close to what Kant calls the schema. Systemic class, logical class, and schematic class are therefore one and the same, and so are empirical, existential, and axiological class.

Classes, whether systemic or existential, are combinations of the extension and the intension of a concept in the sense that without intension extension is not possible and without extension we should not speak of intension. For this reason, concepts — rather than terms — without either extension ("null-classes") or intension (undefined or of contradictory predicates)[6] are not classes in our sense. "Extension" means logical or schematic existence for systemic class-members and actual or empirical existence for existential class-members. There is a systemic but no empirical class of unicorns or proconsuls of France, but there is no class of native aliens or square circles. There is both a systemic and an empirical class of horses. There is a *systemic* class of horses if *a*] there is a definitional concept "horse," and *b*] there are (at least two) things which have the properties of the definition of "horse." (Unless there were at least two such things no *abstraction* of *common* horse properties would be possible.) What is considered in these (two) things, however, is not their individual, total character but merely their common character, the logical schema "horse" inherent in them, the set of minimum characteristics of horsiness. There is an empirical class of horses if *a*] there is an expositional concept "horse," and *b*] there are (at least two) things which have the properties of the exposition of "horse." These (two) things, again, are not considered in their individuality but in the character they have in common, which is that of empirical horses, having maximum common characteristics of horsiness. They are more concrete than the schemata of "horse" but less concrete than an individual horse. Particularity, in other words, is between generality and individuality.

Whereas the systemic class can exist without the empirical, the empirical class cannot exist without the systemic (schematic); all actual horses necessarily contain the schema "horse," but the schema "horse" does not necessitate the empirical existence of horses (otherwise, as Kant showed, we would all be millionaires).[7] The two kinds of horse must not be confused; in particular, it must not be held that a horse whose hooves have been amputated is no longer a horse because it has lost its definitional attributes. These attributes belong to the schema "horse," not the empirical horse. They can be omitted in thought from the schema, but not "amputated" from a horse. The hoofless horse is still an ungulate; even though it has lost its hooves it has not lost its "ungues" which inhere in the schema. And the latter is still with the amputee, just as the blueprint is still with the roofless house.

By "systemic class" we must thus understand two kinds of classes, the synthetic and the schematic class. The former is the class of *constructs,* such as numbers and electrons; the latter the class of *schemata,* such as *Equidae, Canidae,* and *Syringae.* Both kinds of classes, when fulfilled, give rise to systemic value; whereas the concepts "horse," "dog," and "lilac" are *empirical* concepts and give rise to *extrinsic* value.[8] The two kinds of systemic classes correspond to the two kinds of systems previously discussed.

Intrinsic value, or Uniqueness, arises out of the fulfillment of the kind of intensional structure we called Description or Depiction. (We shall use the latter term, in order to avoid confusion with the many uses of "description," and in order to express the Gestalt character of this kind of intension.) Intrinsic value, then, arises from the fulfillment of a depictional intension, as extrinsic value arises from the fulfillment of an expositional, and systemic value from the fulfillment of a definitional or axiomatic one (a construct). Uniqueness, Goodness and Perfection are the three values indicating, respectively, the three different dimensions of value, or the dimensional pattern of value.

In the following we shall examine the pattern of Goodness.

2. *The Logical Pattern of the Value Terms*

Although it is clear that a thing is a "good," "fair," "bad," or "no good" class member depending on the degree in which it fulfills the intension of the class concept, it is not yet clear how two things are valuationally related that have *the same number of properties but different ones,* e.g. two bad chairs; one, let us say, lacking a back and the other a seat. Here we need a pattern which assigns to every property of an intension a definite logical position. We are forced, in other words, to delve into the problem of the structure of analytic intension. This structure may be developed from the relation between definition and exposition.

The definition of an analytic concept is an equality between a *definiendum* and a *definiens.* The latter, the definition in the strict sense, consists of two concepts, the genus and the differentia (which specifies the genus). Any analytic definition, then, consists of four terms: *definiendum* ("man"), *definitional equality* ("is"), *genus* ("animal"), and *differentia* ("rational").

Applying the theory of types,[9] every analytic definition consists of two subjects and a predicate: the definiendum and the genus are the subjects, and the differentia is the predicate. Both subjects are of the same type because of the definitional equality, while the predicate is of a higher type. Thus, in the definition $D = \alpha\beta$, if D is of type one (D_1), the genus α is also of type one (α_1), while the differentia β is one type higher (β_2); hence $D_1 = \alpha_1\beta_2$; or, in terms of type only we write: "$D_1 : 1 = 1,2$" — definition 1 consists of an equality of type 1 and types 1 and 2.

The concepts α and β have, in turn, their own definitions. The genus of α is again of type one and its differentia of type two, while the genus of β is of type two and its differentia of type three:

$$\alpha_1 : 1 = 1, 2$$
$$\beta_2 : 2 = 2, 3$$

Developing this process of specification — intensionally one of differentiation — *ad infinitum*, we find the following sequence:

Types	Levels of Abstraction
1	n
1 2	$n - 1$
1 2 2 3	$n - 2$
1 2 2 3 2 3 3 4	$n - 3$
. . . .	etc.
. . . .	

As is seen, on the highest level of abstraction, level n, there is only one property, which is of type one; on the next level, $n - 1$, there are two properties, one of type one and the other of type two; on level $n - 2$, there are four properties, one of type one, two of type two, and one of type three. From this results Table 8.

The specificatory process arranged in this second form shows something surprising: the table corresponds to Pascal's Triangle, which means briefly that the series of analytic implications corresponds to the series of the natural numbers, continuing ad infinitum quantitatively, and being ordered qualitatively according to the combinatorial calculus. See Table 9.

Thus, there is an order in the analytic structure. And if this structure corresponds to the world of discrete or abstracted properties, then this kind of world (Husserl's *Lebenswelt*) has a logical structure. The analytic pattern represents an order of the world of everyday experience; it assigns to each property not only a place but also a value.

The degree of *differentiation* is the number of properties corresponding to, or contained in, each level of abstraction. It is

8. *Series of Analytic Implications*

Levels of Abstraction	Types							*Degrees of Specification*	*Degrees of Differentiation*
	1	2	3	4	5	6	...		
n	1							0	2^0
$n - 1$	1	1						1	2^1
$n - 2$	1	2	1					2	2^2
$n - 3$	1	3	3	1				3	2^3
$n - 4$	1	4	6	4	1			4	2^4
$n - 5$	1	5	10	10	5	1		5	2^5
....								....	....
$n - m$								m	2^m

the degree of the expositional determination of analytic content. On level n the content is 2^0, (the definiendum), on level $n - 1$ the content is 2^1 (the definition). On level $n - 2$ the content is 2^2 (the definition plus two expositional properties), and in general, on level $n - m$ the content is 2^m. The exponent indicates the level of *specification*. The singular as a continuum is theoretically reached on level $n - \aleph_0$, where the content is $2^{\aleph_0} = \aleph_1$. If in this case $n = \aleph_0$, the level of abstraction is either 0 or any level from

9. *Pascal's Triangle*

1

1 1

1 2 1

1 3 3 1

1 4 6 4 1

1 5 10 10 5 1

1 6 15 20 15 6 1

1 7 21 35 35 21 7 1

1 to $\aleph_0$, since $\aleph_0 - \aleph_0$ may be any value from 0 to $\aleph_0$. In the first case — if the level of abstraction is 0 — the differentiation $\aleph_1$ means the concrete individual. In the second case — if the degree of abstraction is higher than zero — it means the ideal individual: any statement used as Axiom.[10]

The *place* of a property is determined by its position in the degree of differentiation, and by the type to which it belongs. Likewise, degree and type determine the *value* of the thing. In the process of abstraction, the higher degree of differentiation potentially contains, or is the base of, the lower degrees, and its fulfillment is that of more properties. It is that from which the lower degrees are abstracted. Hence, a set of properties of a higher degree of differentiation, if actually differentiated, that is, containing the fullness of its determinations, is worth more than a set of properties of a lower degree of differentiation equally fulfilled. Thus, the set of properties of degree 2^4 is of greater value than the set of properties of degree 2^3, simply because in the former set there are more properties than in the latter. Conversely, the lack of a set of properties of higher degree of differentiation is a greater disvalue than that of a set of lower degree of differentiation. The lack, for example, of a horse's foot is a greater disvalue than the lack of his tail, but the lack of his mouth is a greater disvalue than the lack of his foot.

If of two classes one is the specification of the other, then the objects of greater specification are more valuable, since their properties are actually more differentiated. Thus, the goodness of a horse's tail is a greater goodness than the goodness of just any tail, or the goodness of a restaurant chair is a greater goodness than the goodness of just any chair. On the other hand, the objects belonging to two different classes, for example the classes "horse" and "horse's tail," *possess the same value*, due to the fact that both concepts can be differentiated ad infinitum. In other words, the members of all analytic classes possess the same scale of value, characterized by $\aleph_0$.

A doubt might arise as to how the goodness of a horse's tail can be of the same value as the goodness of a horse. However, this follows from the axiometric structure of the analytic intension. If a horse's tail and a horse, understood as members of two different, unrelated classes, fulfill their respective expositional properties, then this fulfillment has the same extrinsic value, since the number of fulfilled properties in both classes may always be equal. Therefore, the goodness of a horse has the same value, as goodness, as the goodness of a horse's tail. The same occurs in the case of all other extrinsic values, such as "fair," "bad," "no good," etc. Here formal axiology is applied not to valuation but to the value of valuation. The same results are reached if we take into consideration the relation which holds between "horse" and "horse's tail." Everything that we might be able to say about the horse's tail, up to the infinity of its extrinsic properties, is part of the differentiation of the intension of the horse, which also is denumerably infinite. And precisely because of that, the goodnesses of both are the same, since an infinite part of an infinity is equal to the infinity itself.

The situation is different when we consider one of these concepts ("tail") extensionally and the other ("horse") intensionally. In this case, an object or member of one class (*a tail*) — rather than tail properties — is part of the intension, or results from the differentiation, of the other ("horse"), and an extensional part of one concept is an intensional part, or property of the other. In this case, the first is *good for* the second, and the value of this property *for* that which has it is a function of the degree of differentiation which the former represents for the latter. That is, a lesser degree of differentiation gives less value "for," and a greater degree of differentiation gives greater value "for." A tail is of less value for a horse than a foot.

We have here, in the relation between "horse" and "horse's tail," two entirely different kinds of value. In the first case, the class "horse" and the class "horse's tail" have the extrinsic and independent values of *goodness, badness,* etc. In the second case, we do not speak of the extrinsic value of the horse's tail, but rather of the value of the property "horse's tail" within the exposition of "horse"; we speak of the value "good for," but not of the value "good." While the extrinsic "goodnesses" of things are equal and mutually independent, the extrinsic "*goodnesses for*" of things are unequal and mutually dependent. We can state the following theorems: If X is good for U, then a good X is better for U than a bad X. If X is good for U, it may or may not be good for T. A good horse's tail is better for a horse than a bad horse's tail; but nothing can be asserted, on the basis of the goodness of horse's tails for horses, concerning the goodness of horse's tails for things other than horses.

Thus, the relation between "horse" and "horse's tail" is

a relation of *differentiation* which leads to two different kinds of value: either to two independent "good" values or to one dependent value "good for." On the other hand, the relation between "tail" and "horse's tail" is that of *specification:* both are tails, but the horse's tail is a more specific one. In this last case, there are fewer properties in the tail than in the horse's tail. After making the one-to-one correspondence of the properties of "tail" and "horse's tail," there remains a residue in the series of properties of "horse's tail," that is, the specific property "horse's." If both series are infinite, it would be necessary to add to the infinity of "horse's tail" this new property. We have, then, a different order within the two infinities, in accordance with transfinite mathematics: if the infinity of properties of "tail" has the ordinal number ω, that of the "horse's tail" is $\omega + 1$.

The degree of specification, or differentiation, is proportionate to the value "good for" of the set of properties in question, for the thing that has them. The set of properties of maximum specification, that is, of degree m, or of degree of differentiation 2^m, serves as standard for the measurement of all sets, $m - 1$, $m - 2$ etc. If the set of degree m is worth 100 per cent for the thing in question, then the set $m - 1$, having half as many properties — since the former set has 2^m properties and the latter 2^{m-1} properties — is worth 50 per cent for the thing, or 100 per cent/2^1; the set of degree of specification $m - 2$ is worth half again or 25 per cent, or 100 per cent/2^2, and in general, each set of degree $m - p$ is worth 100 per cent/2^p for the thing that possesses it. Since each set has half as many properties as the preceding one in the scale of abstraction — which is the scale of specifications m, $m - 1$, $m - 2$, etc. — and since each set is worth half as much for the thing in question as the preceding one, *each property of the thing is worth as much as each other property, no matter on what level of abstraction or differentiation, but depending on the degree of maximum specification.* Thus, if the level of specification is m, the level of differentiation is 2^m, and if the total set of 2^m properties is worth 100 per cent for the thing in question, then each property is worth 100 per cent/2^m, and this is the value of each property on each level for the thing, since

$$100 \text{ per cent}/2^m = 50 \text{ per cent}/2^{m-1} = 25 \text{ per cent}/2^{m-2}, \text{ etc.}$$

Thus, if $m = 4$, then each property of the thing, no matter on what level of abstraction, is worth 6.25 per cent for the thing in question, if $m = 5$, each property is worth 3.125 per cent, etc. The higher the differentiation of the thing, the less *each* property is worth *for* the thing; but the more properties the thing itself has, that is, the more differentiated it is, the more the thing *itself* is worth *for* something that has it. For this reason the horse's foot is worth more for the horse than its tail, and its mouth is worth more for it than its foot, and so on; any more differentiated part is worth more for the whole than is a less differentiated part.

The standard of the whole set of (horse) properties is, as we have seen, the set of greatest differentiation. Since this set determines the value of every horse property, no matter on what level of abstraction, a set of higher abstraction, that is, of lower differentiation, will have fewer properties of the same value as a set of higher differentiation. Thus, the mouth has more properties than the foot and the foot more than the tail, and all properties are of the same value for the horse. Hence the value of these respective sets of properties for the horse is proportionate to the number of properties contained in each.[11]

As there are many properties in the particular thing, all of the same value for the thing, we need a determination of the place of a property within the thing. Thus, for example, on the level of abstraction $n - 4$ (degree of specification 4 and degree of differentiation 2^4) there are 16 properties which are of 5 different types, including 6 of type 3. All the properties possess the value 6.25 per cent for the thing; as, for example, the well-roundedness of the back of an easy chair or the spring of the chair's seat (presupposing that these are properties rather than sets of properties). Thus, if the spring is lacking we subtract 6.25 per cent from the price of the chair and we do the same if the well-roundedness of the back is defective, possibly too for lack of a spring.

In order to localize logically the place of these two different properties of the same value, we need another rule, especially when it is a matter of properties that are not only of the same degree but also of the same type. The type differentiation gives a primary indication of the place of the property; for example, in the case of a horse defined as *Equus caballus*, the properties connected with "*Equus*" are all of a lower type than the properties connected in the same way with "*caballus*," since the genus is of a lower type than the differentia.

The question then is what is the qualitative rather than the quantitative, the topological rather than axiological difference, of the lack of different properties of the same degree and type. Axiologically, every lack of a property is worth the same, but topologically it is different. Thus, one thing is the back spring and another the seat spring. In order to resolve this problem, we need a *topological*[12] *rule*.

Each property can be expressed by its corresponding judgment: the property α by the judgment "x is α," the property β by the judgment "x is β," etc. Such a procedure can also be followed with properties of the same type and degree, for example with properties of type 2 of level $n - 4$, of which there are 4. Let us call them γ, δ, ε, ζ. The corresponding judgment would then be (1) "x is γ," (2) "x is δ," (3) "x is ε," (4) "x is ζ." These four judgments we shall call, respectively, "p," "q," "r," and "s." The thing x can possess either all or several of these properties, which means that all or several of these judgments "p," "q," "r," or "s" may be true, while if it does not have any of them, they will all be false; and in the intermediate cases, in which the thing possesses some of the properties, some of the judgments will be true and others false.

Since all possible truth values of judgments can be arranged in a table which gives to each judgment its proper place, and since each judgment represents one and only one property, it follows that the table of truth values gives an exact place to each of the properties predicated of x, Table 10, for $m = 4$.

10. *Logical Pattern of Typical Value Terms*

	TYPICAL GOOD *Disvalue:* 0 *Value:* 4 × 6.25 = 25	TYPICAL FAIR *Disvalue:* 6.25 *Value:* 18.75	TYPICAL MEDIAN *Disvalue:* 12.5 *Value:* 12.5	TYPICAL BAD *Disvalue:* 18.75 *Value:* 6.25	TYPICAL NO GOOD *Disvalue:* 2 *Value:* 0
p	T	T T T F	T T T F F F	T F F F	F
q	T	T T F T	T F F T T F	F T F F	F
r	T	T F T T	F T F T F T	F F T F	F
s	T	F T T T	F F T F T T	F F F T	F

The possession of all four properties of the type and degree in question gives rise to the value *Typical Good;* the possession of more than half of the properties, to the value *Typical Fair;* the possession (or lack) of half the properties, to the value *Typical*

Median; the lack of more than half of the properties, to the value *Typical Bad;* and finally, the lack of all properties gives rise to the value *Typical No Good.* If the thing — e.g. the back of a chair — possesses the axiological value *Typical Good,* it means that the thing has a typical disvalue of o per cent; if it possesses the value *Typical Fair,* its typical disvalue will be 6.25 per cent, and so on successively according to what has been said above with respect to quantitative or axiological value.

In the example the value of m was 4, and the number of properties in question was also 4. For other values of m the values and disvalues will be the corresponding multiples of $100/2^m$, and the number of propositions $p, q, r, \ldots$ will correspond to the number of properties in question.

The nature of the value terms themselves can be determined by the truth-value pattern, that is, the truth-values of the entailments belonging to each intensional property. If x is a *good* C and the intension Φ of C contains the properties $\alpha, \beta, \gamma,$ then x is a good C if and only if the entailments "x being a C is α," "x being a C is β," and "x being a C is γ" are all true. Thus, if x is a chair and to be a chair means to be a knee-high structure (α) with a seat (β) and a back (γ) then x is a *good* chair if and only if all these are predicated of it, otherwise x is not a good chair. Let us say that the first entailment, "x being a chair is a knee-high structure," is "p"; the second, "x being a chair has a seat," is "q" and the third, "x being a chair has a back," is "r." The possible truth-values "T" for Truth and "F" for Falsity of these propositions, then, produce a value pattern, as per Table 11.

11. *Logical Pattern of Value Terms*

	"GOOD"	"FAIR"	"BAD"	"NO GOOD"
p	T	T T F	F F T	F
q	T	T F T	F T F	F
r	T	F T T	T F F	F

There is only one way in which a thing can be either good or no good. It is good when all the entailments belonging to its intensional [13] properties are true, and no good if they are all false. But it can be fair and bad in as many ways as there are intensional properties. To our case of the chair that lacks a seat belongs the combination where *q* is false but *p* and *r* are true, hence the *second* column under "fair"; whereas the chair that lacks a back is represented by the *first* column under "fair."

That these chairs are "fair" rather than "bad" is due to the simplicity of our example. Obviously, the pattern can be augmented both horizontally and vertically. Horizontally, it can be applied to any number of sets of intensional properties.[14] Vertically, of course, the value pattern can be applied to the values of the properties within the pattern themselves, so that the relative goodness or badness of a seat or a back can be determined: a chair with a *good* back but a *bad* seat will then have a different axiological position from one with a *bad* back but a *good* seat; and one with a fair seat has a different position from one with a bad seat, etc. The elaboration of all these relations leads to the intensional pattern just given.

As is obvious, we are dealing here with the axiological counterpart of the process of logical division; what we are discussing is the interrelationship of the properties of a subject, and of the properties of these properties, etc. At the same time, we are discussing the nature of the relation ω, which holds between an intension and the properties contained in it. This relation will now be set forth in arithmetical terms. This will lead to an elaboration of Moore's distinction between value predicates and nonvalue predicates, and to further exact distinctions between the value predicates themselves.

3. *The Arithmetical Pattern of the Value Terms*

a) THE RELATIONSHIP BETWEEN VALUE AND NONVALUE TERMS

The gap between value and nonvalue predicates appeared in G. E. Moore as the difference in descriptive power of the two kinds of predicates, called by him the nonnatural intrinsic and the natural intrinsic predicates. The latter "*describe* the intrinsic nature of what possesses them in a sense in which predicates of value never do. If you would enumerate *all* the intrinsic [natural] properties a given thing possessed, you would have given a *complete* description of it, and would not need to mention any predicate of value it possessed; whereas no description of a given thing could be *complete* which omitted any intrinsic [natural] property." [15] This difference in descriptive power of the two kinds of predicates, Moore came to regard as *the* differentia between fact and value; and his incapacity to define it with precision he regarded as his failure to construct the scientific ethics he envisaged.[16] Yet he saw clearly that "*a* solution of the problem is to be found in the different way in which [intrinsic natural properties] are related to one particular sense in which we use the word 'description.'" Once the precise nature of the descriptive properties is known, he believed, it would be clear not only in which sense "good" is "an importantly different *kind*" of property from the descriptive ones, but also in which sense yet it depends upon and *follows from* the descriptive properties.

We have presented "a solution" of this problem and stated with precision both the sense in which value predicates do not "describe" and the sense in which they "follow from" descriptive predicates. We must now turn our attention to the descriptive predicates themselves, that is, again, the structure of the intension, and ask ourselves *in which sense the natural predicates "describe."* What, in other words, is the sense of "description" in question? We shall reach an unexpected result: not only are the value predicates "a different *kind* of predicates" from what the descriptive predicates are thought to be, but the descriptive predicates themselves are a different kind of predicates from what they have been thought to be. They are one particular, and the most significant, set of *value predicates*. And it is precisely this special significance for valuation that makes them both "descriptive" and "factual"!

We shall determine the precise, mutual relationship between the set of descriptive and the corresponding set of value predicates entirely a priori, on the basis of the axiom presented which states that *value properties are sets of descriptive properties*. We shall not need any support in empirical observation; the strength of our procedure lies exclusively in the fertility of the axiom. But we shall find that this formal determination leads to results which empirical observation confirms, and which confirm empirical observation.

The value axiom enables us to determine *with precision all the possible relations in which a thing's value properties stand to its descriptive properties;* and to do so on the basis of nothing but the thing's descriptive properties.

The problem to be solved then is this: Given the set of a thing's descriptive properties, to define, *in terms of these properties,* not only *a*] the thing's value properties but also *b*] the relations between the value properties, and *c*] the relation between the value properties and the descriptive properties. Or, more simply, *if a thing has n descriptive properties, what does it mean for it to be good, fair, bad, and no good?*

According to the axiom, a thing is *good* if it has *all* its intensional properties, *fair* if it has more intensional properties than it lacks, *bad* if it lacks more than it has, and *no good* if it lacks most of the intensional properties. We were able in this way to define the fundamental value terms — the value predicates in Moore's terminology — in terms of the descriptive predicates, by introducing intensional quantifiers, or qualifiers. We shall now express the same matter arithmetically. By "intensional properties," of course, we shall again mean properties of *particular* things, that is, things as members of classes, of at least two things. These properties are signified by the predicates contained in the intension of the class concept, and these predicates are *descriptive* in the sense of the word here used. We do not speak, then, of *unique* things, subject to singular concepts, nor of *constructs,* subject to formal systems. Their value laws — those of intrinsic and systemic value — follow from the laws of extrinsic value here discussed.

Given a set of n descriptive properties, a thing of the corresponding class is *good* if it has n properties, it is *fair* if it has more than $\frac{n}{2}$ properties, it is bad if it has less than $\frac{n}{2}$ properties, and it is *no good,* we shall stipulate, if it has only one, or $\frac{n}{n}$ or n^0 properties. There is no difficulty in the case of an *odd number* such as 9 to determine what is "more than $\frac{n}{2}$" and "less than $\frac{n}{2}$," because $\frac{n}{2}$ always is between two integers, the higher of which is more and the lower of which is less than $\frac{n}{2}$. A 9-propertied thing, therefore, is fair if it has 5 properties and bad if it has 4. *With even numbers,* say 10, there is a difficulty in determining fairness and badness, for $\frac{n}{2}$ itself is a possible number of properties, and it is neither more nor less than $\frac{n}{2}$, but *is* $\frac{n}{2}$. It thus represents neither fairness nor badness nor for that matter, goodness or no-goodness. In these cases, we must introduce a fifth fundamental value term, *average*, ("indifferent," "so-so," "passable," "all right," etc.) which holds when the thing has $\frac{n}{2}$ of its properties. But this, now, introduces a difficulty in odd sets of properties. A thing cannot possess, we assume, halves of properties — although when the pattern here discussed is developed this notion may become feasible. How then shall we determine the value *average* with odd sets of properties? Let us stipulate that *average* holds with sets of odd properties when the thing has $\frac{n \pm 1}{2}$ of its properties. Thus, in the case of a set of 9 properties, a thing has the value *average* if it has either 5 or 4 properties; it is *fair* if it has more than 5 that is 6, 7, or 8 properties, and *bad* if it has less than 4, that is 3 or 2 properties.

In order to determine $\lessgtr \frac{n}{2}$ with precision, we introduce $m < \frac{n}{2}$. Then any number larger than $\frac{n}{2}$ (but smaller than n) is $\frac{n}{2} + m$, and any number smaller than $\frac{n}{2}$ (including 1) is $\frac{n}{2} - m$. Thus, in the case of 10, m has the values 4, 3, 2, or 1, and hence $\frac{n}{2} + m$ the values 9, 8, 7, or 6. These are the values for "fair," so that we define "fair," for even-propertied things, as the possession of $\frac{n}{2} + m$ properties. On the other hand, $\frac{n}{2} - m$ has the values 1, 2, 3, or 4. The first is the value for "no good," the other three are the values for "bad." We thus define "bad," for even-propertied things, as possession of $\frac{n}{2} - m$ properties, except 1.[17] For odd-propertied things we stipulate the range of fairness $\frac{n-1}{2} + m$, the range of badness as $\frac{n+1}{2} - m$, except 1, and that of average as $\frac{n \pm 1}{2}$. This means, in the case of a 9-propertied thing, that its range of fairness is 8, 7, and 6, its range of badness 3 and 2, and its range of average 4 and 5. The range of average of odd-propertied things is thus slightly larger than that of even-propertied things, and the range of fairness and badness of even-propertied things slightly larger than that of odd-propertied things. In odd-propertied things, a small slice of the fairness and badness range is included in the average range. Since the value properties in practice pass into one another, this overlapping is of no practical importance. But it is of importance for the theoretical pattern. The definitions of the value predicates in terms of the descriptive properties n are shown in Table 12.

12. *Arithmetical Definitions of Value Terms*

Value Predicate	Number of Descriptive Properties	
	Even	*Odd*
"good"	n	n
"fair"	$\frac{n}{2}+m$	$\frac{n-1}{2}+m$
"average"	$\frac{n}{2}$	$\frac{n\pm 1}{2}$
"bad"	$\frac{n}{2}-m$	$\frac{n+1}{2}-m$
"no good"	$\frac{n}{n}$	$\frac{n}{n}$

As we see, these definitions not only give the relations between the set of descriptive and that of value properties, but also enable us to state significant relations between the value properties themselves. First of all, we shall determine a notion which is being used over and over in value theory without any significant meaning, that of a *sum of values*. It is very often said, with no justification, that "the value of a whole is more than the value of its parts." Leaving aside just what may be meant by "whole," "part," and the like, we shall simply perform some obvious operations with the values here defined.[18]

By adding the values defined we obtain the sum of all possible fundamental values corresponding to a set n of descriptive properties. As is seen, this sum is not, of course, equal to n;

this is only the value *goodness*. But it is the arithmetical expression, for sets of even properties, $n + \frac{n}{2} + m + \frac{n}{2} + \frac{n}{2} - m = 2\frac{1}{2}n$.

We do not include the factor 1 or $\frac{n}{n}$ which is included in the factor $\frac{n}{2} - m$ when $m = \frac{n}{2} - 1$. For odd-propertied things the corresponding sum is $2\frac{1}{2}n \pm \frac{1}{2}$. Thus, in a thing of 10 properties, the sum of its possible values or its *value sum,* is 25, in a 9-propertied thing it is 23 (or 22), and so on. While, usually, it makes no sense to speak of the "sum of values" of a thing, or of different things, or of the sum of goodness plus fairness, plus averageness, plus badness, plus no-goodness of a thing, or of things, it does make sense to speak of such a sum or sums when the respective values can be arithmetically determined. This they can be in terms of the descriptive properties whose set *defines* them. The arithmetical definitions refine the logical definitions, through axiological quantifiers, mentioned before. In other words, the arithmetical definitions make precise the axiological quantifiers *all, some, not some,* etc. (properties). In doing so they disclose, beside the relations between descriptive and value properties, precise arithmetical relations between the values themselves.

The first such relation, then, is the *value sum* possible in a thing of *n* descriptive properties. We shall denote it by V_s and write,

$$V_s = 2\frac{1}{2}n \qquad \text{I}$$

understanding that this will always represent the nearest whole number.

We now have reached a most significant and peculiar result. If a thing has 10 properties, then the *value sum* of the thing is worth 25 properties of the thing. How a thing of 10 properties can have 25 properties is not immediately clear but will be if we think of the thing's properties as capable of combinations rather than as being a fixed set. But it is not for the moment even necessary to think of them in this manner. The formula makes clear that the sum of the value properties a thing has is greater in descriptive properties than the total of descriptive properties it has. Obviously, the sum of its value properties must be of greater value than any one value property it has, say goodness. Thus, if the thing is good if it has *all* its descriptive properties, and if the sum of its values includes the value Goodness, this sum must be *more* than *all* the thing's descriptive properties. How a thing can have more properties than it has is the secret of valuation. The clue lies in the different ways in which a thing "has" its descriptive properties and "has" its value properties. The "having" of the two *kinds* of properties by the thing is itself of two *kinds*, as will become clear. For the moment, however, let us interpret the formula merely as meaning that whatever the number of descriptive properties of a thing, the sum of its value properties equals two and a half times that number.

We shall now determine the product of fundamental values possible with a thing. We shall only give the formula for even-propertied things, of which the formula for odd-propertied things is a negligible deviation. The product of the fundamental values possible with a thing is $n \times (\frac{n}{2} + m) \times \frac{n}{2} \times (\frac{n}{2} - m)$. The value *product* then is

$$V_p = \frac{n^2}{2}\left(\frac{n^2}{4} - m^2\right) \qquad \text{II}$$

Since m has all the values $< \frac{n}{2}$, there are as many values of this formula as are the values of m. In other words, there is no one value product of a thing but rather a number of *value products*. There is a product for each value of m, which means that there are $\frac{n}{2} - 1$ such products. The total of the thing's value products then is

$$V_{pt} = V_{p1} + V_{p2} + \ldots + V_{p\,(\frac{n}{2}-1)} \qquad \text{III}$$

For example, for the 10-propertied thing, the four possible value products are

m	V_p
1	1200
2	1050
3	800
4	450

The sum of these, or the *total value product*, is 3,500. As is seen, the larger m, the smaller the corresponding value product, and the smaller m, the larger the corresponding value product. There is no reason why the various value products should only be added. They could also be multiplied and thus form a higher and definite value product. This *product of value products* or *secondary value product* of the thing is

$$V_{p_p} = V_{p_1} \times V_{p_2} \times \ldots \times V_{p(\frac{n}{2}-1)} \qquad \text{IV}$$

In the case of a ten-propertied thing, this is 453,600,000,000 or 4.5×10^{11}. As we see, the number of descriptive properties a secondary value product of a thing is worth is already astronomically larger than the number of the descriptive properties, or goodness. A set of 10 descriptive properties can, by some valuational operation, become a set of billions of properties.[19]

b) THE ESSENCE OF VALUATION

This we can understand only if we grasp the fundamental reason for the possibility of valuation: *the mobility of the descriptive properties.* In valuation, the thing's descriptive properties can appear in any kind of combination, the thing becomes fluid. Indeed, this combinatorial arrangement of the thing's properties *is* valuation. Thus, if the ten properties of the ten propertied thing are *a, b, c, d, e, f, g, h, i, j*, the thing is *good* only if it has all ten of these properties. The other fundamental values are subsets of ten. It is *average* if it has any 5, say, *a, b, c, d, e,* or *f, g, h, i, j,* or *a, d, f, g, i,* etc. In all there are 252 sets of 5 properties, or ${}_{10}C_5$ sets, in which the thing can be average. There are other numbers of sets in which it can be fair or bad. In the fairness of having 8 properties, there are 45 possible sets, and in the badness of having 3 properties 120 sets, etc. The totality of sets in which a thing can have value properties, is $2^n - 1$ or, in the case of the 10-propertied thing, 1023. In other words, there are 1023 different value possibilities of the ten-propertied thing; for every set of its properties represents one value, be it "good," "fair" (of different degrees), "average," (of different degrees for odd properties), "bad" (of different degrees), or "no good." The theoretical basis of these combinatorial possibilities is again Pascal's triangle which, we found, ordered the discursive intension. The practical relevance extends over the whole field of extrinsic value. Let us take a few examples.

Suppose a job definition consists of ten properties or job requirements. In how many ways can an employee fulfill or not fulfill the job? In $2^{10} = 1024$ ways (including the value zero). There is 1 way of *good* performance, but there are 385 ways of *fair* performance, 252 ways of *average* performance, and 385 ways of *bad* performance. It is not sufficient, therefore, to say that a worker is doing a fair job. It must also be determined which of the "fair" sets of properties his performance fulfills — 9, 8, 7, or 6 — and which set within these sets. Thus, there are 120 different possibilities of fulfilling 7 out of 10 job requirements. Which of the 120 is he fulfilling? By dividing the possible number of performances on any value level through the possible total of *all* performances one obtains the percentage of performance expectation: 0.098 per cent for *good*, 37.64 per cent for *fair*, 24.64 per cent for *average*, 37.64 per cent for *bad*. The difference between this theoretical expectation and the actual performance in the shop is an objective measure of shop performance. The same procedure can be applied to any other extrinsic valuation, e.g. that of a company's product, say, a refrigerator. If the refrigerator, in the mind of the public, is determined by 10 properties, the theoretical expectation of evaluation of it is $2^{10} = 1{,}024$; there are 385 ways in which the product may appear *fair* or *bad* and 252 ways in which it may appear so-so. These ways may in turn be broken down; of the 385 ways in which the thing may appear *fair*, there are 10 ways in which 9 properties may be accepted, 45 ways in which 8 may be accepted, 120 ways in which 7, and 210 ways in which 6 properties may be accepted. The corresponding percentages of expectation are, respectively, 0.98, 4.4, 11.73, and 20.53. Again, the actual acceptance as against the possible acceptance is an objective measure of the product's success.

The total possibilities of value of an *n*-propertied thing will be called the *total value* of the thing

$$V_t = 2^n - 1 \qquad \text{V}$$

It now becomes clear in which way a thing "has" its descriptive properties and "has" its value properties. The value properties are all possible subsets of descriptive properties; and the descriptive properties are the corresponding set. The value properties are the combinatory possibilities of the descriptive properties; and valuation consists, precisely, in dissolving sets of descriptive properties and rearranging them as value properties.

Dissolution and recombining, as Galileo has made clear, is

the activity of the creative scientist who dissolves the secondary properties into primary properties and rearranges the latter. Thus, creative science is valuation in a profound and specific sense; and the valuer is a scientist in the same sense. While the natural scientist dissolves secondary properties, the valuer dissolves *sets of* secondary properties, that is, intensions; and he recombines their elements into new configurations which *are values*. The valuer, thus, makes the descriptive sets fluid, dynamic; he breaks up intensions into their elements and rearranges the latter to form values. Since these elements are the secondary properties, he uses secondary properties as primary properties of value, and *sets of* secondary properties — intensions — as secondary properties of value. *The secondary properties thus are the primary properties of value.*

The much discussed analogy between natural and moral science now becomes precise and the relation between fact and value is resolved in an unexpected way.[20] If the $2^n - 1$ possible sets of descriptive properties are values, then *the set n of descriptive properties is only one among the value sets; it is that set which serves as norm or standard for the value sets.*

This means that the relation between fact and value becomes inverted: a thing is a certain number of combinatory sets of properties, or values, $P = V_t$. This number, P, is being interpreted as $2^n - 1$, where n is any one set selected to serve as norm for all the rest, and in terms of which the rest is determined. This set n is regarded as the *normal* set in terms of which the thing is considered as such a thing; and it is at the same time the *normative* set for the thing's value: the set is posited as normal by virtue of its normativity and as normative by virtue of its normalcy. As normal, the set determines the thing as *fact,* as normative, it determines the thing as *value. As the former, it is called the description of the thing, as the latter it is called the thing's goodness*. Thus, if the normal set of a thing called "table" is the set of n properties — *a, b, c . . .* — then this set *describes* the thing as a table; and the thing so described is the table in, and as, fact. But the same set, *by virtue of its being the normal or descriptive set,* is also the normative set, for it determines all the other sets as its subsets; and as determining all its subsets, or values, the thing is itself a *value,* namely a good table. In this aspect, it is a norm for all its other values, bad table, no good table, etc.

Thus, the descriptive properties as such determine the

thing as fact; and the same descriptive properties, as representing the set n among $2^n - 1$ possible sets, determine the thing as value. The thing as having n properties is a fact; but as having these same n properties regarded as one set among $2^n - 1$ possible sets, is a value.

In the first case we say that a table simply is *a table,* in the second case, that it is *a good table.* The first expression means that the thing is what it is; but the second means that in being what it is, it is at the same time a part of a number of other possibilities of being what it is. Everything thus is what it is and not another thing; but it can be what it is in different ways. *And each of these ways is one of its values. A value predicate is a subset of a description.*

Thus, the factual set of descriptive properties is a fixation of one set out of the variety of valuative sets. It is the fixation of a most important, indeed, the most important such set: the set that makes order out of the chaos of property combinations. If this chaos has the number p of properties, then the first step in making order out of it is that of regarding p as a totality of sets, $p = P = 2^x - 1$. The second step is to determine x, or solve the equation $x = \log_2(p + 1)$. Given p, the number of properties in question, x, can be determined as the number n of properties to serve as *norm* for the whole collection. This number, then, is the *number of descriptive properties.* Among the totality of properties the *specific* properties must therefore be selected to form this basic set. The resulting set is the set of descriptive properties of the thing, and a thing having these properties is regarded both as such a thing and as a good such thing. It is in this way that perception and conception produce the description and with it the value standard of things.

This not only clarifies the use of the analytic concept as value standard, but also throws light on the process of abstraction, especially the relation of analytic Definition to Exposition and Description in the Kantian sense. Our perception cuts out of the undifferentiated chaos of properties a certain set for differentiation into thinghood.[21] This set is the set of properties Kant called Description. It must be differentiated down to the minimum set of descriptive properties, Definition. We now learn that this minimum set bears a logarithmic relation to the original set. Thus, in the Kantian terminology, the first determination of a set, p, is *Description.* The conception of it as a set of sets,

$2^x - 1$, constitutes *Exposition.* The final crystallization of n, or solution of the equation $x = \log_2(p + 1)$ is *Analytic Definition.* This Definition constitutes the thing as *fact;* but the totality of descriptive properties is not lost; it reappears, now ordered and through the medium of the intension, as the totality of sets $2^n - 1$. Thus, the original setting of Description constitutes not only, as Kant says,[22] the "raw material" for the Definition, but also for the thing's value properties. Valuation is the process of reintegrating the thing, ordered and in a new medium, into the matrix out of which it was originally cut out.[23]

This analytic set of descriptive properties, the *intension,* is the set which language and custom have fixed in order to be able to deal with things and not become diverted by the multiplicity of value forms in which things may appear. It is the common denominator of these forms. Factuality, then, is nothing but the fixation of one value set as normative for all value sets. It is the one value set distinguished among all the others and awarded for its distinction with factuality. Factuality then, far from being the decline and fall of value is, on the contrary, its special mark and significance. The ground for the fixation in factuality of a particular value set is the capacity of this set to serve as common denominator for the multiplicity of value sets, its capacity of ordering these values in an organized manner, of serving as *norm* and *standard* for these values, which become determined by reference to it.

A fact, therefore is a fact by virtue of its fundamental value significance. The *normativity for value* is, from the point of view of value, the very essence of the factual, it is that which defines the factual as factual. Fact is the one specific value, among a totality of values, which is used as value measure.

By measure is meant the isomorphism of a numerical set with material prepared for measurement-by-this-set, through being broken down into, or reduced to primary properties (the Galilean "resolution"). Fact, we said, was the primary property of value. Descriptive properties are primary value properties and their set, the intension, is the value measure. Values, as measurable, are combinations of descriptive properties, just as lengths, as measurable, are combinations of centimeters. In science, secondary properties are broken down into primary properties which serve as units of measurement and are reassembled in terms of a standard of such units in order to measure the original secondary entity. In like manner, in valuation, the totality of value sets, *P*, is broken down into primary value properties (descriptive properties), which serve as units of value measurement and are reassembled in terms of a standard of such units, the set *n* or intension, in order to measure the original secondary entity, the total value *P*, now determined by the formula $V_t = 2^n - 1$, and all its subsets, which are the values. The set *n* or value standard, is called *fact* and happens to be the very set of secondary properties from which natural science has taken its point of departure. Just as all natural science is broken-down factuality, so all factuality is broken-down valuation.

We usually see only the factual nature of fact and not its valuational normativity: only the broken-down aspect but not the reality it measures. Thus, we are living in a broken-down world, a world of primary rather than secondary value properties. We live in the factual world as in the normal but not as in the normative world; and thus we live in the world as Mr. Tompkins does in Wonderland: in a world of reduction in which we are as bewildered as is poor Mr. Tompkins. The real world is the value world which we must build up from the normative set of value properties which we regard as the normal set. Fact, in other words, is our still unrecognized medium for value. We have the magic wand in our hand but we think it merely another stick. Value reality is the gigantic and multifarious domain, $2^n - 1$, of which we only discern, vaguely, occasional subsets. We have a world by the tail, but think the tail is the world.

Formal axiology arrives at a Copernican inversion of fact and value: rather than value being a kind of fact, *fact is a kind of value*; rather than value being the norm of fact, *fact is the norm of value*; rather than fact being real and value unreal, *value is real and fact is unreal.* Value is the reality of which fact is the measure. Fact is to value as a measuring rod is to a mountain. It measures the mountain, but that is all. The geometer is no mountaineer. Fact measures value, but that is all. To live in the world of fact is not living in the world of value. But charting a realm is the first condition for conquering it. The geometer precedes the mountaineer. To recognize the measuring capacity of fact for value is the first condition for conquering the value realm.

The second is to recognize the mutual relativity of fact and value. In a thing with intension $P = n$, the total value $V_t = 2^n$ (dropping "$- 1$" as insignificant in the present context) is the value pattern with respect to which the thing with n properties is a fact. This value pattern itself, however, may be regarded as a *fact*, representing a thing with 2^n properties and intension $P = 2^n$. This thing, in turn, has a value pattern, namely $V_t = 2^{2^n}$, with respect to which, precisely, the thing $P = 2^n$ *is* a fact. Again, there is thinkable a thing with intension $P = 2^{2^n}$ with a value pattern $V_t = 2^{2^{2^n}}$, with respect to which the thing $P = 2^{2^n}$ is a fact, and so on ad infinitum. We have then, within extrinsic value, a hierarchy of intensifications or differentiations which represent levels of the fact-value relationship such that the value of one level is the fact of the next.

This hierarchy parallels that of the theory of types. The total value of one set n is equal to the total possibilities of value on that level, 2^n. Similarly, in the theory of types, "assuming that the number of individuals in the world is n, the number of classes of individuals will be 2^n" and the number of classes of classes of individuals 2^{2^n} and so on ad infinitum.[24] The levels of progressing intensification thus correspond to the types of progressing extensification. We shall call the levels of extrinsic intensification the *extrinsic value types*. They demonstrate the relativity of fact and value within extrinsic value.

This relativity may be pursued beyond the dimensions of extrinsic value. It will then be seen that *a*] the dimensions of value are further types of value, and *b*] that the value dimensions — systemic, extrinsic, intrinsic — bear to one another the relationship of fact to value. In other words, systemic value is fact to extrinsic value, and extrinsic value is value to systemic value; and extrinsic value is fact to intrinsic value, and intrinsic value is value to extrinsic value; while systemic value is a lower type fact to intrinsic value, and intrinsic value a higher type value to systemic value. And since, as will be seen in the next section, the exponential relationship between types of value need not stop anywhere, the same will hold throughout the infinite range of value dimensions. *Any value dimension is fact to the succeeding dimension and value to the preceding dimension.*

The series of extrinsic value types n, 2^n, 2^{2^n} . . . ad infinitum, approaches toward the limit $\aleph_0$. The latter, if we follow the procedure of defining a process by its limit, may serve as theoretical definition of extrinsic value, even though in any given case, the analytic intension — whose fulfillment is extrinsic value — will have some finite number n of properties. Extrinsic value has the properties of $\aleph_0$; it is the actual infinity of discursive (denumerable) properties. Its upper limit, with respect to which it is fact, is intrinsic value. Its lower limit, with respect to which it is value, is systemic value.

Systemic value, which is not one of discursive properties, is that fact whose total value V_t is extrinsic value. If extrinsic value theoretically is regarded as the total intensional range of natural number, and practically as any number n of properties, then systemic value may be defined by a symbol which when used as exponent of 2 results in a natural number (n). This symbol is $\log_2 n$; and this indeed, as we have seen, represents the systemic intension.[25] The definition of the systemic intension, and thus of system — as having the intensional range $\log_2 n$, and the structure of a series of logarithms — shows up from a new side the difference between synthetic and analytic concept, as well as that between synthetic and analytic intension.

The main difference between synthetic and analytic *concepts* we found to lie in the larger operational power of the former, which arose through the process of "intensive abstraction" based on the principle of "convergence to simplicity with diminution of content." The limit of this process was the term which contained the power of the whole series leading up — or down — to it. We now see that the term is to the analytic concept as a logarithm is to its number (or as the number of primes within a number is to the number). This means that the process of intensive abstraction is the stripping down of analytic intensional properties to properties which represent the logarithm of these properties; and that the resulting synthetic properties have the power of the logarithm compared to the analytic from which they were derived; so that extrinsic multiplication of value corresponds to systemic addition, extrinsic division to systemic subtraction, extrinsic exponentiation to systemic multiplication, etc.[26]

The relation between synthetic and analytic *intension* is made clear by the definition of systemic intension as follows: While the analytic intension corresponds to the logarithm of 2^n, or *Description*,[27] the systemic intension corresponds to the logarithm of n, or of analytic intension. Analytic intension whose number of properties $P = n$, insofar as it is not a schema (a definition) but an exposition, is the norm of extrinsic value; synthetic intension whose number of properties $P = \log_2 n$, is the norm of systemic value. Thus, analytic "system" is to synthetic system as extrinsic is to systemic value, that is as a number is to its logarithm.[28]

The relation of extrinsic value to intrinsic value is as that of systemic value to extrinsic value: compared to intrinsic value extrinsic value is fact; or, in other words, extrinsic value bears to intrinsic value the logarithmic relationship that systemic value bears to extrinsic value; for $2^{\aleph_0} = \aleph_1$, which latter, $\aleph_1$, is the cardinality of the singular intension whose fulfillment is intrinsic value.

In sum, the relativity of fact and value stretches throughout the whole value realm, from the level of systemic value, the fulfillment of the synthetic definition, to that of extrinsic value in all its types, from the fulfillment of the analytic definition to that of exposition and description, and beyond to intrinsic value and hence to higher value infinities. Thus, it represents a hierarchy of values, where every lower level is fact to the higher, and every higher level value to the lower.

We may summarize what has been said in the following four rules:

1] *Any given set of properties is a fact, any subset of a given set of properties is a value of the given set. The given set is the norm of its subsets.*

2] *The totality of subsets of a given set of properties is the total value of the given set.*

3] *The total value is the value type succeeding the given set.*

4] *Any value type is value to its preceding and fact to its succeeding type.*

These rules are no mere game, they are an ordering of experience. The relationship between systemic, extrinsic and intrinsic value corresponds to a process of continuous enrichment with definite leaps from one value dimension to the next. Thus, if I buy a package of cigarettes from a saleslady I am in a legal, a systemic relationship with her. If I take her out for dinner I am in an extrinsic relationship, and if I take her to church and marry her I am in an intrinsic relationship with her: my total being is joined with hers in a common intrinsic Gestalt.[29] This Gestalt grew through successive enrichments, out of the first tenuous bond, the original sales contract.

All intrinsic relationships, except those of the family, grow out of systemic and/or extrinsic relations through processes of enrichment; and such processes are as common as is intrinsic value itself. The further growth of richness within intrinsic value, however, leading to higher and higher dimensions of value, is reserved for the few. General human value capacity, at present, does not seem to reach beyond the intrinsic — to experiences where infinities are piled upon infinities, experiences of mystic exaltation, of higher and higher, wider and wider expansion of awareness. In such experiences, as reported by some,[30] the subjects, on progressing from one infinity to the next, find that a previous infinity is limited and shrinks to insignificance when seen from the vantage point of a higher infinity. Thus, subjects experienced the entire world, with its infinity of events, actions, things, relations, and then rose to a higher state of awareness, where the whole world shrank and a new world infinitely richer than the one just left opened up to their marvelling consciousness. One subject experienced the totality of all works of art, tons and tons of sculpture, acres and acres of paintings, miles and miles of lace — only to see all this later from a distance as one great activity of a divine spirit. Such experiences often appear intolerable, for they seem to burst the limited human frame. On the other hand, knowing the hierarchy of intrinsic values and its transfinite symbolism of alephs, which does heap infinities upon infinities in a rational pattern, $\aleph_0, \aleph_1, \aleph_2 \ldots \aleph_n \ldots \aleph_\omega$[31] such experiences do not seem to be so overpowering nor so anomalous. Rather, they seem to confirm a natural disposition of the human spirit toward experiences of ever heightened awareness and consciousness. Thus, we find "in Cantor's work . . . 'the most rigidly exact in science applied to the shadow and spirituality of the most intangible in speculation.' "[32] We might even use his work to develop such experiences, follow along the ever richer path, using the series of alephs as a line of support as Jacob used the ladder to heaven. In any case, for the axiologist there seems to be no other way of ordering such experiences than through the series of transfinite values: transfinite cardinalities of predicates, transfinite intensional structures to be fulfilled by experience.

The combinatorial relationship between fact and value,

thus, reaches to the very limits of experience. It proves as rich as reality itself. It has both the structure and the infinite variety of spiritual life.[33] Experience is ordering sets of properties in infinite variations. To these sets set theory can be applied; and set theory applied to sets of properties is, precisely, formal axiology.

The relationship between fact and value arises in formal axiology from the relationship between a set of properties and its subsets, $V_t = 2^n - 1$. This relationship is so articulate and implies so many details that the old view of this relationship, as being that between *is* and *ought,* becomes obsolete and meaningless. It now becomes one of the axiological fallacies, the normative fallacy. Based on the superficial observation that what ought to be is not yet, "oughtness" was considered as the differentia of value and "isness" as that of fact. Oughtness then was identified with "normativity," and hence value with the norm for fact. All this, methodologically, not historically speaking, was a typically Aristotelian procedure, based on the observations of common sense. A more thorough examination would have shown that *norm* means *measure;*[34] and that in this sense value is not a norm for fact, but that the norms for fact are the primary properties. Axiologists, if they wanted to understand value as scientists understand fact — through the primary properties — ought to have looked for a *measure* of value, that is, *the primary properties of value.* Instead, they repeated *ad nauseam* the Kantian distinction, which is a distinction of secondary value properties and to this day obstructs the true understanding of value.

The primary properties of value, it turns out, are the descriptive properties. They enable us to understand value as the primary properties of fact enable us to understand fact. As natural science by means of the primary properties created a new world of fact, so axiological science by means of the primary value properties — fact as measure of value — can create a new world of value. In regarding the descriptive properties as a fundamental kind of value properties we perform in moral philosophy the very operation Galileo performed in natural philosophy, when he changed the Aristotelian physics into the science of mechanics: we invert the fundamental relation of the philosophy in question.

For Aristotle rest was the "natural" state and motion was defined in terms of rest.[35] Galileo inverted this relation and defined rest in terms of motion—cutting out, in the process, "causes" and with them the technological, psychological, metaphysical, and other paraphernalia of Aristotelian physics. Rest became a specific and significant kind of motion. It was defined as, and became rest by virtue of, being motion: zero motion. This solved the problem and changed the philosophy in question into a science. In a similar way, we invert the relation of fact and value. Value had been defined in terms of fact; and Moore's achievement was to do so in exact terms: value properties *follow from* descriptive properties and *are* nondescriptive properties. We invert this relation and define fact in terms of value—cutting out, in the process, "oughts" and therewith the teleological, psychological, metaphysical, and other paraphernalia of contemporary value theory. Fact becomes a specific and significant kind of value. It is defined as, and becomes fact by virtue of, being value: the measure of value. As Galileo had made fluid, so to speak, the point of rest by inserting it into a matrix of motion, so we make fluid the set of descriptive properties by inserting it into the matrix of value properties. This is "a solution of the problem" and should, as Moore envisaged, change the philosophy in question into a science.

Moore's question, then, in which sense natural properties describe and nonnatural properties do not, has the following answer, which well justifies Moore's lifelong puzzlement: *natural properties describe as primary value properties.*

Natural properties describe insofar as they serve as norm for nonnatural properties. The intension is *one* value property broken down to form a *set* of a different kind of properties. These properties—the elements of the set—are called "descriptive" properties, but their essence is not so much to describe as to serve as units of a standard to measure value properties, to serve as elements of a *set* which is equivalent to *one* specific value property.[36] This value property is called "goodness," and the set as equivalent to, and measure of, it is called "intension." The relation between a thing's descriptive properties and its goodness, therefore, is not so much that the goodness depends on the descriptive properties as that the descriptive properties depend on the goodness: the goodness, namely, as measure or standard of the thing's value. Thus, the goodness of the thing is not the norm for the thing's factuality, but it is the norm for the thing's value possibilities.

The thing's factuality is this norm seen without its normative qualities, the measure seen without its measuring capacity — like a meter rod seen as merely a stick. It is the "crack" in our understanding running between fact and value, seen as merely a crack. The razor sharp and deep crack between fact and value is not really *between* fact and value. It *is fact.* On both sides of it runs value. It is the sharp line that runs through value and measures it. As such it is part of value itself. It thus has a double nature: it is value as the measure and fact as the crack; just as the meter rod is norm as the measure and fact as the stick. Since the norm of value is the intension, mere factuality is the intension seen merely logically and not axiologically.

Mere fact, thus, is a void within value. It is value deprived of its value meaning, and hence of its essential character; it is the shell of value. It is what appears on the surface, what the senses perceive but the mind does not comprehend — like a savage finding a spring scale and bouncing up and down on it. Thus we bounce up and down on the world of fact, uncomprehendingly and giving it a beating. We do not understand its essential meaning, its capacity for measuring value, its *symbolic representation of the value world.* To call the natural properties "descriptive" is calling them by their *obvious* name, their sense characteristics, their aspect of describing the normal appearance of things. Their *essential* nature is not that of being descriptive of the normalcy of fact but normative for the measure of value — just as the essential nature of the scale is not to serve as trampolin but as a measure of weight. It is fact itself, in its symbolic meaning for, and endless recombination as, value by which we must "purge the world of the tedious stuff of the obvious," to speak with Tolstoy.[37]

As "descriptive," the natural properties appear as primary, as "goodness," they appear as secondary value properties. If we call the value properties tertiary, then fact properties are secondary and natural-science properties primary; and the relation of secondary to tertiary is as the relation of primary to secondary. *This relationship expresses in a nutshell the content of formal axiology. It is the essence of valuation.*

Therefore, to base value disciplines — ethics, aesthetics, etc. — on the tertiary value qualities, such as goodness in the analytic sense, or satisfaction, pleasure, preference etc., is like basing natural science on the secondary qualities, such as the yellowness, hardness, and brilliance of gold in the transmutation of elements. It is value alchemy.

As primary value property, each secondary or descriptive property can appear in any of the various subsets of the set V_t, which constitute the value combinations. The total number of ways in which all descriptive properties can be arranged in value sets is what we call the *valency of the descriptive properties*. It is their capacity of forming value combinations, analogous to the valencies of the chemical elements of forming chemical combinations.[38] This valency is

$$V_c = 2^{n-1}n \qquad \text{VI}$$

Thus, the ten properties of the ten-propertied thing can form value combinations in 10×2^9 or 5120 ways. They have the valency 5120, and can appear 5120 times in different value combinations. The properties of a 20-propertied thing can appear so in over five million ways. Thus, the value mobility of the descriptive properties is considerable.[39]

c) THE RELATIONSHIP BETWEEN THE VALUE TERMS

So far we have specified the Moorean relation between descriptive and value properties. We shall now examine the mutual relationships of the value properties themselves. This will further illuminate the Moorean relation. We shall do this by the simple operations of addition and subtraction, multiplication and division, of the corresponding arithmetical expressions. The results are the following, using the notations introduced earlier — "G" for "goodness," "B" for "fairness," "D" for "badness," "T" for "no-goodness" — and adding "M" for "average," signifying "middle," "medium," or "mean" value. Again, we shall discuss only the even numbered sets. The formulae refer to the value properties of one or of several things within the class in question.

Value Addition VII

Goodness plus Fairness	$G + B = n + n/2 + m$ $= 1.5\,n + m$	a
Goodness plus Average	$G + M = n + n/2 = 1.5\,n$	b
Goodness plus Badness	$G + D = n + n/2 - m$ $= 1.5\,n - m$	c

Goodness plus No-goodness	$G + T = n + 1$	d
Fairness plus Average	$B + M = n/2 + m + n/2 = n + m$	e
Fairness plus Badness	$B + D = n/2 + m + n/2 - m = n$	f
Fairness plus No-goodness	$B + T = n/2 + m + 1$	g
Average plus Badness	$M + D = n/2 + n/2 - m$ $= n - m$	h
Average plus No-goodness	$M + T = n/2 + 1$	i
Badness plus No-goodness	$D + T = n/2 - m + 1$	j

Value Subtraction VIII

Goodness minus Fairness	$G - B = n - (n/2 + m)$ $= n/2 - m$	a
Goodness minus Average	$G - M = n - n/2 = n/2$	b
Goodness minus Badness	$G - D = n - (n/2 - m)$ $= n/2 + m$	c
Goodness minus No-goodness	$G - T = n - 1$	d
Fairness minus Average	$B - M = n/2 + m - n/2 = m$	e
Fairness minus Badness	$B - D = n/2 + m - (n/2 - m)$ $= 2m$	f
Fairness minus No-goodness	$B - T = n/2 + m - 1$	g
Average minus Badness	$M - D = n/2 - (n/2 - m) = m$	h
Average minus No-goodness	$M - T = n/2 - 1$	i
Badness minus No-goodness	$D - T = n/2 - m - 1$	j
Fairness minus Goodness	$B - G = n/2 + m - n = m - n/2$	k
Average minus Goodness	$M - G = n/2 - n = -n/2$	l

Badness minus Goodness	$D - G = n/2 - m - n = - n/2 - m$	m
No-goodness minus Goodness	$T - G = 1 - n$	n
Average minus Fairness	$M - B = n/2 - (n/2 + m) = - m$	o
Badness minus Fairness	$D - B = n/2 - m - (n/2 + m) = - 2m$	p
No-goodness minus Fairness	$T - B = 1 - n/2 - m$	q
Badness minus Average	$D - M = (n/2 - m) - n/2 = - m$	r
No-goodness minus Average	$T - M = 1 - n/2$	s
No-goodness minus Badness	$T - D = 1 - n/2 + m$	t

Value Multiplication IX

Goodness times Fairness	$G \times B = n(n/2 + m)$	a
Goodness times Average	$G \times M = n \times n/2 = n^2/2$	b
Goodness times Badness	$G \times D = n(n/2 - m)$	c
Goodness times No-goodness	$G \times T = n$	d
Fairness times Average	$B \times M = n/2(n/2 + m)$	e
Fairness times Badness *	$B \times D = (n/2 + m)(n/2 - m) = n^2/4 - m^2$	f
Fairness times No-goodness	$B \times T = n/2 + m$	g
Average times Badness	$M \times D = n/2(n/2 - m)$	h
Average times No-goodness	$M \times T = n/2$	i
Badness times No-goodness	$D \times T = n/2 - m$	j

* $G \times M \times B \times D = \frac{n^2}{2}\left(\frac{n^2}{4} - m^2\right) = V_p$ (Formula II.)

	Value Division	x
Goodness/Fairness	$G/B = n/(n/2 + m) = 2n/(n + 2m)$	a
Goodness/Average	$G/M = 2$	b
Goodness/Badness	$G/D = n/(n/2 - m) = 2n/(n - 2m)$	c
Goodness/No-goodness	$G/T = n$	d
Fairness/Average	$B/M = (n/2 + m) : 2/n = 1 + 2m/n$	e
Fairness/Badness	$B/D = [(n/2 + m)/(n/2 - m)] = [(n + 2m)/(n - 2m)]$	f
Fairness/No-goodness	$B/T = n/2 + m$	g
Average/Badness	$M/D = [n/2/(n/2 - m)] = [n/(n - 2m)]$	h
Average/No-goodness	$M/T = n/2$	i
Badness/No-goodness	$D/T = n/2 - m$	j
Fairness/Goodness	$B/G = (n/2 + m)1/n = 0.5 + m/n$	k
Average/Goodness	$M/G = n/2 \times 1/n = 0.5$	l
Badness/Goodness	$D/G = (n/2 - m)1/n = 0.5 - m/n$	m
No-goodness/Goodness	$T/G = 1/n$	n
Average/Fairness	$M/B = [n/2/(n/2 + m)] = [n/(n + 2m)]$	o
Badness/Fairness	$D/B = [(n/2 - m)/(n/2 + m)] = [(n - 2m)/(n + 2m)]$	p
No-goodness/Fairness	$T/B = [1/(n/2 + m)] = [2/(n + 2m)]$	q
Badness/Average	$D/M = [(n/2 - m) : 2/n] = 1 - 2m/n$	r
No-goodness/Average	$T/M = 2/n$	s
No-goodness/Badness	$T/D = [1/(n/2 - m)] = [2/(n - 2m)]$	t

What do these formulae mean?

1] It is easy to see what value *addition* means, namely, a situation in which the value statements added are made together. To take the simplest example, let us say that a thing is being judged by four people as being good, fair, so-so, and bad. Let us take a girl being judged by four young men, all four having agreed on the definition of "a girl." The first says: "What a girl!" meaning she has all the girl properties, that she is n. The second says: "I really don't think she's so hot," meaning she is so-so, she has $\frac{n}{2}$ girl properties. The third says: "I think she is pretty good," meaning $\frac{n}{2} + m$, and the fourth says; "I don't know what you fellows see in her," meaning that, though she is still a girl, she does not have many of the girl qualities, she is $\frac{n}{2} - m$. If we ask ourselves what is the value of the situation of the four young men saying this about the girl, or what their judgments add up to, we would have value addition; we would have to add up the judgments in question, that is, $n + \frac{n}{2} + \frac{n}{2} + m + \frac{n}{2} - m = 2\frac{1}{2}n$. This would be the value of the total situation in question. The girl, as seen by the four young men, does have according to formula I above, two and a half times the properties she has.

Value measurement, thus, is not necessarily the same as intensive measurement by degree. If we have four bodies of different heat, one 10°, one 20°, one 30°, and one 40°, we cannot say that the warmth of the total is $10° + 20° + 30° + 40° = 100°$. Degrees do not add up, they can only be averaged. We can determine what the temperature of the room would be, having these four bodies. Again, if in a class of students one has the grade 60, another the grade 70, another 80, and another 90, then it makes no sense to say that the total grade is 300. We can, however, speak of the average grade which is 300 divided by 4, or 75. With value addition it *does* make sense to say that the total value of the situation is $2\frac{1}{2}n$, for value judgments do not fuse with each other; and equal judgments, e.g. of "fair," usually mean different sets of properties. But even if they do not, the value situation may be cumulative. A girl regarded as "good" by four young men is regarded as four times good, or $4n$, and not just as once good (similarly it may be said that a room regarded as hot by four people is, so far as the people are concerned, four times hot). Four girls regarded as "so-so" by one young man are each regarded as so-so, or $4 \times \frac{n}{2} = 2n$. This does not mean that the situation is twice good, anything beyond n is beyond the term "good." It means that the situation has accumulated $2n$ properties of girlhood, similar to the way that a team acquires points, which are points of excellence, depending on the definition of the particular sport, or as a worker acquires points in job fulfillment.[40] By the same reason, the girl seen by four young men acquired $2\frac{1}{2}n$ points. Value addition, thus, may be compared to, or actually be, *scoring*, the "points" in question being descriptive properties.

On the other hand, there may be average value of a situation; the average value of the one girl in the above situation is $2\frac{1}{2}n : 4 = \frac{5}{8}n$, which means "fair," and that of the four girls is $4n : n = n$. If it should be held that, at least in the case of value judgments concerning identical or overlapping value properties, only averaging should be permitted, it would mean a prohibition of "scoring" which would not seem to be justified.

2] The interpretation of value *subtraction* is more difficult. Value subtraction cannot mean the subtraction of value judgments, for any value judgment, even a negative one, once it is made is made and cannot disappear. Any value judgment thus means addition, although it may be a negative one. It cannot mean silence, for in this case no judgment is uttered and the addition is ± 0. Rather, value subtraction must have a very definite axiological meaning: It is *ought value.*

Let us look at the first four values in VIII: $G - B = \frac{n}{2} - m$, $G - M = \frac{n}{2}$, $G - D = \frac{n}{2} + m$, $G - T = n - 1$. These results represent the values of "ought" of a thing that is fair, average, bad, and no-good, respectively. Since a thing ought to be good, that is, ought to be n, its value of ought is n less that which the thing is. Hence, the value of ought is determined by the formula $o = n - x$, where x is the value the thing has at the moment of valuation. The formula means: "x ought to be n through o." For example, *John ought to be a good student* means that John ought to have an average of 90 rather than 70, as he does. His value of ought is $90 - 70 = 20$; he ought to raise his marks by 20 points. *John ought to be a good student through a raise of 20 points.* Or, another example, everyone who promises (x) ought to fulfill his promise (n) by fulfilling his obligation (o), $n - x = o$. The obligation is the promise fulfilled (n) less the promise made (x); the promise fulfilled is the norm for the promise made, that is, for him who makes it. The distance between the norm and the actuality is the obligation, that which is "owed."

The four formulae then mean the following:

> $G - B = \frac{n}{2} - m = D$ *means* B ought to be G through D.
>
> $G - M_1 = \frac{n}{2} = M_2$ *means* M_1 ought to be G through M_2.[41]
>
> $G - D = \frac{n}{2} + m = B$ *means* D ought to be G through B.
>
> $G - T = n - 1 = G - 1$ *means* T ought to be G through $G - 1$.

D, M_2, B, $G - 1$ are the ought-values of B, M_1, D, and T with respect to G.[42] John who is a fair student (70) ought to be a good student (90) through raising his grades by 20 (which would be the grades of a bad student). Inversely, George, who is a bad student (20) ought to be a good student (90) by raising his grades by 70 (which would be the grades of a fair student). Similarly, if I promise a person to pay him $100.00 in a month, and I pay him $50.00 in two weeks then I have to pay him another $50.00 two weeks later; my promise of the whole sum, $G - o = G$ is now $G - M_1 = M_2$.

Thus, whenever it is said that x ought to be p there is the formula $p - x = o$, where p is the norm in question, usually n; hence $n - x = o$. Depending on whether the value x is "fair," "average," "bad," or "no-good," the ought-value — that through which the subtrahend ought to be the minuend — is, respectively, $\frac{n}{2} - m$, $\frac{n}{2}$, $\frac{n}{2} + m$, $n - 1$. In a fair thing its ought-value is bad, in an average thing it is average, in a bad thing it is fair, and in a no-good thing it is good less 1 (practically good). All these are axiologically synthetic oughts, that is, the thing is not what it ought to be. The valuc of synthetic ought of a thing is, logically, the contrary (or sub-contrary) value of the thing, and, arithmetically, the complementary value of the thing, $n - x$. The *analytic* ought value is o since $n - n = o$. In a good thing the value of ought is o; analytic ought means that the thing is what it ought to be.

From $n - x = o$ it follows that $n - o = x$, which means that o ought to be n through x; in other words, x is the ought-value of its ought-value. Tom, who is a fair student (70) (Tom's value equals George's ought-value) ought to be a good student (90) through raising his grades by 20 (George's value equals Tom's ought-value). In terms of the promise, the promise fulfilled, n, less the obligation to fulfill it, is the promise made. When I have paid $50.00 on account of my debt of $100.00 ($n$) then the obligation is $50.00 ($o$), and the actual promise, my debt outstanding, (x) is $50.00; $100 - 50 = 50$.

The ought-value of x is that through which x ought to be n; it is the *complement* which makes x into n. Of three values in the relation $V_1 - V_2 = V_3$, V_3 is the ought-value of V_2, and V_2 is the ought-value of V_3, both with respect to V_1; $V_2 + V_3 = V_1$. If we substitute in the formula $n - o = x$ for o its value $n - x$, according to the last formula, $o = n - x$, then $n - (n - x) = x$, which means that $(n - x)$ ought to be n through x: x is the ought-value of $n - x$, which latter is the ought value of x, namely o. This formula, peculiar as it may seem, will help us to understand value multiplication.

The formulae of subtraction point to a generalization of ought-values. If we call *all* subtractions ought-values, then *all* values under VIII are ought-values. This means that the norms in question are not merely G, but all the values possible; the formula then is the general one, mentioned above, $o = p - x$, where p is the norm in question. Thus, if the norm is "no good" and the thing is good, the ought-value would be $1 - n$ (VIII,n). Here "ought" is applied to "good" with respect to "no good"—which it ought not, by the definition of "ought." It is an illegitimate use of "ought." A no-good thing ought to be good, but a good thing ought not to be no-good. The norm "no-good" for a good thing is a no-good norm. It is *bad* advice by John's no-good friend to tell him that, being a good student, he ought to be a bad one. It is *bad* advice to be told that promises ought not to be fulfilled. On the other hand, the norm "average" could be a good norm for a no good thing since "average" lies in the positive direction of "ought," from "no good." A no-good thing ought to be (at least) average. If I haven't paid my debt it would be good to at least pay half of it. The ought-value of a no-good thing with respect to average is $\frac{n}{2} - 1$ (VIII,i) and is a legitimate use of ought since it is positive in the sense defined. The same would be true of the ought relation between a no-good and a bad thing. In general, any step from worse to better is in the direction of "ought" and is *improvement*. The ought-value of a good thing with respect to bad is $-(\frac{n}{2} + m)$, since ought-value $o = p - x$, where the norm $p = \frac{n}{2} - m$, or bad, and $x = n$. Hence $\frac{n}{2} - m - n = \frac{n}{2} - m$ (VIII,m) or $-(\frac{n}{2} + m)$. This is the (negative) ought-value of a good thing with respect to "bad" — and it is the negative value of "fair." A good thing ought not to be bad; if it ought to be bad then the ought-value is the negative of the positive ought-value of a bad thing (VIII,c). In general, if a thing with value x ought not to be p, then the corresponding ought-value is the negative of the positive ought-value of p; $o = p - x = -(x - p)$.

The notion of ought-value helps us to understand some of the relations implied in the formulae VII — X — relations not only between descriptive and value properties, but also between the value properties themselves. As is seen,

$$
\begin{aligned}
n &= B + D && \text{VII f} \\
&= G \times T && \text{IX d} \\
&= \frac{G}{T} && \text{X d}
\end{aligned}
$$

This means that the *number of descriptive properties* equals the sum of the values Fairness and Badness, the product of the values Goodness and No-goodness, the quotient of the values Goodness and No-goodness. The relations are obvious from the definitions of Goodness (the set of descriptive properties, n used as value measure), of No-goodness (1), and of Fairness and Badness $(\frac{n}{2} + m, \frac{n}{2} - m)$.

The following relations are not so obvious but are quite beautiful:

$$n = \frac{G \times M}{G - M} \qquad \text{XI (VIII b, IX b)}$$

$$= \frac{G \times B}{G - D} \qquad \text{XII (VIII c, IX a)}$$

$$= \frac{G \times D}{G - B} \qquad \text{XIII (VIII a, IX c)}$$

As we see, the number of descriptive properties corresponds to artful combinations of the value properties, Goodness and Average, and Goodness, Badness, and Fairness. Since $G - M$, $G - D$, $G - B$ are the ought-values of, respectively, an average, a fair, and a bad thing, the *goodness of the thing, n,* equals the quotients of Goodness times Average, and the Ought-value of an average thing; Goodness times Fairness, and the Ought-value of a bad thing; Goodness times Badness, and the Ought-value of a fair thing.

We have, by definition,

$$n = G \qquad \text{Def.}$$

Furthermore,

$$\frac{n}{2} = M \qquad \text{Def.}$$

$$= \frac{M \times D}{G - B} \qquad \text{XIV (VIII a, IX h)}$$

and

$$n^2 = 2GM \qquad \text{XV (IX b)}$$

From the above equations follow, among others, the following relations between value properties:

$$\frac{G \times B}{G \times M} = \frac{G - D}{G - M} \qquad \text{XVI (XI, XII)}$$

$$\frac{G \times D}{G \times M} = \frac{G - B}{G - M} \qquad \text{XVII (XI, XIII)}$$

$$\frac{G \times D}{G \times B} = \frac{G - B}{G - D} \qquad \text{XVIII (XII, XIII)}$$

The product of Goodness and Fairness is to the product of Goodness and Average as the difference of Goodness and Badness is to the difference of Goodness and Average. The product of Goodness and Badness is to the product of Goodness and Average as the difference of Goodness and Fairness is to the difference of Goodness and Average. The product of Goodness and Badness is to the product of Goodness and Fairness as the difference of Goodness and Fairness is to the difference of Goodness and Badness.

Since $G - D$, $G - B$, and $G - M$ are the ought-values of, respectively, a fair, a bad, and an average thing, the formulae mean that the ought-value of a bad thing is to that of an average thing as the product of a good and a fair thing is to that of a good and an average thing, the ought-value of a fair thing is to that of an average thing as the product of a good and a bad thing is to that of a good and an average thing, the ought-value of a fair thing is to that of a bad thing as the product of a good and bad thing is to that of a good and a fair thing. In general, *the ought-values of values are proportionate to the products of good and the complements of the values* [the ought-value of, say B, is the complement $G - B$, which is D, and $D : B = GD : GB = (G - B) : (G - D)$].

3] Let us now turn to the interpretation of *multiplication*. If the ought-values of a thing are subtractions of value, what is multiplication of values? According to equation XI, $G \times M = n\ (G - M) = G\ (G - M)$, that is, the product of Goodness and Averageness equals the product of Goodness times the Ought-value of Averageness. According to XII, the product of Goodness and Fairness equals the product of Goodness and the Ought-value of Badness, $G \times B = G\ (G - D)$. In general, the product of n and x equals the product of n and the ought-value of x, $n \cdot x = n(n - o)$, and this, as we have seen, means $n \cdot x = n[n - (n - x)]$. In words, $x = n - (n - x)$ meant that $(n - x)$ ought to be n through x. The product of this by n, $n \cdot x = n[n - (n - x)]$, merely adds the factor n on both sides, meaning on the left side that there *is* n and there *is* x, and on the right side that there is n, and $n - x$ ought to be n through x. A value multiplication, thus, is a more emphatic value subtraction or ought-value. The emphasis is on the norm which ought to be fulfilled through the ought-value. Any statements of the form "there is a norm n and there is an ought value x" is a value multiplication $n \cdot x$, and it means that $n - x$, or the ought value *of* x, namely o, ought to be n, and that there is n: *John ought to raise his average of 70 to 90, and he will do it; You have to fulfill your promise and of course you will.*

In terms of V_1 and V_2, the product of V_1 times V_2 equals the product of V_1 times the ought-value of the complement of V_2 with respect to V_1. The ought-value of the complement of V_2 is, by the definition of ought-value, V_2. Hence, with respect to V_1 (that is, if the norm of the ought-value of V_2 is V_1) $V_2 = V_1 - (V_1 - V_2)$; V_2 is the ought-value of $(V_1 - V_2)$ with respect to V_1; $(\frac{n}{2} + m)$ or Fairness is the ought-value of $[n - (\frac{n}{2} + m)]$ with respect to n, that is, of $\left(\frac{n}{2} - m\right)$, or Badness, with respect to Goodness. Any value product, then $V_1 \times V_2 = V_1\ [V_1 - (V_1 - V_2)]$, — where $V_1 - V_2$ is the ought value *of* V_2, which latter ought to be V_1. If $V_1 = G$ and $V_2 = D$, then $G \times D = G\ [G - (G - D)]$; the product of Goodness and Badness is equal to the product of Goodness and the Ought-value of the complement of Badness, which is the ought-value of Fairness, that is Badness. If $V_1 = D$, $V_2 = G$, then $D \times G = [D - (D - G)]$; the product of Badness and Goodness is the product of Badness and the Ought-value, with respect to Badness, of the Ought-value of Goodness with respect to Badness, which is the Ought-value of negative Fairness (VIII, m) with respect to Badness, or Badness minus negative Fairness, that is, plus Fairness, which is Goodness. In other words, the complement of Goodness with respect to Badness is negative Fairness, and the latter's ought-value with respect to Badness is again Goodness. As is seen, value multiplication, in this interpretation, is not commutative. It is, however, commutative in the arithmetical interpretation.[43]

If we signify the complement of V by "V_d" (from Spanish "*debe*" for "ought"), and the ought-value of V by "V_D," then $V_D = V_d$. Thus, $B_d = D$, $D_d = B$, $M_d = M$, $G_d = 0$. The ought-value of a value is the complement of the value; that is, $B_d = G - B$, $D_d = G - D$, $M_d = G - M$, $G_d = G - G$. Hence $V_1 \times V_2 = V_1(V_1 - V_{2(d)})$, whence follows that $V_2 = V_1 - V_{2(d)}$ and $V_{2(d)} = V_1 - V_2$.

The product of two values, V_1 and V_2, then, *relates the two values with respect to the latter's ought-value which ought to be the former value through the latter,* the product $V_1 \times V_2$ relates V_1 and V_2 with respect to the ought-value of V_2 which ought to be V_1. The first factor V_1 is the norm which the complement of the second factor $(V_1 - V_2)$ ought to fulfill $[V_1 - (V_1 - V_2)]$. But the norm V_1 is also the norm of the second factor itself, for whereas $(V_1 - V_2)$ is to fulfill V_1 through V_2, V_2 is to fulfill V_1 through $[V_1 - (V_1 - V_2)]$, which means $n - (n - x) = n - o$, which is the ought-value of x. From this follows the *rule of value multiplication: The only permissible multiplier of a value is the norm of the value.* Thus, if I want to multiply V by K, then K *becomes the norm of the value* V, and the resulting formula is $KV = K(K - V_d)$. Since the *norm* of V is the totality of which V is a subset or a *part*, (as the norm of $B = \frac{n}{2} - m$, is $G = n$, of which $\frac{u}{2} - m$ is a subset), this norm, say K, is of a higher logical order than is V; and the rule of value multiplication states that *I may multiply a value only with the set of which it is a subset,* or with the intension in which it is contained; in logical terms, I may multiply a set of properties only with the intension of which it is a part but not with another set of properties.[44]

What then does multiplication of two sets of properties mean? It means logically, that the first set must serve as intension for the second, as the set of which the second is a subset; or, axiologically, that the first set is the norm for the second which the second set ought to fulfill. Thus, $B \times D = \frac{n^2}{4} - m^2$ means, logically, that the first factor (the multiplier) here is the set of which the second (the multiplicand) is the subset; or, axiologically, that the multiplier is the *norm* the multiplicand is to fulfill; such that $B \times D = B\ [B - (B - D)]$. Hence, it implies that a bad thing ought to be fair. Similarly, all the formulae under IX imply that a thing with the value of the multiplicand ought to have the value of the multiplier. Value multiplication, thus, *implies* oughtness. Every value product is a product of a norm and a lower value, even if the latter is absolutely higher, as in $B \times G$. This implies the ought-value of G with respect to B (VIII, k).

Hence, if value subtraction is called ought-value (in $G - D = B$, B is the ought value of D), a value product may be called a *normative ought-value,* meaning that $G \times D$ is the normative ought-value of which G is the norm which is to be fulfilled both by D and its ought-value, or complement, $G - D$; for $G \times D = G\ [G - (G - D)]$. The second factor of the product, D, may be called the *value-ought,* meaning a value which ought to fulfill a norm but is not an ought-value (the ought-value in question is $G - D = B$). John, who is a bad student ($D = 20$) ought to be a good student ($G = 90$) by a grade that in itself would be that of a fair student ($B = 70$). B is the value-ought; and D is the ought-value; at the same time, it is the value-ought of $G - B = D$, that is of a fair student who ought to be a good student by grades which in itself would be those of a bad student. A *normative ought-value, thus is a product consisting of a value norm and a value-ought.* It is equal to the product of the norm and the ought-value of the value-ought's complement [the complement of D is B, and B's ought-value is $G - B = G - (G - D) = D$]. A value-ought, thus, is a complementary value's ought-value; which means simply that it is its complement's complement. D, as a value-ought, is the ought-value of B, just as B, as a value-ought, is the ought-value of D.

4] *Division* is the opposite operation, the splitting up of the normative ought-value by the corresponding value-ought, $\frac{G \times D}{D}$. The result is the norm which makes the value-ought into the normative ought-value, $\frac{G \times D}{D} = G = \frac{G \times D}{G - B} = n$ (XII). Thus $\frac{D}{M} = 1 - \frac{2m}{n}$ (x, r), which is equivalent to $D = M(1 - \frac{2m}{n})$. Here the set D is split up into two sets, M and $(1 - \frac{2m}{n})$ whose multiplication results in D. Hence, D is the normative ought-value of which M is the norm and $(1 - \frac{2m}{n})$ the value-ought (or vice versa). M is what $(1 - \frac{2m}{n})$ ought to fulfill in order to become D. Similarly, all the quotients in x are norms to which the divisors are the value-oughts.[45] For example, $(1 + \frac{2m}{n})$ is the norm which M ought to fulfill in order to become B (x, e). A value quotient, then is a norm which the divisor ought to fulfill in order to become the dividend; it is a norm which is a division of a normative ought-value by a value-ought. In our example of the student, the normative ought-value is John's normative or ideal Goodness as a student ($G = 90$) multiplied by his actual badness ($D = 20$), $90 \times 20 = 1800$, which divided by the corresponding value-ought

$(D = 20)$ is $\frac{1800}{20} = 90$ or his normative Goodness. In terms of the promise, the normative ought-value is the promise fulfilled (n) times the promise made (x), divided by the corresponding value-ought (x) which results in n. Since the norm is a means for the value-ought to realize, by multiplication, the normative ought-value [e.g., the norm M is a means for $(1 - \frac{2m}{n})$ to realize D], the norm may be called the *multiplicative complement* of the value-ought with respect to the normative ought-value. Thus, M is the multiplicative complement of $(1 - \frac{2m}{n})$ with respect to D, for $M(1 - \frac{2m}{n}) = D$; and G is the multiplicative complement of D with respect to $G \times D$.

While value multiplication is emphatic value subtraction, or ought-value, that is the emphasis of the positive in the lack of value through the complement, the meaning of value division is value diminution down to value cancellation. Thus, if $x = n - 0$, then $\frac{x}{n - 0} = 1$; if $B = G - D$, then $\frac{B}{G - D} = 1$. The pure number 1 has the lowest value dimension "no-good," and value < 1 means even less, down to 0, value neutrality or valuelessness. To understand this we must turn to another value operation, exponentiation.

The axiological interpretations of the value operations so far appear in Table 13.

13. *Axiological Interpretation of Value Operations*

VALUE OPERATION	NAME	FACTORS
Addition	Scoring	Values
Subtraction	Ought-Value (Value Complements)	Norm; Value
Multiplication	Normative Ought-Value	Norm; Value-Ought
Division	Norm (Multiplicative Value Complement)	Normative Ought-Value; Value-Ought

5] Value *exponentiation,* in its systematic import, means the following. A value product always is a norm multiplied by a non-fulfilled value-ought. An exponentiation of a value means its multiplication by itself so that $G^2 = G \times G$. This would mean, accordingly, that the value would be split into two different kinds: logically, the multiplier would be the set of which the multiplicand is a subset; axiologically, the former would be the norm of which the latter is the value-ought. The former would be the set of Goodness, the latter a specific goodness. This would mean that the various goodnesses of things are combined to the set of Goodness, and that the latter is the norm of which the particular goodnesses of things are the value-oughts, that is the norm to which they logically strive. The same is true of all other values. The exponentiation of Badness, D^2, means that the various badnesses of things are combined to form the set of Badness, and that the latter is the norm of Badness of which the particular badnesses of things are the value-oughts, and to which they logically strive. Value in *general* is, then, the set of all these sets of which specific values are subsets.

An exponentiation, thus, is *a value as a normative ought-value.* Values can, potentially, be exponentiated ad infinitum. Thus, there is possible G^3, G^4, . . . G^n, . . . G^∞; B^2, . . . B^n, B^∞, or, in general, V^2, V^3, . . . V^n, . . . V^∞, which means that there is always possible a higher fulfillment of the value in question. Only the actual infinity of an exponentiated value, $G^{\aleph_0}$, $B^{\aleph_0}$, or in general, $V^{\aleph_0}$, where $V > 1$, is the *fulfillment* of the value — and such fulfillment makes it another value (since $n^{\aleph_0} = \aleph_1$), a non-denumerable continuum, whereas the values here in question are denumerable. Thus, "$G \times G = G^2$" means that the actual good-value of a thing is an *ought-value* with respect to a higher good-value which includes all the good-values of things. But this higher good-value, in turn, is an ought-value with respect to an even higher good-value $G \times G \times G = G^3$, and so on ad infinitum. In other words, value fulfillment is never final. One value potentiality realized, or fulfilled, in turn implies another higher potentiality, which when fulfilled implies another higher potentiality and so on. We have here the axiological basis of *Teleology.* Even at "infinitum" the process does not stop, for here we have $G^{\aleph_0}$, which is $\aleph_1$ and is an ought-value with respect to the value $\aleph_2$. The exponents "2", "3" etc. signify the level of self-fulfillment of the value in question, the fulfillment of its nature of being this value, *its level of perfection as this value*. Thus D^2 is a more perfect bad-value than G^1 is a good-value. While the latter is n, the former is $(\frac{n}{2} - m)^2$ which may be more than n. Note that *all* values > 1 denumerably infinitely potentiated, are equal to the new value dimension $\aleph_1$, since any number > 1 with exponent $\aleph_0$ is $\aleph_1$. The value $\aleph_1$ thus is a first "highest value," a *summa virtus* rather than a *summum bonum*, and a first link in an infinite series of such "transfinite" values.[46]

A value to the exponent 1 is the value itself, yet the product in question is not without significance, since the expression $V \times 1$ would mean that V is the norm of which 1 is the ought-value, and the expression $1 \times V$ that 1 is the norm of which V is the ought-value. In the first case, $V \times 1 = V[V - (V - 1)]$, in the second case $1 \times V = 1[1 - (1 - V)]$. The question then arises what besides no-good, or having only one property, signifies "1". A clue may be found in the exponential meaning of "1." "1" signifies exponentiation by zero, since $V^0 = 1$. This means that a value multiplied zero times by itself is a no-good value: a stagnant value which does not develop is no good (mathematically, exponentiation and radication are "involution" and "evolution"). Similarly, the product $V \times 0 = 0$ means that the value is the norm of nothing and nothing is the value-ought of it. Here the lack of development means that the value is no value. Thus, we have a scale between 0 and 1, signifying the scale from no value to no-good value. This scale is that of fractions which are between 0 and 1, and which therefore, represent values of no good or less than no-good. Such values are found in value *division*, e.g. IX, 1, X, n, and X, *s*. They signify *transpositions* of value, that is, combinations of values which result in a value less than no-good. An example would be the combination of coffee and sawdust, which is less than no-good coffee and less than no-good sawdust. It is noncoffee, and nonsawdust. Yet, it is not no coffee or no sawdust; for there is both coffee and sawdust. Both have existence, but they hardly have value, not even the value *no good*. Such non-values but existences are expressed in language by privative prefixes, or by the prefix "non." These prefixes concern the intension (value) but not the extension (existence) of the thing in question. Thus, *Time* calls a visit of De Gaulle to Bonn "a kind of unvisit — no parades, no crowds, none of the pageantry so dear to the heart of De Gaulle," a state visit without the properties of a state visit; Bruno Walter calls the music of Schönberg "Unmusik," and the Mexico City *News* speaks of "the roads and non-roads" of Mexico. In all these, the things are *what* they are supposed to be but they are not *as* they are supposed to be — they are non-such-things, their properties are non-properties, *Uneigenschaften*, either not present or, if present, in one way or another degenerated, "fouled up," "mixed up," "*verkorkst*," "*verkehrt*." They have lost their interconnection with respect to their norm *n*.

What does the exponent mean in such cases? It means, as we said, the level of perfection of the value as this kind of value. A value to the exponent zero has zero perfection as this kind of value, and the symbol of such "zero perfection" is "1." A value of "zero perfection" as this kind of value is no good to such a degree as hardly to be said to be that kind of value. Yet, it is not 0, that is, nothing. On the other hand, a value to the exponent "1" always remains that kind of value. Thus, "1" is a symbol which as an exponent signifies value stagnation and as result of exponentiation zero perfection of the value as such.

There is a third value operation which always results in 1. Exponentiation is possible not only as self-multiplication of a value, that is, in terms of a pure-number exponent, but also in terms of a value as an exponent. Thus G^G means the level G, or good, of the perfection of a G-value, the valuation of goodness in terms of goodness — attribution of goodness *to* goodness — approval of goodness as good. G^D means the level D, or bad, of the perfection of a G-value, the valuation of G in terms of *D*: disapproval of goodness as bad. This does not mean lack in goodness. Rather, it means more than goodness, it means an abundance beyond goodness. And though it is a disapproval of goodness it is still a better value than a simple good-value. It recognizes G as G and in addition disapproves.[47] In other words, $n^n > n^{\frac{n}{2} - m} > n^1$, since $m < \frac{n}{2}$. Any exponentiation by a positive value — even if it is smaller than the base value — enhances the value. In other words, any value valued in terms of a positive value becomes a higher value. Such a higher value we call a *value composition* or *abundance value*. A diamond is beautiful, but even more so when set properly in a unique setting; and even if the setting is not beautiful but functional, such as a cutter's tool, the original value of the diamond may be enhanced. On the other hand, exponentiation by a negative value devaluates the value. Thus, valuing good coffee in terms of good sawdust is a devaluation of the good coffee; and instead of the exponentiation G^G we have the exponentiation $G^{-G} = \frac{1}{G^G}$, which is a fraction between 0 and 1 and, as we have just called it, a *transposition.*

Now we are ready to have another look at the value symbol "1." As is seen, composition and transposition with the same values are inversely proportionate, and their product always is 1. Thus $G^G \times G^{-G} = 1$; or, in general $V^V \times V^{-V} = 1$. This means that the product of a composition and its corresponding transposition cancel each other, that is, cancel their respective value elements. The result is a situation of no-good value, or hardly any value. This we call a *mere fact,* and the symbol of *mere fact* is "1." Our theory shows that the *value of mere fact* is exactly between composition and transposition, between approval and disapproval. It is neither the one nor the other. *Mere fact* is value with the exponent 0. And a value to the exponent 1 is the *mere fact of a value,* a stagnant value.

The nature of Composition and Transposition as Approval and Disapproval, respectively, shows up a relation between value exponentiation and value subtraction. Approval expresses addition — exponentially — to an existing value, oughtness expresses subtraction — nonexponentially — from a value to be fulfilled (or addition to a nonexisting value).[48] Thus approval is axiologically opposed to *oughtness.* A good chair to which — by a second value judgment — goodness is added, is axiologically opposed to a chair which *ought to be* good and hence *is not* good. Any approval of a good thing exponentially adds value to the goodness, since, in addition to *stating* the goodness of the thing — which is *attribution* of goodness — this goodness is in turn called good; goodness is attributed to goodness: "*x* is good and it is good that *x* is good." *Approval thus is attribution of goodness to the attribution of goodness.* Since approval is opposed to oughtness, or value lack, approval of a person as good is opposed not only to disapproval as bad, but to admonishing the person that he ought to be good — which latter *is* disapproval of him as bad. Disapproval is attributing badness to the attribution of badness. "X ought to be good" implies that *x* is bad and that it is bad that *x* is bad. "Ought" here is synthetic. Approval, on the other hand means that *x* is good and that it is good that *x* is good. This may also be expressed in terms of analytic ought, since analytic ought is equivalent to "good": "*x* is good" equals "*x* is as *x* ought to be." Hence, approval may be expressed in the following eight ways (where "is good" equals A, and "is as ought to be" equals B):

1] X is good and it is good that *x* is good (A A A)
2] X is good and it is good that *x* is as *x* ought to be (A A B)
3] X is good and it is as it ought to be that *x* is as *x* ought to be (A B B)
4] X is as *x* ought to be and it is as it ought to be that *x* is as *x* ought to be (B B B)
5] X is as *x* ought to be and it is as it ought to be that *x* is good (B B A)
6] X is as *x* ought to be and it is good that *x* is good (B A A)
7] X is as *x* ought to be and it is good that *x* is as *x* ought to be (B A B)
8] X is good and it is as it ought to be that *x* is good (A B A)

As is seen, approval and disapproval, disclose a relation between "it is good that" and analytic "ought." "It is good that *x* is good" may be taken as a definition of analytic "ought," and "it is bad that *x* is bad" as a definition of synthetic "ought."

The generalizations of ought-value discussed above, e.g. "*x* is good and ought to be bad," are equivalent to disapprovals of goodness and approvals of badness, that is, to uses of analytic "ought" combined with negative uses of synthetic "ought." "X is good and ought to be bad" is equivalent to "*x* is as *x* ought to be and *x* *ought to be* as *x* ought not to be." The underlined "ought" is synthetic, the others are analytic. On the other hand, "*x* is bad and ought to be good" is equivalent to "*x* is not as *x* ought to be and *x* *ought to be* as *x* ought to be" which is a positive use of synthetic "ought." Or, the same expressed by uses of synthetic "ought not," "*x* is bad and ought not to be good" equals "*x* is as *x* ought not to be and *x* *ought not to be* as *x* ought to be," the illegitimate use; and "*x* is good and ought not to be bad" equals "*x* is as *x* ought to be and *x* *ought not to be* as *x* ought not to be," the legitimate use.

The legitimate forms of *Disapproval* can be arrived at by denying all eight of the above forms: $\overline{A}\,\overline{A}\,\overline{A}$ ("*x* is not good and it is not good that *x* is not good") or "*x* is bad and it is bad that *x* is bad"), $\overline{A}\,\overline{A}\,\overline{B}$ ("*x* is bad and it is bad that *x* is not as *x* ought to be") $\overline{A}\,\overline{B}\,\overline{B}$, $\overline{B}\,\overline{B}\,\overline{B}$, etc. Illegitimate forms arise when positive and negative forms are mixed. These forms are of two kinds, contradictions and nonsense. Contradictions arise when there is a symmetric mixture, that is, the central form is different from the outer forms ($A\bar{A}A$, $\bar{A}A\bar{A}$), while nonsense arises in all other cases. Thus, it is axiologically contradictory to say "*x* is good and it is bad that *x* is good" ($A\bar{A}A$) or "*x* is bad and it is good that *x* is bad" ($\bar{A}A\bar{A}$) but nonsense to say "*x* is good and it is bad that *x* is bad" ($A\bar{A}\bar{A}$).

Value composition thus reveals a logical pattern for Approval and Disapproval. These, in turn, are certain interpretations of Value Compositions revealing relationships between value exponentiation and value subtraction.

The latter are two of the operations possible when, as a solution of Moore's problem, value properties are regarded as subsets of descriptions.

We have examined in this chapter some of the many formal relations within the system of formal axiology, and only those belonging to extrinsic value. Yet, what has been said may suffice to show that an axiological system with theoretical import is possible, and that Moore's suggestion can be developed in a strict way. We have offered a very simple model of valuation, pertaining to things in classes. Unique things of intrinsic value have, of course, a different pattern (in some ways more complex, in others more simple), but it is based entirely on the simple model discussed. Operations with transfinite values follow logically from those with finite values. The differences between the two kinds of operations characterize the differences between extrinsic and systemic values on the one hand, and intrinsic values on the other. These relationships may best be understood by reference to the vast field of value phenomena. Their explication by the axiological pattern constitutes the empirical import of formal axiology.

7

THE EMPIRICAL IMPORT OF FORMAL AXIOLOGY

Much is already gained if we can bring a number of investigations under the formula of a single problem.

— IMMANUEL KANT

The most exact of all measures is the Good. — ARISTOTLE

Let it be clearly understood that there is nothing irrational or unscientific in the conception of transfinite values, provided it is developed by strictly logical means.

— EDWIN T. MITCHELL [1]

1. *The Hierarchy of Values*

The empirical import of formal axiology arises, as does the systematic import, from the axiometric structure of the intension. While the systematic import arises from the *axio*-metric nature of intension, its value structure, the empirical import arises from its axio-*metric* nature, its capacity to measure value. The value structure is the structure of "*value*," that is, of the value form; the value measure is the measure of *value*, that is, of the value phenomenon. The value form is capable of measuring the value phenomenon, because the former is isomorphous with the latter.

In their capacity of measuring value, intensions are not different from any other kind of measure, say meter rods which measure lengths. Just as meter rods measure lengths by being "fulfilled," either wholly or partly, by phenomenal lengths, and express this fulfillment by numbers such as ½, ¼, etc. meters of length, so the intensional structures measure value by being "fulfilled," either wholly or partly, by phenomenal values, and express this fulfillment in axiological numbers, axiological quantifiers, or qualifiers, such as "good," "so-so," "bad," which correspond, respectively, to "1 intension," "½ intension," and "¼ intension" ($n, \frac{n}{2}, \frac{n}{4}$). Just as some actual phenomena of length may "go the whole length" or part of it, so some actual phenomenal value may "go the whole intension" or part of it. And just as the meter rod contains units, the centimeters, so the intension contains units, the predicates. Finally, just as each object requires its particular kind of meter (rod, tape, etc.), so each object requires its particular kind of concept (the value of an apple cannot be measured by the concept "pear"). But all concepts are concepts as all meters are meters, and all concepts contain predicates as all meters contain centimeters.

Also, as there are different dimensions of natural measurements — length, time, weight — with different units each, so there are different dimensions of value measurement — systemic, extrinsic, intrinsic — with different units each. Just as the standard of time has as units not centimeters but seconds, so the standard of systemic value, the synthetic intension, has as units not predicates but terms. And just as the standard of weight has as units neither centimeters nor seconds but grams, so the standard of intrinsic value, the singular intension, has as units neither predicates nor terms but a third kind of unit, Gestalt elements, the linguistic representative of which is the metaphor.[2]

Intensional structures, thus, are *axiometric* in the same fundamental sense that extensional structures are *physiometric*: the former measure value or "nonnatural" phenomena in the same sense that the latter measure physical or natural phenomena — with respective standards fitting, and appropriate to, each realm. Since natural science measurement is that of systemic value, or rather of a species of systemic value, the application of system to extensions, and systemic value is itself only one among the three value dimensions and thus in turn a species of value, *measurement* is a term of much wider meaning and extension than usually considered. There is measurement in all three value dimensions; and although all these kinds of measurement are based on the structure of intension, they have the greatest differences among themselves — precisely because the structure of the intension itself may have these differences. Systemic measurement is the abstract, formal kind of measurement used by the scientist, objective and detached. It is the way axiology itself measures value, or rather "value" — it measures the value measure.[3] Extrinsic measurement is by the value terms "good," "fair," etc. — and it is more "subjective" in application although as objective in theory as is systemic measure. This kind of measurement is acted out, thought through, judged, in short, is lived in everyday life. Intrinsic measurement is exalted experience, enjoyment, involvement in the thing valued, indeed, *the stages of involvement* are *this measurement* — yet, theoretically, it is as objective as are the other kinds of measurement.[4]

Thus, our view of measurement is similar to the classic view, founded by Plato and revived by Husserl. It is, at bottom, an elaboration of the Platonic conception of measure, especially as found in *Philebus* and *The Laws*. "The rationality of the world seems [to Plato] to lie in its being measurable, rhythmical, articulate; to understand any part of it is to find its unit of measurement or to trace the lines of its structure. . . . *The measure of a thing is its reality, its true self;* . . . to fulfill its own measure, to be entirely what it is meant to be, and neither to exceed nor to fall short of its place in the great whole of which 'God is the measure' (*Laws*, 716 C). This is to obey the law of its existence both for itself and for others." [5] The "scientific" view of measurement is only a partial — the formally extensional — fulfillment of the Platonic vision. The intensional aspect still awaits formal development, although informal developments have been frequent and of great historical importance.[6]

The values measured can be the referents of any concept whatsoever, and any concept whatsoever can appear in any of the three *logical* dimensions, synthetic, analytic, or singular, and hence in any of the three *axiological* dimensions, systemic, extrinsic, and intrinsic. Here it is where the Oxford Analysts' famous doctrine of "use" acquires axiological relevance and precision: it becomes a value dimension. For, axiological practice consists in nothing else but determining in which *axiological use*, that is, in which value dimension, a concept is employed. For example, in the three propositions, "my soul is filled with God," "the Christian God is different from the Buddhist and the Mohammedan God," and "the Trinity of God is well above mathematics; the *iota* that made all the difference in *homoiousion*, was extirpated as heresy," [7] the concept "God" is used as an intrinsic, an extrinsic, and a systemic value, respectively, that is as a concept in mysticism, comparative religion, and theology. In the three propositions "I ought to be myself," "I ought to be the best streetcar conductor in Chicago," and "I ought to move KKt-KB_3," "ought" is used respectively intrinsically, extrinsically, and systemically, in this case ethically, socially, and according to the rules of a game. The art of applying axiology is to determine such uses; and the fundamental difference between the axiological determination of uses and that of the Oxford school is that the axiological determination is precise and exactly defined within a formal system, giving rise to definite *value sciences*, whereas the determination of the Oxford school is a hit-and-miss affair based on contextual intuition rather than axiomatic foundation.[8]

As we have seen, the axiometric structures are connected with the level of abstraction of the intension in question. The synthetic concept is on the highest level of abstraction — indeed beyond that, it is construction — and has the minimum of intensional structure (or plenitude); the singular concept is on the lowest level of abstraction — that of immediate experience — and has the maximum intensional structure (or plenitude). As is seen, for the synthetic and singular concept, abstractive level and axiometric structure or plenitude are in inverse proportion. The minimum abstraction has the maximum structure and the maximum abstraction has the minimum structure. Only for the analytic concept do the degrees of structure and of abstraction coincide. It is probably for this reason that almost all our articulate valuation is extrinsic. Intrinsic valuation is inarticulate, and systemic value rarely recognized. Extrinsic valuation is that of everyday and works well as such. But our capacity for valuation all but breaks down as soon as systemic or intrinsic valuation is required. Then there appear confusions both in practice and in theory. The most fundamental, most consequential, and most prevalent such confusion is that between systemic and intrinsic value — both equally unknown theoretically — which appears, in practice, in the confusion of systemic value with "spiritual" value, in the hypostatizations of national social, theological, and other ideologies as demanding man's supreme loyalties,[9] and, in theory, in the denial of the logical nature of intrinsic value and the fallacy of method. Recently, recognitions of systemic value and its conflict with intrinsic value have begun to appear,[10] but academic ethics has not yet taken cognizance of this issue, let alone made the necessary distinctions. Yet, the formal recognition of systemic value is easy in ethical theory, and is indicated by the uses of "ought." In systemic value, the thing in question, by virtue of being, is as it ought to be. "Ought" here is analytic. The thing is perfect; its very existence means its being in its essence (as construct or as schema). When it is not as it ought to be it *is* not. In the case of extrinsic and intrinsic value, "ought" may be analytic, but also hypothetical or synthetic. The thing is as it ought to be not by its mere being but only if and when it actually is as it ought to be. These formal distinctions have had, and are having, profound influence in ethics, metaphysics, and politics. In *ethics* we have Kant's "perfectly good will," which is different from the will that is "good without qualification" in that it needs no "ought," does not "stand between" the rational and the empirical, and, according to its own subjective constitution, can be determined to act "through the mere conception of the good." This "holy will" contains no imperfection, hence (synthetic and hypothetical) "ought" is "out of place here." [11] In Hegel's *metaphysical politics*, the rational is the real, the real is as it ought to be, the state is the rationally universal, and the citizen as the particular of this state is always rational, real, and as he ought to be, that is, moral. This has led, on the side of the political right, to the totalitarian states of our days. On the side of the political left, a similar construction by Rousseau has led to a similar kind of state. Rousseau's General Will, "for the simple reason that it is, is always everything it ought to be." [12] The General Will is a most thoroughly elaborated synthetic concept, or construct, and its fulfillment a most thoroughly elaborated systemic value. As any construct, "it can no more be modified than alienated; to limit it is to destroy it" — just as Aristotle's "number." [13] As any synthetic concept, "the general will is always right." [14] Its members are like the members of a synthetic class or system — terms within the whole, lieu-tenants, placeholders for formal relations. They have no extension and no intension of their own, but only through and by the system. In themselves they are zeros — as the zero in arithmetic is nothing but a placeholder. "Each citizen should be in perfect independence of the others, and excessively dependent on the State. . . . For it is only the power of the State which makes the freedom of its member." [15] This freedom is systemic freedom, derived from, and precisely circumscribed by, the system. "The social pact gives the body politic an absolute power over all its members"; [16] and this power extends also over the organs of the State who *"are not and cannot be its representatives, they are only its commissars."* [17] Democracy, for Rousseau, is only for the best, for the most perfect, the gods themselves, but not for the people.[18] At the cradle of the great revolution against the divine right of kings there stood, as an ungodly godmother, the fiction of a new mystic body, the People, which assumed without change these same divine rights. These fictions, from the right and the left, led straight to certain political systems of our own times. The "State," the "Party," becomes "perfection," and what serves it *is good.* The analytic "ought," which is an attribute of God, reappears in the theory of sovereignty, of totalitarian as well as democratic states, of theistic as well as atheistic ones, driving both, but especially democracy, "into intolerable self-contradictions," and spawning the devastating wars of our era.[19]

Formal axiology tells us that what is by its mere existence as it ought to be is a systemic value, a construct or fiction. The democratic as the totalitarian citizen, thus, dies for the same kind of fiction, of "sovereignty," that did his medieval ancestors. *"Ce sont toujours les mêmes qui se font tuer."* [20]

Since the axiometric structure determines, by its being fulfilled, the value connected with this structure, value and abstractive level are in inverse proportion for the synthetic and singular concept, and in direct proportion for the analytic concept. This means that the value scale connected with concepts in general is the inverse from the abstractive scale connected with them. For, if value is defined as the fulfillment of conceptual intension, then intrinsic value — the fulfillment of the singular intension — is more of a value or, according to the definition of "better," a *better* value than the fulfillment of the synthetic concept, that is, systemic value. There arises, then, a *hierarchy of values* due to the different axiometric structures of the respective intensions, and connected with the application of the relation "better" to the concept of value itself. This hierarchy is opposed to the hierarchy arising from abstraction. The universal has the lowest, the unique the highest value. To the "Copernican inversion" of the relationship between fact and value in formal axiology thus belongs an equal inversion in the scale of values, the valuation of valuation. Formal axiology confirms the radical value reversal of existentialism, in particular Kierkegaard: its highest value is the individual, its lowest the system, with classes — of individuals or things — in the middle.[21]

Depending on the nature of the concepts used, there arise three different languages, which are axiologically systemic, extrinsic, and intrinsic, respectively. These languages may use different expressions for different value dimensions of the same thing, as "lilac" and "Syringa vulgaris," or "brother" and "male sibling" — expressions which in a one-dimensional logic are regarded as synonymous; or they may use the same expression for the different dimensions, and in this case axiological error can be avoided only when the rules of the value dimensions are strictly observed. Thus, since extrinsic valuation presupposes a class of at least two things, and two things presuppose empirical coexistence in space or succession in time, the subjects of extrinsic valuation include all spatio-temporal objects — objects, that is, within empirical, rather than systemic, space and time, *as such*. The star over the city can be "a good star" in the extrinsic sense of having all the empirical star properties, but not the star of astronomy. The latter is not in empirical space and time but in the physical space-time continuum. On the other hand, neither of the two can be "my good star," for this is neither in empirical space and time nor in the physical space-time continuum, but in intrinsic space-time — the *space* which includes paradise, utopia, the Kingdom of God, and heaven (not the sky), "theological space";[22] and the *time* which includes eternity, the millennium, the "fullness of time," *Hochzeiten*, honeymoons, vacations, gestations and births, Bergsonian durations. It is neither a systemic nor an extrinsic but an intrinsic value, and the fulfillment of a singular concept. Similarly, of the two sets of three propositions mentioned above, those using "God" and those using "ought," the first proposition in each set belonged to intrinsic, the second to extrinsic, the third to systemic value language.

Now let us consider the following three propositions: 1] "I am in pain," 2] "They are both in pain, and she in worse than he," 3] "The patient in eighteen has a referred pain in the sternocleidomastoid." Here, the word "pain" means three different things. In 1] it is an intrinsic (dis)value, in 2] it is an extrinsic (dis)value, and in 3] it is a systemic (dis)value. In (1) a sufferer *expresses* his pain; and nobody will understand his words unless he suffers the same pain — which he never will for he is another person.[23] This is a statement in a unique, a private language, and "pain" a singular concept; its intension is a Gestalt. This means, according to our definition, that it is an intrinsic (dis)value and that the statement is one in the language of intrinsic value. In (2) two pains are judged as members of the class of pains and compared — our definition of extrinsic valuation. This is a statement in extrinsic value language. In (3) pain is stated not of a person but of a certain physiological and medical entity, a unit in a certain hospital room with a certain pathological symptom. Here pain is precisely determined within a network of relations and belongs to systemic value language. As is seen, the wholly private language — which Wittgenstein regarded as impossible — is the language of intrinsic value; and its relation to the other languages is determined by formal axiology.[24]

Let us now ask which of the three kinds of pain is the "best" — that is, from the sufferer's point of view, the worst. Obviously, this question is meaningless unless there is a scale by which the three kinds of pain can be measured. For unless there is, there is no connection whatever between the pain I have, the pain somebody else has, and the pain described in pathology texts. This scale must be objective, that is, all three kinds of pain must appear with equal status, namely, as subjects of the scale. The scale must not repeat the three kinds of knowing pain by three kinds of measuring them; they must all be measured subject to the same scale. Formal axiology happens to be such a scale. In it the three kinds of pain appear as three kinds of value: the pains are all (dis)values; and their differences in kind are determined by the scale, formal value.

According to the scale, the pain *I* suffer is the worst, for it is an intrinsic value. It is not the worst pain because *I* have it; if that were all there were to it there would be no *tertium comparationis* with the other kinds of pain. These become "other kinds" only by and through the axiological measure. The second worst pain, axiologically, is the one the others have, being an extrinsic value; and the least bad pain, axiologically, is the one constituting, or constituted by, a system.

But all this can be known only by a system. The axiological system presents the scale by which the pains can be known and the best kind of knowledge, according to this scale, is the most systematic. This kind of knowledge tells us with precision that the first kind of pain — the intrinsic — is the worst and the third kind — the systemic — is the least bad, axiologically; it is just a term in a system (and so is the patient). The systemic kind of knowledge thus determines with precision that the non-systemic, the intrinsic kind of pain, is the worst. Obviously, if knowledge itself were intrinsic, then I could only say by intuition and compassion what the pains are, and my compassion would be the same for him who says he suffers, him who is said to suffer, and him who is defined to suffer. Intrinsically, there is no knowledge *scale,* at least not in the same cognitive sense as above; and hence no way of comparing the three kinds of pain. Neither is there such a scale extrinsically, for the three kinds are incomparable analytically, they do not belong to one genus. Only synthetically, by axiological *science,* is there possible any knowledge, and indeed a precise knowledge, of the three *types,* or *dimensions,* of pain.

Formal value, thus, is a synthetic concept, whose intension is formal axiology and which systemically contains three measures, n for systemic value (in its schematic rather than systematic aspect, which latter is of the order log n), $\aleph_0$ for extrinsic value, and $\aleph_1$ for intrinsic value. Each of these measures belongs to the system of numbers, to which also belongs the relation "greater than." Hence, measuring value phenomena, or values, by these measures, we can demonstrate that an intrinsic value is a greater value than an extrinsic value, and an extrinsic value a greater value than a systemic value. The value phenomenon measured is not the everyday phenomenon consisting of secondary properties but the axiological phenomenon as determined by the value measure. The measure imparts to it primary axiological properties. These properties "depend," in the peculiar Moorean sense, on the secondary properties in such a way as to be elements of the sets of the secondary properties; which means that they *are* the secondary properties, but rendered more fluid, combinatorially arranged and rearranged. In a way, therefore, "value" may also be regarded as an analytic concept, a genus consisting of three species, systemic value, extrinsic value, and intrinsic value. The extension of the genus would be all the value phenomena, the intension the definition of value regarded as analytic: Fulfillment of an intension. The extensions of the species would be the respective value phenomena; their intensions, respectively, fulfillment of a synthetic intension, fulfillment of an analytic intension, fulfillment of a singular intension. To these analytic concepts there could be applied the rules for extrinsic value, especially the definition of "better than" as "having more properties of the intension than." That value would be a better value than another which has more properties of the intension of "value." The latter is fulfillment of an intension. The more fulfillment of an intension there is, the better is the value. Since the intrinsic value fulfills a singular intension ($\aleph_1$) it fulfills more of an intension than an extrinsic value which fulfills an analytic intension ($\aleph_0$) and an extrinsic value fulfills more of an intension than a systemic value which latter fulfills a synthetic (schematic, n) rather than an analytic intension. Hence an intrinsic value is a better value than an extrinsic value and an extrinsic value a better value than a systemic value. But this "analytic" derivation of the hierarchy of value is not quite so rigorous, and hence not quite so exact as the synthetic. It comes down to the synthetic only if the logical terms are interpreted synthetically and not analytically. That "value" can be regarded as both synthetic and analytic is due, in part, to the methodological ambiguity of logic, which is partly analytic and partly synthetic, and in part to the subtlety of the distinction between fact and value, value being a set of secondary properties regarded as primary.

The hierarchy of value follows from the axiometric sequence of value dimensions. Theoretically, it results from the progressive structuring of the intensional sequence of predicates. Practically, it is recognized through the sense of axiological proportion — the sense which corresponds to intensional structurization. It sees the given property-richness of a situation, its *value texture*, in relation to other possible such textures of the same situation, the axiological variations of the situation. For the value sense, any situation is a theme to be developed; the value view of the world is, in the widest sense, a harmonious view of the world. The world for him who values is a great harmony, a texture of situational themes; and the sense of values is a talent similar to the musical which sees in a given harmonic structure its possible variations; only that the sense of values seems to be the *general* talent of which the musical — and other artistic talents — are species.[25] The whole theory of formal value is one of axiological harmony.

The detailed laws of this harmony will be given later when the *elements* of the harmonics will be developed, and it will be shown how one and the same given value aspect or harmonic element may appear in many different value situations. Here we shall illustrate the harmony itself and show how one and the same given situation may appear in many axiological variations.

We shall use two examples, one by Ortega y Gasset, the other by Eddington.

1] Ortega y Gasset calls his example "A Few Drops of Phenomenology";[26] and indeed, we remember Husserl's saying that fact is only one of the possibilities of varying the given in imagination. Husserl does not say that the other possibilities are values, or that fact itself is both the primary property of value and the prime value. But the whole phenomenological modulation of experience can be systematized in axiological form. This is Ortega's example:

> A great man is dying. His wife is by his bedside. A doctor takes the dying man's pulse. In the background two more persons are discovered: a reporter who is present for professional reasons, and a painter whom mere chance has brought here. Wife, doctor, reporter, and painter witness one and the same event. Nonetheless, this identical event — a man's death — impresses each of them in a different way. So different indeed that the several aspects have hardly anything in common. What this scene means to the wife who is all grief has so little to do with what it means to the painter who looks on impassively that it seems doubtful whether the two can be said to be present at the same event.
>
> It thus becomes clear that one and the same reality may split up into many diverse realities when it is beheld from different points of view. And we cannot help asking ourselves: Which of all these realities must then be regarded as the real and authentic one? The answer, no matter how we decide, cannot be arbitrary. Any preference can be founded on caprice only. All these realities are equivalent, each being authentic for its corresponding point of view. All we can do is to classify the points of view and to determine which among them seems, in a practical way, most normal or most spontaneous. Thus we arrive at a conception of reality that is by no means absolute, but at least practical and normative.

Ortega analyzes the viewpoints of the four persons. The wife "is drawn into the scene," she is an intrinsic part of it, "it becomes one with her person." In our terminology, she values it intrinsically. The doctor "is involved in it, not with his heart but with the professional portion of his self"; he values it intrinsically-systemically. The reporter "observes it with a view to telling his readers"; his valuation is extrinsic-systemic. And the painter's is purely systemic, all he sees are "color values, lights, and shadows." Ortega *measures* these aspects of the situation by a common denominator: "*the emotional distance between each person and the event they all witness.*" This measure coincides with our value hierarchy: the intrinsic valuation involves the valuer completely; "for the wife of the dying man the distance shrinks to almost nothing." The intrinsic-systemic valuer, "the doctor, is several degrees removed. For him this is a professional case." Yet he carries the responsibility for it, and "hence he too, albeit in a less integral and less intimate way takes part in the event." The reporter, the extrinsic-systemic participant, is "a long way from the tragic event." His profession requires him to stay aloof. "To him the event is a mere scene, a pure spectacle," which he is to communicate. And "the painter, in fine, completely unconcerned, does nothing but keep his eyes open. What is happening here is none of his business; he is, as it were, a hundred miles removed from it. His is a purely perceptive attitude; indeed, he fails to perceive the event in its entirety." All he sees are lines, shapes, colors, systemic aspects of an event whose intrinsic meaning escapes him and indeed does not interest him. "In the painter we find a maximum of distance and a minimum of feeling intervention."

These "drops of phenomenology," can be formalized in terms of our symbolism of composition and transposition: the woman intrinsically values a whole life the loss of which is still too new for her to be understood and her situation is the exponentiation I^I, the intrinsic valuation of an intrinsic value. The doctor values the moving scene professionally, I^S. The reporter values an everyday event professionally, E^S, and the painter values professionally a constellation of spatial and optical elements, S^S. Applying the laws of exponentiation to these compositions, with $I = \aleph_1$, $E = \aleph_0$, and $S = n$, we find the wife's attitude to have value intensity $\aleph_2$, the doctor's $\aleph_1$, the reporter's $\aleph_0$, and the painter's n.[27] Thus, the axiological variations of the situation are expressible in a symbolism which faithfully represents the "measure" given by Ortega y Gasset. The value symbolism is isomorphous with the value reality.

Ortega not only shows the axiological variations of the situation, but also points out that these variations are part of a larger number of aspects which, in their totality, are variations of one and the same theme. In our terminology, the combinations of intensional subsets are all variations of a given set.

> Among the diverse aspects of reality we find one from which all the others derive and which they all presuppose: "lived" reality. If nobody had ever "lived" in pure and frantic abandonment a man's death, the doctor would not bother, the readers would not understand the reporter's pathos, and the canvas on which the painter limned a person on a bed surrounded by mourning figures would be meaningless. The same holds for any object, be it a person, a thing, or a situation. The primal aspect of an apple is that in which I see it when I am about to eat it. All its other possible forms — when it appears, for instance, in a Baroque ornament, or on a still life of Cezanne's, or in the eternal metaphor of a girl's apple cheeks — preserve more or less that original aspect. A painting or a poem without any vestiges of "lived" forms would be unintelligible, i.e. nothing — a discourse is nothing whose every word is emptied of its customary meaning.
>
> That is to say, in the scale of realities "lived" reality holds a peculiar primacy which compels us to regard it as "the" reality.

"The" reality is none other than the "factual" aspect of the event, where "factual" is taken in the relative or type sense discussed above. It is its "typical" constellation of properties — typical both as to type of value, and to definition of concept.

2] Eddington's example shows the "Generation of Waves by Wind" in all three value dimensions.[28] This is the systemic view:

> One day I happened to be occupied with the subject of "Generation of Waves by Wind." I took down the standard treatise on hydrodynamics, and under that heading I read
>
> "The equations (12) and (13) of the preceding Art. enable us to examine a related question of some interest, viz. the generation and maintenance of waves against viscosity, by suitable forces applied to the surface.

If the external forces p'_{yy}, p'_{xy} be given multiples of $e^{ikx+\alpha t}$, where k and α are prescribed, the equations in question determine A and C, and thence, by (9) the value of η. Thus we find

$$\frac{p'_{yy}}{g\rho\eta}=\frac{(\alpha^2+2\nu k^2\alpha+\sigma^2)\,A-i\,(\sigma^2+2\nu km\alpha)\,C}{gk\,(A-iC)},$$

$$\frac{p'_{xy}}{g\rho\eta}=\frac{\alpha}{gk}\cdot\frac{2i\nu k^2A+(\alpha+2\nu k^2)\,C}{(A-iC)},$$

where σ^2 has been written for $gk+T'k^3$ as before. . . ."
And so on for two pages. At the end it is made clear that a wind of less than half a mile an hour will leave the surface unruffled. At a mile an hour the surface is covered with minute corrugations due to capillary waves which decay immediately the disturbing cause ceases. At two miles an hour the gravity waves appear. As the author modestly concludes, "Our theoretical investigations give considerable insight into the incipient stages of wave-formation."

The extrinsic, sensorial dimension is described thus:

> "Aethereal vibrations of various wave-lengths, reflected at different angles from the disturbed interface between air and water, reached our eyes, and by photoelectric action caused appropriate stimuli to travel along the optic nerves to a brain-centre. Here the mind set to work to weave an impression out of the stimuli. The incoming material was somewhat meagre; but the mind is a great storehouse of associations that could be used to clothe the skeleton. Having woven an impression the mind surveyed all it had made and decided that it was very good."

As John Dewey rightly says,[29] this statement includes ordinary objects, water, air, brain, which have to be reduced to give the scientific account. The latter consists of the mathematical functions of the systemic aspect, relating physical constants with no counterpart in ordinary perception. On the other hand, these functions when related to perception give the extrinsic aspect, though prepared, as it were, for scientific treatment. A purely extrinsic statement of the "Generation of Waves by Wind" would be that of a sea captain surveying the windy surface of the ocean, sensing the coming of the storm, as in this account of Captain MacWhirr in Joseph Conrad's story "Typhoon":

> He was conscious of being made uncomfortable by the clammy heat. He came out on the bridge, and found no relief to this oppression. The air seemed thick. He gasped like a fish, and began to believe himself greatly out of sorts.
>
> The *Nan-Shan* was ploughing a vanishing furrow upon the circle of the sea that had the surface and the shimmer of an undulating piece of gray silk.
>
> . . . The lurid sunshine cast faint and sickly shadows. The swell ran higher and swifter every moment, and the ship lurched heavily in the smooth, deep hollows of the sea.
>
> "I wonder where that beastly swell comes from," said Jukes aloud, recovering himself after a stagger.
>
> "North-east," grunted the literal MacWhirr, from his side of the bridge. "There's some dirty weather knocking about. Go and look at the glass."

The intrinsic aspect of the situation finally appears in this account from Eddington:

> On another occasion the same subject of "Generation of Waves by Wind" was in my mind; but this time another book was more appropriate, and I read —
>
> There are waters blown by changing winds to laughter
> and lit by the rich skies, all day. And after,
>   Frost, with a gesture, stays the waves that dance
> And wandering loveliness. He leaves a white
>   Unbroken glory, a gathered radiance,
> A width, a shining peace, under the night.
>
> The magic words bring back the scene. Again we feel Nature drawing close to us, uniting with us, till we are filled with the gladness of waves dancing in the sunshine, with the awe of the moonlight on the frozen lake. . . . The critical faculty was lulled. We ceased to analyze and were conscious only of the impression as a whole. The warmth of the air, the scent of the grass, the gentle stir of the breeze, combined with the visual scene in one

> transcendent impression, around us and within us. Associations emerging from their storehouse grew bolder. Perhaps we recalled the phrase "rippling laughter." Waves — ripples — laughter — gladness — the ideas jostled one another. Quite illogically we were glad; though what there can possibly be to be glad about in a set of aethereal vibrations no sensible person can explain. A mood of quiet joy suffused the whole impression. The gladness in ourselves was in Nature, in the waves, everywhere. That's how it was.

It is obvious from this example that what is seen is one and the same situation, but that there are three different ways of seeing it. There is more precision in the mathematical formulae of Lamb's *Hydrodynamics* than in the observations of Captain MacWhirr or the images of the poem. On the other hand, there is more experience in Conrad's account, and more awareness, more allusion and texture in the poem. The opposition of a book about hydrodynamics and the experience of a storm is vividly presented by Conrad:

> He had been reading the chapter on the storms. When he had entered the chartroom, it was with no intention of taking the book down. Some influence in the air — the same influence, probably, that caused the steward to bring without orders the Captain's sea-boots and oilskin coat up to the chartroom — had as it were guided his hand to the shelf; and without taking the time to sit down he had waded with a conscious effort into the terminology of the subject. He lost himself amongst advancing semicircles, left- and right-hand quadrants, the curves of the tracks, the probable bearing of the centre, the shifts of wind and the readings of barometer. He tried to bring all these things into a definite relation to himself, and ended by becoming contemptuously angry with such a lot of words and with so much advice, all head-work and supposition, without a glimmer of certitude.

This opposition is one of the main themes of the story, exemplified by that of the sophisticated chief mate Jukes and the single-minded captain. The story ends with words written by Jukes: "The skipper remarked to me the other day, 'There are things you find nothing about in books.' I think that he got out of it very well for such a stupid man."

As we see from these examples, the hierarchy of values is not static but dynamic. The value dimensions follow each other in experience in any order. But they can be recognized only when their theoretical order is known, the progression of the intensional cardinalities from n to $\aleph_0$ to $\aleph_1$, and beyond, to ever higher cardinalities — the order of cardinal intensification, as we may call it. In theory, this order is static as is the orderly progression of numbers. In practice, the value dimensions do not follow each other in this hierarchical order but interweave in experience, as do the numbers. Experience continuously oscillates between the dimensions; and each experience is a pattern in which these dimensions are arranged in different ways. In other words, *the relation between the hierarchy of values and the value dimensions is equal to that between an intension and its predicates.* Just as the intension of a concept is a theoretical structure which in value reality is varied by the flow of its constituent predicates, so the total intension of all predicates, the progression of intensional cardinalities, is itself a theoretical structure which is varied in value reality by the flow of its constituting cardinalities. The hierarchy of values is a theoretical order that must be recognized in its dynamic variations in experience.

Just as the predicates of the intension are variations, according to the combinatorial laws, of the intensional set, so the value dimensions themselves are variations, according to these same laws, of the total set of intensional cardinalities. The application of the combinatorial laws to the value dimensions constitutes the calculus of value.

2. *The Calculus of Value*

The calculus of value is the application of exponentiation to the value dimensions. Each typical intension has its characteristic number — n, $\aleph_0$, $\aleph_1$, etc. — and these numbers can be combined in arithmetical operations. While it is clear that the intensional numbers n and $\aleph_0$ represent series of predicates, or discursive intensions — due to the denumerability of the sets in question — it is not so clear what a nondenumerable intensional set may mean. How can a continuum be a *set of predicates*? How can the structure of the singular intension be expressed in *logical terms?*

The intension of the singular ought to be discursive, that is, it should be constituted by individual words. Yet, the set of such words ought to form a continuous nondiscursive whole. This means that the singular intension must be constituted by discursive elements which are understood nondiscursively, or by nondiscursive elements which are understood discursively, where by "discursive" we understand denumerability of logical elements. In sum, the *singular intension is a discursive set of nondiscursive elements, or a nondiscursive set of discursive elements.*

This logical formulation corresponds to the mathematical formula $2^{\aleph_0} = \aleph_1$; the nondenumerable (nondiscursive) set $\aleph_1$ consists of all possible arrangements of denumerable (discursive) sets $\aleph_0$. Logically this means that there must be concepts with nondiscursive meanings which come about by all possible arrangements of concepts with discursive meanings. Such concepts, as we have already seen, are metaphors.

A metaphor is a word of the ordinary discursive language used as a predicate with a nondiscursive meaning. Its meaning is an *exposition without definition and without extension.* Thus, the expression "a peach" when used metaphorically, does not mean "*malum persicum*" (definition of "peach"), nor does it refer to the fruit of the Amygdalus tree which we eat in natural slices and after peeling; the set of predicates of the metaphorical concept (peach) does not refer to and express the set of properties of the corresponding things (peaches). What then does it refer to and express?

Such a set, since it lacks definition as well as extension, refers to *nothing and anything.* This means that it can replace any other word of the language which does express complete extension and intension. A peach can mean a woman or a dog ("peach of a woman," "a peach of a dog") just as a dog can mean a peach ("a dog of a peach") or a woman ("a bitch"). A metaphor, as we said before, is a *set of predicates used as a variable.* Hence it can in principle replace every other word of the language — and even itself as an ordinary word rather than as a metaphor, for example, "a peach of a peach." As the totality of all possible ordinary languages has $\aleph_0$ words, *each one* of which signifies as metaphor $\aleph_0$ senses or meanings, the total meaning of the metaphorical language is of $2^{\aleph_0} = \aleph_1$ meanings. The metaphorical language therefore is a nondenumerable infinity, a continuum. Since an element of a continuum may, itself, be a continuum, it follows that a metaphor may, itself, be a continuum.[30] A metaphor, as part of a metaphorical language, is a Gestalt containing the meanings of all possible meanings of the language.

The logical analysis of the value dimensions showed that to each of them — systemic, extrinsic, intrinsic — belongs a specific language: the technical, the ordinary, and the metaphorical, respectively; and that each meaning or connotation of the language shares the characteristic number of the language. Thus, a systemic connotation has the characteristic number n, (or $\log_2 n$), an extrinsic connotation the characteristic number $\aleph_0$, and an intrinsic connotation the characteristic number $\aleph_1$. Value being the fulfillment of these respective connotations, systemic, extrinsic, and intrinsic value can be characterized by these same respective numbers.

The hierarchy of value was based on the fact that, since value is defined as the fulfillment of a connotation, the more of a connotation there is to be fulfilled the higher is the value. A systemic value fulfills a connotation of at most n elements, an extrinsic value one of at most $\aleph_0$ elements, and an intrinsic value one of $\aleph_1$ elements. Thus, the logic of value applies the calculus of transfinite numbers to value theory, through the analysis of the structure of intension. Since this structure, through the axiom of value, mirrors value reality, it is no wonder that combinations of such structures do the same. The isomorphism of formal value and phenomenal value is due to the correctness of the axiom. Therefore, the value calculus here presented is not to be compared with calculi in moral philosophy as those proposed by Bentham, Hutcheson, and others. All these are implicative combinations of analytic concepts put into the form of letters; they have no synthetic basis. The application of arithmetical symbols to these letters, such as Hutcheson's $M = B \times A$, "The moral Importance of any Agent . . . is in compound Ratio of his Benevolence and Ability," [31] is an illegitimate mixing of analytic and synthetic procedures, and actually an example of the metaphysical fallacy.[32] Our value dimensions, on the other hand, represent *numbers of properties*, and hence arithmetical operations with their symbols, "S," "E," "I," are not only legitimate but demanded by the axiom itself.

The calculus of value arises by combining the three value

dimensions, or types, S (systemic), E (extrinsic), and I (intrinsic) and their respective arithmetical values n, $\aleph_0$, and $\aleph_1$. As we have seen, combinations of these three value types can be either *compositions* or *transpositions*. A *composition* of values is a positive valuation of one mode of value by another, while a *transposition* is a negative such valuation. The terms "positive" and "negative" were defined by means of the term "ought." The positive value of a thing is that which the valued thing ought to be, and the negative value is that which it ought not to be. As we have seen, a thing ought always to be good and ought not to be bad; it ought always to fulfill and ought never not to fulfill the intension of its concept: it ought, in other words, always be as valuable as possible. The same is true when the "thing" in question is a value. The most valuable value, that is, the value that fulfills the concept "value" most fully, is intrinsic value. Any value, thus, ought to be intrinsic value. Intrinsic value is the positive value of a value, that which the value ought to be. Any direction away from intrinsic value is negative, any direction toward or beyond it, positive. In the positive direction lie goodness and its intensifications, in the negative direction badness and its intensifications. The former, we might say, is the direction of enhancement, the latter that of debasement.[33]

As we have seen, the bad is always a transposition of values. Hence the general form V^{-V} or V_V indicates badness. This implies, as we have seen, that no thing as such is bad; all that is, is good the way it is, it is V-value. But the combination of things can be bad; and bad is indeed nothing but the incompatibility of things, or things in transposition, which means the *lack of a positive concept covering the things in question;* of a concept, that is, which fulfills or contains the expositions of both. Thus, any good thing can turn into the constituent of some bad, that is, of some complex that is a transposition.[34] A good Buick and a good Ford transpose each other when they collide; and the wreck may be called a transposition in the literal sense of the word. The result is both a bad Buick and a bad Ford, or rather, a Buick disvalued in terms of a Ford and vice versa, $E_{1_{E_2}} + E_{2_{E_1}}$. The wreck, however, is a good wreck, fulfilling the definition of "wreck," which in turn means a combination of two bad cars. On the other hand, a good Buick and a good Ford in a show room form a composition of values, E^E, and this complex is a whole whose concept contains the expositions of both automobiles.[35]

A good man, a good rope, and a good Christmas tree are transposed when the man is hanged by the rope on the Christmas tree. As a transposition, this is axiologically bad; but it is an axiologically good bad, that is, a good transposition, namely, a good hanging. On the other hand, if a man pulls the Christmas tree behind him on a rope we have a composition of values, of the same elements, and the complex value "man-pulling-Christmas-tree" or "Christmas-tree-puller," the concept of which contains all three connotations in question. The actual person fulfilling this concept is "a good Christmas-tree-puller."

Since the transposition comes about by the lack of a concept combining its elements, the *transpositional intension* is no true intension. It is a collection of predicates without a uniting tie. While the synthetic, the analytic, and the singular concepts were all *rational,* that is, their intensions signified or indicated something meaningful — even though it may have referred to something unreal, ideal, or imaginary — the transpositional concept is an *irrational* concept. Its intension signifies or indicates *lack of meaning.* Yet, it is a legitimate concept, with its own intension and extension. Its logic is that of the senseless and the absurd, of paradoxes and sophisms, of the language of dreams and neuroses, of jokes, jabberwocks, and jinnees.[36] The bad is only *one* aspect of this general transpositional logic.

The following propositions are examples of conceptual transpositions: "The number 3 is masculine"; "a thing can be and not be at the same time"; "there is no iron on Mars." In these three examples, the attributes predicated of the subjects "number 3," "thing," and "Mars," constitute — in various ways — transpositions of concepts. They are examples of three types of transposition within the transpositional concept: *Nonsense, counter-sense* (*or contradiction*), and *falsity.*[37]

In the first proposition, "The number 3 is masculine," the predicate and the subject are incoherent, they do not fit within the whole in question. What is predicated of the number 3 is an analytic concept of ordinary life, unsuitable to that number, which is a synthetic concept of the mathematical system. Such an assertion, we must say, is not really a proposition, for its nonsense precludes any union between subject and predicate. It is neither false nor true. A nonsensical proposition is beyond the true-false opposition, and hence beyond the law of excluded middle. "The number 3 is not masculine" and "the number 3 is feminine" are equally nonsensical. Other examples are the following: "the Mediterranean Sea is hypocritical"; "Robert is $\alpha - 13$"; "the mathematics of the transfinite is black and red"; and "green is or." This kind of transposition is intravaluational, that is, the three kinds of value are transposed.

In the statement that "a thing can be and not be at the same time" we have a new transposition: *counter-sense* (*Widersinn*). In this kind of transposition subject and predicate are coherent; the addition of "not" to the proposition would give it perfect sense. Similar contradictions are committed in the following examples: "a pentagon has twelve sides"; "physical being is immeasurable"; "two bodies may occupy the same place at the same time";[38] "houses cannot be lived in"; "bachelors are married"; "war is peace." In all these propositions the predicates contradict the *meaning* of the subject. It can be *demonstrated* that they are false, they are *formally* false. The transposition is *systemic;* what is transposed are systemic values.

Finally, in the proposition "on the planet Mars there is no iron" we have the third type of transposition: *falsity*. Falsity is not formal but material. It could well be that there is no iron on Mars. In these propositions which err from falsity, the predicate contradicts *facts*. This kind of transposition is extrinsic; what is transposed are extrinsic values.

A *false* proposition can be refuted or corrected by *showing* the correct state of affairs. A *contra-sensical* proposition can be corrected by *demonstrating* the consequences of the subject's meaning and uniting it coherently with the predicate, by means of an affirmation or a negation. A *non-sensical* "proposition" cannot be corrected, for the respective meanings of the subject and the predicate prevent any union of the two, nor can they even be thought under any common aspect. In the Kantian terminology of analytic and synthetic judgment, it may be said that *transpositions are ana-synthetic;* synthesis itself is dissolved, *aufgelöst,* in an *analysis* of even the tenuous synthetic connection between subject and predicate. Transpositional propositions are so "synthetic" that there is either *no possible* connection between subject and predicate (nonsense) or an *opposition* between them, either logically (contradiction) or epistemologically (falsity). In the case of *nonsense,* the predicate is not only removed from the subject, it is removed from the total range or context of the subject. Thus, a nonsense proposition is really two unconnected enthymematic contexts "connected" in one proposition. Such a proposition is a malformation, like a monster — and nonsensical intensions can only be "fulfilled" by malformations. Transpositions, thus, may also be called *parathetic*, subject and predicate put side by side without connection, or *teratic*, malformed.

In the case of *opposition* between subject and predicate, we have in contradiction and falsity, either *analytic* or *synthetic* opposition. In both cases, the subject and predicate, although, and because of, belonging to the same context, repel each other within it. In both cases, therefore, we have a looser connection between them than in synthetic judgments, indeed, a negative such "connection." In all transpositional propositions, thus, we have an *analyein* of synthesis, which means *a fortiori* (actually *a debiliori*) one of analysis. We may, therefore, call such propositions *anathetic judgments*, an expression which not only shows the privative nature of the transposition but also conveys its meaning of badness. (Similarly, singular propositions may be called *synlytic judgments.*) [30]

As synthetic, analytic and singular (synlytic) concepts lead, respectively, to systemic, extrinsic, and intrinsic value, so the transpositional (anathetic) concept leads to transpositional value. Transpositional value is the pattern of disvalue. Its character is negative.

In the field of *logic*, the transpositional concept, as an *irrational* concept, is opposed to the synthetic, analytic and singular concepts, all of which are *rational.* In the field of *value* something analogous happens: systemic, extrinsic and intrinsic values are *positive* values distinguishing themselves from transpositional value which is *negative value* or *disvalue*. The relation between positive and negative value is the axiological counterpart of the logical one between rationality and irrationality.

The logic of transpositional value, as the rule of disvalue, renders a service analogous to that rendered by Husserl's morphology as pure logical grammar. The logic of disvalue delimits fields and clarifies disordered situations just as Husserl's morphology clarifies irrational propositions. Thus, even though this value type is destructive or negative, its knowledge is constructive; it prepares the field for every positive valuational judgment. By classifying disorder it prepares for order.

In the transpositional concept, the totality of intensional attributes do not constitute a whole. There is not, strictly speaking, a transpositional intension; the attributes are fragmentary and do not succeed in forming an intensional unit. This means that

14. Secondary Value Combinations Nos. 1–9 Compositions

RANK	SYMBOL	EXAMPLES
1	I^I	A Baby; Mystic Experience; Creative Act
2	E^I	My New Car; "Peach"; Creative Engineer
3	S^I	Corporate Personality; Morale of Army, Shop, etc.; Maxim; Creative Thinker; Hypostatization
4	I^E	"Mr. Republican"; Materialist God; Selling Favorite Painting
5	I^S	"Elizabeth II"; Axiological Value; Philosophy of Creativity
6	E^E	Ice Cream Sundae; Binding of Book; Easy Chair
7	S^E	"By This Ring I Thee Wed"; Application of a System; Popular Science
8	E^S	Production Line; Game; "Legal Tender"; Abstraction; Corroborating Witness
9	S^S	Technical Improvement; Deduction; Corroboration

such an intension *cannot, strictly speaking, be fulfilled.* Leibniz called this the "impossibility" of a concept. Since value is intensional fulfillment, the transpositional intension does not lead to value. Its "fulfillment" is disvalue. Any negative exponentiation of values is such a disvalue.

We have already seen that the intensional range of a transposition is between 0 and 1. Let us now examine this range from another aspect. In logic, meaning is represented by the number "1" and falsity by the number "0." Disvalue has no meaning; but neither is it all falsity. On the other hand, value is more than 1; disvalue is less. Extrinsic value is represented by the range of numbers from 2 — the minimum number of definitional properties — to infinity. *One* property does not constitute a meaning and hence cannot be fulfilled (except by an action "fulfilling" an imperative, e.g., "red!" shouted to a painter). A transpositional set of properties is *no set* of properties or is a set of *no properties*; and hence cannot be fulfilled. The numbers "1" and "0" thus

Nos. 10–18 Transpositions

RANK	SYMBOL	EXAMPLES
10	S_S	Puzzle; Existentialist Depreciation of Thought; Jabberwock; Logical Paradox; "Red Tape."
11	E_S	Uniform; Pedant; Policeman Stopping My Car
12	S_E	Bad Popularization; Bribed Judge; "Egghead"; False Application of System
13	E_E	Chocolate and Sawdust; Inkblot on Book; Chair Smashed by Hammer
14	I_S	Paranoia (Systemic "Self"); Color Line; Menotti's *The Consul;* The Metaphysical Fallacy; Killing in War
15	I_E	Person as Function ("Alienation of Self"); Idol; Jesus Tempted by Satan (Matt. 4:1); Metaphor Taken Literally; Christmas Shopping and Merchandizing; Act of Killing
16	S_I	Argumentum ad Personam; Burning Heretics; Strategy in War; "*Intelligenzbestie*"; Rationalization
17	E_I	Lovesick Truant; Building an Ugly Bridge; Train Running over Suicide
18	I_I	"We'll Always Be Friends" (Friendship used to terminate love); Nazi Irma Greese (tied women's legs in labor, used life to kill)

represent a specific fulfillment and no fulfillment, and hence a specific value and no value, respectively. The specific value, as we have seen, is that of fact. An action "fulfilling" the kind of imperative in question has the value of fact. We, therefore, let disvalue be represented by the numbers between one and zero, that is, the fractions. The transpositional intension is, so to speak, a fragmented intension. These fractions, since they can be systemic, extrinsic or intrinsic, give rise to the characteristic number of transpositional value $1/n$, $1/\aleph_0$ and $1/\aleph_1$, with zero as the limit.[40]

The interpretations of the value compositions and transpositions are fairly obvious. Thus, examples of S^S are our treatise on axiology, or any systematic treatment of a systematic subject. Examples of E_E are any combination of incompatible empirical elements, such as hot coffee with sawdust or the collision of two automobiles. Compositions and transpositions containing intrinsic value are, for example, I^I or I_S. The former is the intrinsic valuation of intrinsic value, such as a mystic experience, where the intrinsic value of human experience is intrinsically experienced; the second the systemic disvaluation of intrinsic value, for example a depreciation of faith by a system, be it juridical, philosophical, scientific, or any other. The first kind of depreciation — juridical — has been fought by Jesus, the second — philosophical — by Kierkegaard, the third — scientific — we have met as the metaphysical fallacy. As we have already seen, the axiological fallacies are transpositions of value. The reciprocal of the transposition I^{-S} or I_S is the composition I^S, the systemic valuation of a person (Newton as the founder of his system). The inversion of I^S yields the composition S^I, the intrinsic valuation of a system (the intrinsic appreciation of logic by a logician, or the embodiment in a person of a juridical system, like a state, as "Elizabeth II"). The reciprocal of S^I is S^{-I} or S_I — which is the reciprocal of the inverse of I^S or the reciprocal inverse, or inverse reciprocal of I^S — and is the intrinsic disvaluation of a system; a fanatical anti-intellectualism, as was evinced in the Nazi term "*die Intelligenzbestie*," the beast of intelligence, the irrational fanatically arrayed against the rational. A minor form of this is S_E, the extrinsic disvaluation of thought, as in the kinder American term "egghead." The least disvaluating such notion is S_S, the systemic depreciation of thought, as found in the axiological and the existential hierarchy of values.

There are nine compositions and nine transpositions of the three value categories, so that the positive and the negative permutations of the three types of value are eighteen. These combinations of value are found in Table 14 in the order of their axiological rank with a few examples for each.

The axiological rank or value of each composition or transposition arises from the values of each of the values which make it up, as deduced from the nature of the systemic, extrinsic, or intrinsic intension, respectively. Since systemic value — that is, the fulfillment by an object of the schematic intension of its concept — has the axiological value n, being the fulfillment of a minimum number of intensional attributes, while extrinsic value has the value $\aleph_0$ and intrinsic value the value $\aleph_1$, the compositions and transpositions consisting of these values can in turn be exactly valued. In this way the hierarchy of values, which originally only comprises the three dimensions S, E, and I, can now be expanded into one of combinations of the three. While the former hierarchy is atomic, the latter is molecular.

The rank of the value combinations is as follows, with the arrow indicating the scale from the highest composition (I^I) to the lowest transposition (I_I), and the position of *fact* in the hierarchy of values, between S^S and S_S (n and $1/n$),[41] showing that fact is the intermediary between a complete and a fragmentary intension, that is, between the rational and the irrational: any transition from valuation to disvaluation must pass through factual cognition, and so must any transition from disvaluation to valuation. *Comprendre, c'est pardonner.* The first step from prejudice to estimation is a factual look at the situation. See Table 15.

15. *Numerical Values of Secondary Value Combinations*

COMPOSITIONS			TRANSPOSITIONS	
I^I E^I S^I	$\aleph_2$	↑	S_I E_I I_I	$\frac{1}{\aleph_2}$
I^E I^S E^E S^E	$\aleph_1$		S_E E_E I_S I_E	$\frac{1}{\aleph_1}$
E^S	$\aleph_0$		E_S	$\frac{1}{\aleph_0}$
S^S	n	1 (Fact) →	S_S	$\frac{1}{n}$

Although some compositions and transpositions have the same axiological value, for example S_I and I_I both have the value $\frac{1}{\aleph_2}$, this axiological value covers a difference,

$$S_I = n^{-\aleph_1} = \frac{1}{n^{\aleph_1}} = \frac{1}{\aleph_2} \text{ whereas } I_I = \aleph_1{}^{-\aleph_1} = \frac{1}{\aleph_1{}^{\aleph_1}} = \frac{1}{\aleph_2}$$

As is obvious, the combinations of value can in turn be combined. Thus, arise tertiary, quaternary, etc., compositions and transpositions of value, and the above combinations appear as only the secondary value combinations. A tertiary composition, containing three primary values, for example, is $(S^S)^E$, which would be the empirical value — or price — of our treatise on axiology, or the extrinsic valuation of any of the examples in number 9 of Table 14; a tertiary transposition is $(I_S)_S$, a systemic disestimation of the consul's behavior in Menotti's opera, or of the metaphysical fallacy, or of any of the examples under number 14 of Table 14. On the other hand, $(S_I)^I$ would be the intrinsic valuation of rationalization, or of strategy in war, or of any of the examples in number 16 of the Table — a certain perverted enjoyment of systemic-intrinsic transposition, of the devaluation of thought in terms of intrinsic value. Here belongs especially *Fanaticism*, the enjoyment of a system which, by being fanatically enjoyed, is disvalued in terms of the valuer's personal needs. Thus, a person who values a system not for its inherent systemic merit S but out of his own personal needs disvalues its systemic value in terms of his own person. He is really an anti-systemic person, an anti-intellectual posing as an intellectual. He uses the system to build up his own ego, that is, he uses it as rationalization, (S_I), and then values intrinsically, again out of personal need, this rationalization or building up of his own pseudo-self, $(S_I)^I$. Here belong fanatics of all kinds. *Fetishists* behave analogously toward things, *E*. The formula for fetishism then is $(E_I)^I$. Whenever there appears a *disvalue valued*, the conjunction "sub-super," such as "*E* sub *I* super *I*" $(E_I)^I$, we have a perversion of values — disvalue posing as value. This is worse, as the axiological calculus makes clear, than straightforward disvaluation — the disvaluation of value, which is easily recognized in life. Axiology thus helps to expose the real evils — the disvalues posing as values — of our civilization, from the horrors of war disguised as virtues to the hypocrisies in sex and other relations — e.g. the pretense of the Savior's birth for arranging merchandizing orgies — which are chronic diseases of the so-called Christian world and which arise from its inverted hierarchy of values.

Quaternary value combinations are, for example, $((S^S)^E)_E$, the extrinsic disestimation of the price of our treatise on axiology, as being too high for one to buy it, or $((I_S)_S)^S$, the systemic positive criticism of the negative estimation of the metaphysical fallacy. In other words, this formula expresses the following situation: Professor A explains the metaphysical fallacy as a transposition of intrinsic and systemic value, for example, Freud's application of the scientific framework to faith. Freud says that religion has not stood the test of science. Professor A says that Freud confuses the intrinsic value of faith with the systemic value of science and applies science to something to which it must not be applied, thus deprecating faith, without justification, and thus Freud commits the metaphysical fallacy. So far we have situation $(I_S)_S$. Professor B agrees with Professor A and writes an article in turn, praising Professor A's treatment of Freud's treatment of religion. This gives us $((I_S)_S)_S$.

Complicated though it sounds, this is one of the simpler value situations. The difference between its exposition and its symbolization shows the power of the symbol. This power becomes even more striking when we consider that the formula $(I_S)_S)^S$ covers not only the one situation described, but any situation where a systemic disvaluation of a systemic disvaluation of an intrinsic value is systemically praised, that is, systemically valued. The intrinsic value may be any such value whatsoever, a human or a thing or a thought, that is, a moral, an aesthetic, or a metaphysical, or any other kind of value, and the systemic values in question may be any system, whether philosophical, scientific or juridical. Thus, to combine any such values at random, if "*I*" represents a painting and "S" a juridical value, then the situation would be the disestimation of the painting's intrinsic value by a judicial decision, the reversal of that decision by the higher court, and the confirmation of the reversal by the highest court. Cases like this were decided in Holland and Germany shortly after the Second World War, where falsifications of paintings were discovered, with ensuing legal proceedings. Again, if "*I*" means the intrinsic value of a painting and "S" art criticism, the formula means the adverse criticism of the painting, the adverse criticism of that criticism, and the positive criticism of the latter. In other words, a critic has panned a painting, another critic pans the critic and lauds the painting, and a third critic lauds the second critic's praise of the painting and his panning of the first critic — all in all a situation over which the painter may feel complacent.

Thus, like a mathematical formula in the natural sciences, *a value formula is capable of infinite interpretation* — it is a set of *variables.* It represents the law of form governing a large class of value phenomena. Formal axiology is the science of the logical interrelationships of these forms.

Just as secondary value combinations have their exact axiological values so do tertiary, quaternary, etc. combinations. Thus, if $E^E = \aleph_0^{\aleph_0} = \aleph_1$, then $(E^E)_I = \aleph_1^{-\aleph_1} = \frac{1}{\aleph_2}$. In other words, the value of the tertiary value combination $(E^E)_I$ is equal to that of the secondary value combination I_I — the worst possible value in Table 14. We may interpret $(E^E)_I$ as the fanatical aversion not against one thing (E_I) but against an order, and implicitly *the* order, of things. Again, $(E^E)_E$ would be the extrinsic disvaluation of such an order, it would be a physical or mental disorder, as an allergy or the capacity for making the worst of a good situation — a *schlemihl.* The value of this tertiary combination is $\aleph_1^{-\aleph_0} = \frac{1}{\aleph_1}$, or the same as that of the secondary transposition I_E. Thus, to regard people as functions (I_E, see number 15 in Table 14) has the same axiological value as making the worst of a good situation. While this is a fitting axiological relation between two interpretations of axiological values, it seems less fitting to say that to regard people as functions has the same axiological value as having an allergy — even though this statement is as true axiologically as the former. Here axiology outruns ordinary understanding, perhaps in the same way as did the Copernican or Einsteinian theories in natural science. It may well make sense to say that a person who treats people as things is also likely to have allergies to orderly combinations of things, but at this point we do not understand this.

What we do understand is that this is axiologically true, just as it is arithmetically true that the value of two pears added to two horseshoes is four, even though pears and horseshoes have nothing with each other to do. Arithmetically, what is true is the addition of two *numbers*, and it is of no importance what these numbers are *interpreted* to be, whether pears or horseshoes or what not. Similarly, what is axiologically true is that $(E^E)_I$, or the intrinsic disvaluation of an extrinsic value composition, is of equal value as I_E, or the extrinsic disvaluation of an intrinsic value, no matter what these extrinsic and intrinsic *values* may be *interpreted* to be, people regarded as functions, allergies, or what not.

However, these interpretations point to an extreme sensitivity of the value calculus. Thus, it is axiologically of the same value to commit the metaphysical fallacy and to kill in war. The former is, for example, to degrade faith by systematization, the latter to extinguish systematically a human life. Can faith be equated with human life? The only value scale equally sensitive is that of Jesus. With Jesus, the *thought* of a deed already was the deed — and a life without spirit was *no life*.[42] Thus, we may not understand certain consequences of formal axiology simply because our value sensitivity is not subtle enough.

Secondary, tertiary, quaternary, etc., value combinations can be calculated down to definite axiological values, with interesting possibilities for the axiological scientist.

The tertiary value combinations arise by adding the exponents *S*, —*S*, *E*, —*E*, *I*, —*I*, to each of the secondary combinations of Table 14. Since there are 18 secondary combinations, each of which can take the six exponents, we get $6 \times 18 = 108$ tertiary value combinations. Again, each of these can take six exponents, and thus we have $6 \times 108 = 648$ quaternary combinations, and so on. In general, the number of possible value combinations on a level n is $3 \times 6^{n-1}$, that is to say, on the secondary level there are $3 \times 6^{2-1} = 18$ combinations, on the tertiary $3 \times 6^2 = 108$, on the quaternary $3 \times 6^3 = 648$ combinations, and so on. The number of possible value forms rises rapidly. There are 3,888 quinary forms and 23,328 senary forms. A quinary value combination is one with five elements, for example $(E_S)^I)_I)^S$. This may mean that a policeman stops my car (Table 14, number 11); an element of a system, the policeman, S, disvalues my driving, *E*, producing situation, E_S. My wife who felt all along that I was driving too fast, thoroughly enjoys the situation, $(E_S)^I$; I thoroughly dislike her enjoying it $(E_S)^I)_I$, and the policeman systemically appreciates the whole situation, $(E_S)^I)_I)^S$, and gives me a ticket, both in order to fulfill the systemic requirements, to please my wife and to discourage me. There are 3,888 different forms of such quinary value situations, and each form may mean an infinity of such situations. Thus, in a different situation E_S may mean my uniform as an army private, which I dislike as the systemic disvaluation of a good piece of clothing (some other private may feel differently and thus regard it as E^S, the systemic valuation of an ordinary piece of clothing). My girl friend likes my uniform greatly, $(E_S)^I$, her father dislikes her liking it $(E_S)^I)_I$, and the commanding officer likes the situation, in the name of the army and of a soldier's love, $(E_S)^I)_I)^S$. Here we have the stuff of an operetta in our little formula.

Of senary value forms there are 23,328, each again applicable to an infinity of situations. Thus, to add the disvaluation —E to the above, we get $(E_S)^I)_I)^S)_E$, which means that I go on driving very much slower — I disvalue the whole situation in terms of the extrinsic value of driving — (from my wife's point of view this is, of course upgrading the situation, that is, adding the exponent E rather than —E. An event on which both of us would agree as representing —E would be a nail in the tire). In the second interpretation, the addition of —E would be, for example, my throwing an ink bottle on my uniform or at the father.

The sum of all value forms on all levels is $3 \times (6^1 + 6^2 + \ldots + 6^n)$. Thus, on the secondary through the quaternary level there are $3 \times (6 + 36 + 216) = 774$ forms. The total of all forms up to the senary level is 27,990.

The totality of all possible combinations of value constitutes the *calculus of value* or the *axiological calculus.* With the present electronic computers a table of the value forms can be produced, similar to a logarithmic or other mathematical table, in thirty minutes. This would give the axiological scientist an opportunity to apply the calculus to all possible situations, to create prototypical situations, and thus create a catalogue of literary plots, metaphysical possibilities of thought, juridical cases, moral conflicts and solutions, in short, a genuine axiological science. As is clear from the little that has been said, even the most complicated axiological arguments and situations can be analyzed by means of this calculus.[43]

We shall illustrate this in two ways, first, by showing how one and the same value aspect can appear in many different situations (just as we showed the inverse in the preceding section, that namely one and the same situation can appear in many value aspects); secondly, by analyzing a famous philosophico-literary text. The first illustration will be from Thomas Mann's *Magic Mountain,* the second from Plato's *Euthyphro.*

1] Thomas Mann varies one particular value relationship, that between systemic and intrinsic value, in a series of situations throughout his novel. It is the relationship between the hero,

Hans Castorp and one of his fellow patients at the Sanatorium Berghof, Frau Clawdia Chauchat. The story weaves back and forth between Hans Castorp's systemic and intrinsic, and occasionally extrinsic, exposures to his beloved. Ghostly and fascinating scenes contrast the two value dimensions which are neglected in everyday valuation. The crassest episode, probably, is one in which the head of the sanatorium, Hofrat Behrens, gives Hans an X-ray plate of Clawdia's skeletal structure which Hans lovingly keeps on his chest of drawers. But the whole relationship between Hans and Madame Chauchat is a variation on this theme. In a first encounter Hans meets her in the waiting-room of the X-ray laboratory.

> Frau Chauchat had crossed one leg over the other again, and her knee, even the whole slender line of the thigh, showed beneath the blue skirt. . . . Hans Castorp recalled, suddenly, that she too was sitting here waiting to be X-rayed. The Hofrat painted her, he reproduced her outward form with oil and colours upon the canvas. And now, in the twilighted room, he would direct upon her the X-rays which would reveal to him the inside of her body. When this idea occurred to Hans Castorp, he turned away his head and put on a primly detached air; a sort of seemly discretion presented itself to him as the only correct attitude in the presence of such a thought.

In this passage the extrinsic, the systemic and the intrinsic are transposed, the description of the woman, the thought of her X-ray, and the peculiar intimacy which the latter represents. In the X-ray room the transpositions become sharper. Hans sees his cousin's pulsating heart, "Good God, it was . . . Joachim's honour-loving heart, that Hans Castorp saw!" He sees his own skeletal hand and thus "what is hardly permitted man to see . . . : he looked into his own grave. The process of decay was anticipated by the powers of the light-ray, the flesh in which he walked disintegrated, annihilated, dissolved in vacant mist. . . . The Hofrat said: 'Spooky, what? Yes, there is something distinctly spooky about it.' " As Hans leaves the X-ray room, "ushered in by the technician, Frau Chauchat was entering the laboratory" — to be according to Hans, intimately investigated by the Hofrat, but actually to be reduced to what Eddington calls the shadowy world of science, the primary properties of her physical existence.

The Hofrat not only X-rays her, he also paints Madame Chauchat. This occasions another scene in which the Hofrat discusses with Hans and his cousin the interplay of two value dimensions, the aesthetic and the medical, while Hans burns with love for the subject of the Hofrat's detached exercises. The Hofrat has painted Frau Chauchat's picture:

> "You might think she would be easy to capture, with those hyperborean cheekbones, and eyes like cracks in a loaf of bread. Yet, there's something about her — if you get the detail right, you botch the ensemble. Riddle of the sphinx. Do you know her? It would probably be better to paint her from memory, instead of having her sit. Did you say you knew her?"
>
> "No; that is, only superficially, the way one knows people up here." "Well, I know her under her skin — subcutaneously, you know: blood pressure, tissue tension, lymphatic circulation, all that sort of thing. I've good reason to. It's the superficies makes the difficulty. Have you ever noticed her walk? She slinks. It's characteristic, shows in her face — take the eyes, for example, not to mention the complexion, though that is tricky too. I don't mean their colour, I am speaking of the cut, and the way they sit in the face. You'd say the eye slit was cut obliquely, but it only looks so. What deceives you is the epicanthus, a racial variation, consisting in a sort of ridge of integument that runs from the bridge of the nose to the eyelid, and comes down over the inside corner of the eye. If you take your finger and stretch the skin at the base of the nose, the eye looks as straight as any of ours. Quite a taking little dodge — but as a matter of fact, the epicanthus can be traced back to an atavistic vestige — it's a developmental arrest."
>
> "So that's it," Hans Castorp said. "I never knew that — but I've wondered for a long time what it is about eyes like that."

From the eyes the Hofrat turns to the skin.

> "Now look at the skin — the epidermis. Do you find I've managed to make it lifelike, or not?"
>
> "Enormously," said Hans Castorp. "Simply enormously. I've never seen skin painted anything like so well.

You can fairly see the pores." And he ran the edge of his hand lightly over the bare neck and shoulders, the skin of which, especially by contrast with the exaggerated red of the face, was very white, as though seldom exposed. Whether this effect was premeditated or not, it was rather suggestive.

And still Hans Castorp's praise was deserved. The pale shimmer of this tender, though not emaciated, bosom, losing itself in the bluish shadows of the drapery, was very like life. It was obviously painted with feeling; a sort of sweetness emanated from it, yet the artist had been successful in giving it a scientific realism and precision as well. The roughness of the canvas texture, showing through the paint, had been dexterously employed to suggest the natural unevennesses of the skin — this especially in the neighbourhood of the delicate collar-bones. A tiny mole, at the point where the breasts began to divide, had been done with care, and on their rounding surfaces one thought to trace the delicate blue veins. It was as though . . . were one to press one's lips upon this surface, one might perceive, not the smell of paint and fixative, but the odour of the human body. Such, at least, were Hans Castorp's impressions, which we here reproduce — and he, of course, was in a peculiarly susceptible state. . . .

Hofrat Behrens rocked back and forth on his heels and the balls of his feet, his hands in his trouser pockets, as he gazed at his work in company with the cousins.

"Delighted," he said. "Delighted to find favour in the eyes of a colleague. If a man knows a bit about what goes on under the epidermis, that does not harm either. In other words, if he can paint a little below the surface, and stands in another relation to nature than just the lyrical, so to say. An artist who is a doctor, physiologist, and anatomist on the side, and has his own little way of thinking about the under sides of things — it all comes in handy too, it gives you the *pas*, say what you like. That birthday suit there is painted with science, it is organically correct, you can examine it under the microscope. You can see not only the horny and mucous strata of the epidermis, but I've suggested the texture of the corium underneath, with the oil- and sweat-glands, the blood-vessels

> and tubercles — and then under that still the layer of fat, the upholstering, you know, full of oil ducts, the underpinning of the lovely female form. What is in your mind as you work runs into your hand and has its influence — it isn't really there, and yet somehow or other it is, and that is what gives the life-like effect."
>
> All this was fuel to Hans Castorp's fire. His brow was flushed, his eyes fairly sparkled, he had so much to say he knew not where to begin.

And now Hans says in his own way what Husserl and Ortega have said in theirs: that the various aspects are all variations of one theme, and that this theme can be understood as a formal structure.

> He eagerly began: "Yes, yes indeed, that is all very important. What I'd like to say is — I mean, you said, Herr Hofrat, if I understood rightly, you said: 'In another relation.' You said it was good when there was some other relation besides the lyric — I think that was the word you used — the artistic, that is; in short, when one looked at the thing from another point of view — the medical, for example. That's all so enormously to the point, you know — I do beg your pardon, Herr Hofrat, but what I mean is that it is so exactly and precisely right, because after all it is not a question of any fundamentally different relations or points of view, but at bottom just variations of one and the same, just shadings, so to speak, I mean: variations of one and the same universal interest, the artistic impulse itself being a part and a manifestation of it too, if I may say so . . . and I find it wonderful, I find it a simply priceless arrangement of things, that the formal, the idea of form, of beautiful form, lies at the bottom of every sort of humanistic calling. It gives it such nobility, I think, such a sort of distinterestedness, and feeling, too, and — and — courtliness — it makes a kind of chivalrous adventure out of it. . . ."

There are many more transpositions between the various value dimensions, especially between the systemic and the intrinsic, as when "our young adventurer, supporting a volume of embryology on the pit of his stomach, followed the development of the organism from the moment when the spermatozoon, first among a host of its fellows, forced itself forward by a lashing motion of its hinder part, struck with its forepart against the gelatine mantle of the egg, and bored its way into the mount of conception, which the protoplasm of the outside of the ovum arched against its approach." And then having fallen asleep, "he beheld the image of life in flower, its structure, its flesh-borne loveliness. She had lifted her hands from behind her head, she opened her arms. On their inner side, particularly beneath the tender skin of the elbow-points, he saw the blue branchings of the larger veins. These arms were of unspeakable sweetness. She leaned above him, she inclined unto him and bent down over him, he was conscious of her organic fragrance and the mild pulsation of her heart. Something warm and tender clasped him round the neck; melted with desire and awe, he laid his hands upon the flesh of her upper arms, where the fine-grained skin over the triceps came to his sense so heavenly cool; and upon his lips he felt the moist clinging of her kiss." Finally, in a rapturous but elusive moment he falls on his knees before her real life person and stammers in French both his love and his anatomical knowledge, a scientific apotheosis of ecstasy. "*Oh, les douces régions de la jointure intérieure du coude et du jarret, avec leur abondance de délicatesses organiques sous leurs coussins de chair! Quelle fête immense de les caresser, ces endroits delicieux du corps humain! Fête à mourir sans plainte après! Oui, mon dieu, lasse-moi sentir l'odeur de la peau de ta rotule, sous laquelle l'ingénieuse capsule articulaire sécréte son huite glissante! Laisse-moi toucher dévotement de ma bouche l'Arteria femoralis qui bat au front de ta cuisse et qui se divise plus bas en les deux artères du tibia! Laisse-moi ressentir l'exhalation de tes pores et tâter ton duvet, image humaine d'eau et d'albumine, destinée pour l'anatomie du tombeau, et laisse-moi périr, mes lèvres aux tiennes!*" She dismisses him, "*Adieu, mon prince Carnaval!*" — a carnival scene beyond the flatlands of worldly events, on the Magic Mountain, where the timeless and spaceless, the systemic and the intrinsic, have their domain and the extrinsic is excised — a poetic epochē. Much later, when an uncle visits Hans, "in Hans Castorp's room, he lifted from its easel on the chest of drawers a black glass plate, one of the small personal articles with which the owner adorned his cleanly quarters. He held it toward the light; it proved to be a photographic negative. He looked at it — 'What is that?' he said. He might well ask. It showed the headless skeleton of a human form — the upper half, that is — enveloped in misty flesh; he recognized the female torso. 'That? Oh, a souvenir,' the nephew answered. To which the uncle replied: 'Pardon me,' and hastily replaced the picture on its easel." [44]

In this example we see not only the masterly play upon a fascinating and unusual value theme,[45] but also the power of formal axiology to lay bare the formal structure of the novelist's art.[46]

2] The second example is the value analysis of the opening statement of Plato's *Euthyphro*: "*What surprising thing has happened, Socrates, that you have left your pastime at the Lyceum and are passing your time at the portico where the King Archon resides? For it cannot be that what brings you here is an action before the king, as it does me.*" As will be seen, this sentence contains 22 value combinations, consisting of 53 values. The entire first page consists of 169 value combinations containing 493 values. And the entire dialogue of some 3000 value combinations containing some 8000 values. All these can be combined into *one* value, as will be the 22 value combinations of the opening statement.

The following Commentary should be compared with Burnet's.[47] He explains who Euthyphro is, that his "surprise is natural for Socrates had never yet appeared before a court," that "the Lyceum was one of the three great gymnasia outside the walls of Athens, the other two being the Cynosarges and the Academy," where these were located, how to get to them, etc.; that the king's stoa was "the first building to the right as you entered the Agora from the Ceramicus," etc. As the comparison clearly shows, Burnet's is a factual and not a valuational commentary. The present analysis is as detailed valuationally as is Burnet's factually, that is, etymologically and historically. The comparison gives a good illustration of the difference between fact and value.

The value analysis of a text as the following demands a great deal of concentration by the axiologist and his reader. The result may at times not seem to be worth the trouble taken. But an analysis as the following is to the art of the axiologist as are finger exercises to that of the pianist. Without them the pianist's hands would soon stiffen. With them he develops the agility to comprehend and interpret the multiform harmony of a piece of music. The same is true *mutatis mutandis* of the axiologist. He must continuously practice his value sensitivity in every detail; and he is rewarded, after exercises done long enough and well enough, with a profound and satisfying view of unsuspected depths and variety in a seemingly well-known text supposedly exhausted by centuries of close study.

I. — EUTHYPHRO. — "What surprising thing has happened, Socrates, that you have left your pastime at the Lyceum . . ."

(a) An extrinsic valuation of an intrinsic valuation. I^E

Calling "pastime" (*diatribē*) that which Socrates does at the doors of the Lyceum, his philosophizing, which for Socrates is his life's vocation (I) and which is taken by Euthyphro approvingly as a kind of social entertainment (E).

(b) An extrinsic disvaluation of an intrinsic value. I_E

Euthyphro considers the questions and discussions of Socrates at the door of the Lyceum (I) as a *mere* social entertainment (E). While (a) is a social valuation of Socrates' doing, (b) is a social disvaluation of it.

(c) A systemic disvaluation of an extrinsic value. E_S

Euthyphro depreciates as a loss of time (S) what Socrates does and what for Euthyphro is a social entertainment (E). Or

(d) A systemic disvaluation of an intrinsic value. I_S

Euthyphro depreciates as a loss of time (S) the philosophizing of Socrates which for him is his life's vocation (I).

(e) In calling the philosophical investigations of Socrates "pastime" Euthyphro charges the word with irony. The ironic transposition is a contradiction between the valuation expressed in the proposition (V^V) and the disvaluation implicit in the underlying judgment (V_V).

(1) *The valuation expressed in the proposition*

Extrinsic valuation of a systemic value. S^E

In calling the philosophical discussions of Socrates, directed to the clarification and systematization of truth (*S*), "pastime" (*E*) Euthyphro values these discussions socially. He may think entertainment better than futile discussion.

(2) *Disvaluation in the underlying judgment*

(α) Systemic disvaluation of an intrinsic value. I_S

Euthyphro depreciates as futile discussion (*S*) Socrates' profound philosophical disquisitions (*I*).

(β) Systemic disvaluation of a systemic value. S_S

Socrates' disquisitions, directed toward the clarification and systematization of truth, may also be regarded as systemic values (*S*), disvalued systemically as futile (*S*).

(γ) Extrinsic disvaluation of a systemic value. S_E

Here the disvaluation of the Socratic disquisitions directed toward clarifying and systematizing truth are debunked as social pastime (*E*).

(f) Extrinsic disvaluation of an extrinsic value, and systemic disvaluation of an extrinsic value. $E_{(1)}\, E(2)$ E_S

Euthyphro believes that the mode of life — E_2 — of Socrates is inadequate within Athenian society — E_1 — since as a member of the class of citizens he should live in accordance with the customs of the Athenians and use his time in something useful both to him and to society, as for example, earning his living or in some professional office. As is known to all, Socrates has no resources and thus is not justified in losing his and his community's (*E*) time in talks and discussions (*S*).

16. *Summary I* *

a	*b*	*c*	*d*	*e*				*f*	
				1	2			1	2
					(α)	(β)	(γ)		
I^E	I_E	E_S	I_S	S^E	I_S	S_S	S_E	E_E	E_S

Value Sum: $a + b + c + d + e + f = I;$

Value Product: $a \times b \times c \times d \times e \times f = \frac{1}{I}$

II. — "and are now passing your time at the portico where the King Archon resides?"

(a) An extrinsic disvaluation of an intrinsic valuation. I_E

The disvaluation by Euthyphro in observing that Socrates leaves his philosophizing (*I*) in the Lyceum for the portico of the king, perhaps only in order to change his auditorium (*E*).

---

* The rules to remember for value operations are:

(1) $S = n$; $E = \aleph_0$; $I = \aleph_1$; $J = \aleph_2$; $Y = \aleph_3$

(2) Value Composition is positive exponentiation (V^V)

(3) Value Transposition is negative exponentiation ($V^{-V} = V_V = \frac{1}{V^V}$)

(4) $a^{\aleph_n} = \aleph_{n+1}$; $\aleph_n^{\aleph_n} = \aleph_{n+1}$; $n^{\frac{1}{\aleph_n}} \rightarrow 1$ (1 = mere fact).

All formulae must be read backward to follow the explicatory text, e.g. "I^S" means "the *s*ystemic valuation of an *i*ntrinsic value."

(b) An intrinsic disvaluation of the extrinsic disvaluation of an intrinsic value. $(I_E)_I$

Euthyphro may disapprove in his innermost mind (I) of Socrates' use for his talks (E) of so exalted a place as the king's portico (I).

(c) An extrinsic valuation of the extrinsic valuation of an extrinsic value. $(E^E)^E$

The estimation (E) by Euthyphro of the satisfaction (E) which he supposes Socrates feels on changing the auditorium or scenario (E) for his inquiries, moving to the portico of the King certainly in order to continue with his discussions in a more outstanding place.

(d) An extrinsic valuation of an extrinsic value. E^E

Euthyphro still may suppose that if none of the two previous motives have induced Socrates to change the place or auditorium for his inquiries (E) it may be some social motives (E).

17. *Summary II*

a	*b*	*c*	*d*
I_E	$(I_E)_I$	$(E^E)^E$	E^E

Value Sum: a + b + c + d = J;
Value Product: a × b × c × d = J;

III. — "For it cannot be that what brings you here is an action before the king, as it does me."

(a) 1. An intrinsic disvaluation of an intrinsic disvaluation of systemic values. $(S_I)_I$

Euthyphro depreciates in his innermost mind (I) the innermost depreciation of Socrates (I) for the laws (S).

2. An intrinsic disvaluation of the systemic disvaluation of an intrinsic value. $(I_S)_I$

Euthyphro depreciates in his innermost mind (*I*) the depreciation of Socrates of the laws as merely systemic (*S*). For him the laws are intrinsic values since they serve to fulfill justice (*I*) even against his father.

(b) An intrinsic valuation of systemic values. S^I

The appreciation (*I*) which for Euthyphro the laws (*S*) merit since he takes recourse to them even to accuse his father.

(c) A systemic valuation of a systemic value. S^S

The respect (*S*) which for Euthyphro the laws merit (*S*).

(d) 1. An intrinsic disvaluation of the systemic disvaluation of a systemic value. $(S_S)_I$

Euthyphro condemns (*I*) the disapproval by Socrates (*S*) of the laws (*S*) as not adequately representing justice. Or

2. An extrinsic disvaluation of a systemic disvaluation of a systemic valuation. $(S_S)_E$

The same case as before, with the difference that the disapproval of Euthyphro is extrinsic and not intrinsic.

3. An extrinsic disvaluation of an intrinsic disvaluation of a systemic value. $(S_I)_E$

In this case the depreciation by Socrates of the laws is intrinsic, regarding their insufficiency to express justice as an innermost failure of the existing laws, which he disapproves in his innermost being.

(e) A disvaluation of a systemic value in terms of the intrinsic disvaluation of the systemic disvaluation of an intrinsic value. $S_{(I_S)_I}$

The depreciation (*I*) by Euthyphro of Socrates' (*I*) incapacity to bring legal action (*S*) which incapacity disvalues the laws (*S*). The efficiency of the law (*S*) is disvalued by Socrates' legal incapacity (I_S). His legal incapacity (*S*) disvalues Socrates (*I*).

18. *Summary III*

a		*b*	*c*	*d*			*e*
1	2			1	2	3	
$(S_I)_I$	$(I_S)_I$	S^I	S^S	$(S_S)_I$	$(S_S)_E$	$(S_I)_E$	$S_{(I_S)}\ _1$

Value Sum: $a + b + c + d + e = \mathrm{J}$
Value Product: $a \times b \times c \times d \times e = J$

Summary I–III

Value Sum: $\mathrm{I} + \mathrm{II} + \mathrm{III} = J$

Value Product: $\mathrm{I} \times \mathrm{II} \times \mathrm{III} = \frac{J}{I}$

As we see, the application of formal axiology to value situations and arguments follows definite rules as does the application of any science. So far, the rules for the application of a science have not been stated clearly, one reason being that they presuppose axiology. In Table 14, the application of a science is seen to be an axiological matter (Table 14, No. 7, 12, 3).

The rules of application of axiology exist besides and in complementation of the formal rules of axiology. They are either *a*] formally normative or *b*] materially normative, the latter in turn being either *α*] facultative or *β*] obligatory. The *formal rules* belong to formal specificity, the *formal norms* to theoretical specificity, and the *material norms* to material specificity. The *formal rules* of axiology are the formulas, theorems, etc., which make up the axiological system. The *formal norms* are those which regulate the application of the system to fields of situations and arguments. They establish the procedure one must follow in order to analyze axiologically. The *material norms* are those which regulate actual applications.[48] While the formal rules and the formal norms are systemic values, the applications themselves are either extrinsic or intrinsic values. An application is an *extrinsic* value if the systemic rule is applied in general or by anybody — if, in other words, it is a matter of the class of applications (Table 14, No. 7); it is an *intrinsic* value if the application of the systemic rule is a singular event, for example, if the application is made an intrinsic part of a life.

To use an example, a *formal rule* of axiology says that each thing ought to be good. The corresponding *formal norm* explains what is the particular thing in question, how to find its exposition or description, how to fulfill this exposition or description, etc., and thus, how in actuality such a thing ought to be good. The corresponding *facultative material norm* refers to the application of the rule in general, for example, the application to a person, by himself. The formal norm says that a person, like anything, ought to be good, that is to say, ought to fulfill his concept. In the formal norm, "person" is an element of an applied axiological science, to wit, ethics. The material norms are directed toward the user of axiology applying the formal rules and formal norms about persons. Axiology does not establish that a certain person must be good, but directs itself to all those who apply axiology, saying that *if* they are going to apply axiology then they must apply it correctly; but it does not say that anyone *has* to apply axiology. Hence, if a user wants to apply axiology to himself and regards himself as a person, then the *facultative* material norm says that he has to use the axiological rules and formal norms concerned with persons, which means, that he has to be morally good, which in turn means that he has to fulfill his concept of himself. It guides him in determining his own concept of himself, its intension in his particular case, the mode of fulfilling this intension, and the like. The *obligatory material norm* comes into play when the person decides to make out of the facultative and formal norms as well as the formal rules an intrinsic *maxim* of his life (Table 14, No. 3). In this case, the facultative axiological norm converts itself into an obligatory norm for the individual. "I must be morally good" is then a material obligatory norm, or a maxim, of axiology. The obligatory norms, thus, are singular, while the facultative norms are general. The rules of application of the dimensions, the hierarchy, the logic, and the calculus of value are the formal normative rules of axiology. All the rules of application we have given so far are such formal norms. The material norms, both facultative and obligatory, concern specific applications, that is, the applied sciences of axiology.

3. *The Value of a Value Theory*

The reader of the present study has the task of evaluating the value theory presented. Hence, value theory becomes for him a subject of valuation, and he needs a value theory to evaluate value theory. He may, following so many empirical value "theories," simply say that he likes or dislikes what he has read. On the other hand, he may want to be more analytical and test his own value theory against the one presented and, vice versa, the value theory presented against his own. We shall therefore apply our own theory to the valuation of any value theory.

Such a theory is a mental construct. Hence, in our terms it is a systemic value. Since, however, anything can be considered in all three value dimensions, a value theory can also be regarded as an extrinsic and as an intrinsic value. In the former case, it is seen as a tool functioning in the space-time world among similar such tools, or a member of the expositional class of value theories. In the latter case, it is regarded as unique and incomparable, and some person as fully involved with it.

a) A VALUE THEORY AS A SYSTEMIC VALUE

A value theory regarded as a systemic value cannot be either a good or a bad value theory, but it either is theory of value or is not. It *is* if it fulfills the definition of a value theory, and it is *not* if it lacks an element of that definition. We must therefore first of all define a value theory and determine the criterion or criteria. A value theory, obviously, must be a *theory of value,* that is to say, it must be a theory which accounts for the value world, which in turn is the totality of all the value phenomena. Hence, the one criterion that makes or breaks a value theory is that of *universal applicability to values.* A theory not universally applicable to values is not, according to this criterion, a value theory.

As the proof of the pudding is in the eating, so the proof of a value theory is in the application. There is an elegant way and there is a less elegant way of making this proof. The elegant way is to investigate the level of abstraction. Since a theory is the more applicable the more abstract it is, the most abstract theory is the most applicable. The "best" value theory, then, will be a purely formal one and, indeed, no value theory that is not formal will, by this test, be a value theory.

But this elegant test is only available to minds trained in philosophy and science. The less elegant way is to enumerate all the value phenomena to which a value theory is applicable and cross off from membership in the systemic class of value theories any which does not account for all of the enumerated value phenomena. If there is only one value phenomenon which a value theory A does not account for while a value theory B does, then value theory A is not a value theory in the systemic sense of the word.

In our case, there is indeed one value feature which is new in our theory as against any other value theory: that of systemic value, both in conception and application. On the other hand, for those who may not accept this feature we may, by way of summary, enumerate some of the themes and subjects for which our value theory accounts. We shall divide the features of the theory into methodological features and material features, and subdivide both into formal and material aspects.

α) Methodological Features

Methodological features of a value theory are those contained in or implied by the theory as such.

i) FORMAL METHODOLOGICAL FEATURES Formal methodological features are those which constitute, or relate to, the theory *as such.* Among them are the following, which in our theory have been systematically interrelated in a consistent pattern: The definition of moral science as against natural science; the definition of philosophy as against science, category as against axiom, analysis as against synthesis; the definitions of the specific moral sciences; the definitions of the fundamental terms of these sciences; the definition of value in general and value in particular; the definition of the axiological terms "perfect," "good," "unique," "bad," "fair," "no good," and their interrelationships; the explanation of the logical use of these terms in quantification; the definition of the value relations "better," "worse," "good for," "bad for," "worse for," "it is good that," "it is bad that," "it is better that," "it is worse that," "ought"; the difference between *x is good* and *x is a good C,* and that between the latter and *x is a C;* the structure of axiological propositions and judgments; the truth values of axiological propositions; the dimensions of value; the distinction of value languages; the hierarchy of values, both atomic and molecular; the calculus of value and its first 30,000 systematic forms; the rules of valuation as applied axiology; the difference between rule and norm, between formal and material norm, between facultative and obligatory norm, between law and maxim.

ii) MATERIAL METHODOLOGICAL FEATURES Material methodological features of a value theory are the deductions made from the formal methodological features in order to characterize the formal subject matter of the theory, "value," "fact," etc. Here belong, in our theory, the following: The relation between fact and value; the exact position of fact in the hierarchy of values; the exact position of value in a world of facts; the analysis of traditional value theories by means of the axiological fallacies; the features of value: its universality or particularity, its absoluteness or relativity, its rationality or irrationality, its objectivity or subjectivity; the nature of the value world; its goodness or badness, its optimistic or pessimistic aspect, the realistic or positivistic approach to it; the nature of the "moral law"; the nature of axiological agreement and disagreement; the relationship between the various kinds of values, aesthetic and economic, ethical and ecological, metaphysical and technological, etc. Again, all these material methodological features are systematically interrelated within a consistent pattern.

β) Material Features

Material features of a value theory are its applications to its subject matter. This subject matter is the value world, consisting of formal and material features. The former are value arguments (value statements), the latter value situations.

i) VALUE ARGUMENTS Some of the arguments discussed in our theory as subject matter of valuation follow. The arguments of the Church against Galileo; the arguments of the naturalists against the Church; the arguments of modern scientists such as Jeans or Eddington on value; the arguments of ethicists about nature; the relation of the principle of indeterminacy to free will; the relation of relativity theory to the ideality of the world; the relation of the survival of the fittest to moral goodness; the arguments of modern philosophers on moral science; the relation of music to astronomy in Pythagoras and others; the relation of astronomy and theology in Plato and others; the relation of theology and chemistry in alchemy; the confusion between ethics and other sciences; the confusion between sciences and their subject matter; statements such as "to be good is to do God's will"; "to be good is to feel satisfied"; "to be good is to be a proletarian"; Moore's paradox of "goodness"; the value nature of Leibniz's metaphysics and that of Spinoza's; the ontological proof of God's existence; the axiological nature of transcendentals; the value meaning of "Ideals," "Causes," "Rules," slogans; Castellio's observation on Servetus' burning by Calvin; Lord Alanbrooke's observation on General Eisenhower's conduct of World War II; Thomas Mann's *Magic Mountain*; Menotti's *The Consul*; Plato's *Euthyphro*; Kierkegaard's *The Sickness Unto Death*; Rousseau's "General Will"; Dante's notions of Pope and emperor; The Devil's notion of man in Shaw; Dr. Teller's notion of nuclear war; Ortega y Gasset's phenomenology; art criticism; juridical decisions and appeals to higher courts; the value equivalent of the second law of thermodynamics; rationalizations; expressions such as "egghead," "*Intelligenzbestie*," "By this ring I thee wed"; the value nature of Karen Horney's and Abraham Maslow's psychology; the value nature of a value theory. All these applications are systematically interrelated through the formal pattern applied to them.

ii) VALUE SITUATIONS Among the value situations mentioned in our theory are the following: the goodness of an automobile; of a jalopy; of a chair; the value nature of prejudice; of fetishism; a wreck; a hanging; a man pulling a Christmas tree; two automobiles in a showroom; a passport; a birth certificate; love; slavery; personality; individuality; self-alienation; indifference; joke; metaphor; arsenic as medicine for horses; electrons; geometrical circles; the Kingdom of Ends; the Kingdom of God; war; the Tower of Babel; destruction; confusion; hot coffee with sawdust; the mystic experience; a baby; corporate personality; Irma Greese; a policeman stopping my car, my wife's feeling about it, my own feelings about my wife's feelings, the policeman's feelings about the whole situation; jabberwock; Christmas shopping; Jesus' fight with scribes; Kierkegaard's with Hegel; *Schlemihl*; anti-intellectuals posing as intellectuals; allergies; the generation of waves by wind; and many others. Again, these value phenomena are systematically interrelated and thus constitute a value cosmos.

In the systemic comparison of our theory with any other value theory all these features must be taken into consideration, and the other value theory must be shown to account for them, and to do so systematically and consistently. On the other hand, our theory must be shown to account for and interrelate all the features presented and exhibited by the other value theory. In such a comparison, our value theory will exhibit one great lack. We speak nowhere of purpose, goals, and other teleological features of value theories, such as rightness of acts. The reason is that we deal with axiology and not with teleology. The latter is one of the applied axiological sciences, and arises when axiology, or to be exact, extrinsic value, is applied to time. Therefore, it is a special axiological science and must be discussed in its own right.[49]

b) A VALUE THEORY AS AN EXTRINSIC VALUE

A value theory can also be regarded as an extrinsic value. As such it is one of the tools for the understanding of the value world. In this case, again, universal applicability can be used as a criterion; that value theory will be the *better* value theory which is applicable to *more* value phenomena. In extrinsic valuation the *degrees* of value of the things valued are being taken into consideration, and therefore a value theory lacking some of the criteria of a good value theory will not, for that matter, be disqualified as a value theory, as is the case in systemic valuation. Thus there can be better and worse value theories. The extrinsic valuation of value theories, in other words, will admit more theories with the name of value theories into the class of value theories.

Not only the extension, but also the intension of that class will be greater, as we have seen in our analysis of extrinsic value. Therefore, in addition to the criterion of applicability, other extrinsic criteria for a good value theory may be admitted. Among two equally *applicable* theories, the better one will be the more *consistent;* and among two equally consistent theories the one may be preferred which is more *elegant*, that is to say, achieves its result with the smallest and simplest means (principle of parsimony). In natural science, one had to take recourse to this last criterion in deciding the relative merits of the Copernican and the Ptolemaic theories, since at the time both were equally applicable and consistent. In value theory, as we do not yet have many equally consistent and applicable theories, we do not yet have to have recourse to this last criterion.

c) A VALUE THEORY AS AN INTRINSIC VALUE

A value theory may, finally, have an intrinsic value. In this case all that counts is the involvement of the valuer in the theory. Here we must distinguish two cases. The involvement may either be positive or negative, that is to say, either an intrinsic valuation or an intrinsic disvaluation of the theory — either S^I or S_I. In the first case, we have the personal appreciation of the theory. It is based on the appropriate valuation of the theory *as a theory,* that is, its systemic features, and the valuer *gives himself to* the theory as a *rational* being, and is as such fully involved. In the second case, the personal involvement is not based on these objective criteria but the personal needs of the valuer. Thus, the creator of a value theory may be so much in love with it that he is blind to all other theories. This, of course, would imply that his own theory is incapable of dealing *as a theory,* that is to say, systematically and rationally, with other theories, and hence the creator must deal with them unsystematically and irrationally — epithetically rather than analytically. The same is true for the advocate of any value theory who is so much taken by this theory that he does not consider any other. Such an attitude again, and for the same reason, proves the insufficiency of the theory held.

The difficulty, in this case, arises as much from the value theory held as it does from the valuer holding it. A valuer who clings to an insufficient value theory clings to insufficient valuation: he really refuses to value and is guilty of the intrinsic evil of indifference. Indifference to a theory means indifference to intellectual endeavor. Thus, we have here the case discussed above, of the pseudo-intellectual, if not the anti-intellectual, posing as intellectual. Cases of this kind may be called intellectual prejudice. Prejudice, we remember, is the systemic disvaluation of the intrinsic, I_S; intellectual prejudice is the intrinsic disvaluation of the systemic, S_I. As we have seen, this is the formula for rationalization. Hence, we have here the peculiar case of a philosophical system serving as a rationalization, that is, as the crutch for a self. A person who thus clings to philosophy does so out of self-preservation; the theory is part of the system which constitutes his pseudo-existence. No new theory can be written for him nor can he write one; the development of philosophy passes him by, as does life itself. Philosophical literature is full of complaints against and characterizations of such pseudo-philosophers posing under the name of "philosophical scholars" and confusing what has been done in philosophy with what is to be done — "to whom the history of philosophy is philosophy itself," to speak with Kant.[50]

A value theory, of course, must be capable of accounting for a value phenomenon as close to home and as common as this.

The applicability of value theory to philosophy in general, and to value theory in particular, is an important criterion for the validity of a value theory. The application of a value theory to a value theory, therefore, is a touchstone for both theories.

This application is only one more illustration of the empirical import of the theory presented. The examples will have made it clear that axiological terms, relations, propositions, arguments — "value," "good," "bad," "ought," "paradoxes," etc. — can be defined with precision and with both theoretical and empirical import. Formal value theory thus is a genuine synthetic system, and not different in this respect from other axiomatic systems such as mathematics. Indeed, the procedure of defining "value" in formal axiology is exactly the same as that of Whitehead and Russell when they define "number" in mathematics in a strict logical manner. As we have seen, our definition of value is the intensional counterpart of Russell's definition of number. "I call," says Russell, " 'mathematical logic' any work which has as its aim the analysis and deduction of arithmetic, as well as of geometry, through concepts which ostensibly belong to logic." [51] Paraphrasing Russell we call axiological logic (axiologic) any work which has for its aim the analysis and deduction of ethics, as well as of the other moral sciences, through concepts which belong ostensibly to logic. As we have seen, the logical definition of value and that of number have this formal similarity: they are both definitions on the conceptual rather than the empirical level.

The system of value logic is to the moral sciences as the system of mathematics is to the natural sciences. The Galilean method in the natural sciences consisted of defining certain material relations, e.g. those of actual movement, in terms of certain formal relations, as arithmetical division, and then letting the formal relations develop according to their own logic. By representing the formal relations $v = \frac{s}{t}$ in geometrical form ($v \times t = s$, where $v \times t$ becomes the formula of a rectangle with sides v and t) Galileo invented a miniature geometrical system from which he could read off results which pertained to space, time, velocity, acceleration, etc.; even though, once the premisses were determined, no thought other than geometrical — of space, time, velocity, etc. — needed to enter the purely geometrical argument. In this sense Galileo used mathematics as the logic of discovery: the mathematical system is left to its own devices; but, by having defined a fundamental relation in it as one of material phenomena, the algorithm carries on its back, so to speak, like an invisible load, the material relations between natural phenomena; and its formal result is at the same time a material result.

Our procedure is formally just the same. We define certain material value relations — the Moorean relations between the natural and the nonnatural properties of a thing — in terms of a certain formal relation: the structure of *logical* relations determining the correspondence between a set of qualities possessed by a thing and the set of predicates contained in its concept, and then operate logically with this structure of relations.

From these operations we can read off results which pertain to value — goodness, badness, ought, etc. — even though, once the premisses are determined, no thought other than logical — no thought of value, goodness, badness, ought, etc. — need enter the purely logical argument. In this sense we may call formal axiology the logic of valuational discovery. The logical system is left to its own devices; but by having defined a fundamental relation in it as one pertaining to material value phenomena, the algorithm carries on its back, so to speak, like an invisible load, the material relations between value phenomena; and its formal result is at the same time a material result.

Thus, the systematic import of formal axiology carries with it its empirical import. This can only be understood by one who understands the *paradox of knowledge:* the paradox "now fully established that the utmost abstractions are the true weapons with which to control our thought of concrete fact," as well as concrete value. For, as Whitehead continues to remind us, the reconstruction now necessary in the value realm "must seek truth in the ultimate depths" into which, so far, only the constructions of mathematics have led the human spirit.

We have tried to plumb those depths with the help of a master chart found in value theory itself, and to develop a map: a formal pattern representing the structure of value. In the examples of this chapter we have shown that this pattern has both systematic and empirical import. The result is that a formal science of value is possible and that there is no reason to believe that the method of knowledge which proved to be correct in the realm of fact should not prove the same in the realm of value: *that only in utmost abstraction will the phenomenon of value be concretely understood.*

The deepest reason, then, why value theory has not so far become a truly empirical science is the very desire of value philosophers to remain empirical; their failure to grasp the paradox on which Whitehead insists so emphatically for both the fact and the value realm. Not until, but as soon as, formal synthetic concepts have replaced material analytic ones will the empirical science of value, with both systematic and empirical import, be launched. A genuinely *formal* value theory will, precisely by virtue of its formality, be the genuinely *empirical* one.

SUMMARY AND OUTLOOK

We are at the end of our investigation. Let us cast a glance backward from the plateau we have reached and look toward the horizon before us. We have attempted to lay the *foundations* of scientific axiology. Behind us is the climb from moral philosophy up to moral science. We have put aside analytic procedures, secondary properties, definitions by words leading to more words, abstractions leading to higher abstractions, the philosophical pursuit of a "truth" that forever escapes until, in the end, "the philosopher knows nothing of everything." We have discarded the procedure of categorial analysis.

The plateau we have reached is that of axiomatic synthesis. The rock we are standing on is the solid core of the phenomenal essence; its grains are the primary properties of value. The essence of the value phenomenon has been defined in terms of *formal* relations — as in natural science the essence of natural phenomena is being defined in terms of formal relations, e.g. the essence of a ray of light in a homogeneous medium as a straight line. By this process of axiomatic identification, the hooking on, as it were, of the phenomenon to a central notion of an axiomatic system — in the case of optics that of geometry — the phenomenal realm receives structure and precision, and the science is born.

The phenomenal manifold of sense impressions, in order to be prepared for its transfiguration into systematic form, must be broken down, dissolved — resolved, in the Galilean term — and its parts reconstituted. It must lose its empirical life in order to gain its systematic life. Its dissolution is a creative act; the scientist steps through the looking glass; and the phenomenal realm, resolved into primary properties, not only can but *must* be put together again. The resulting scientific object does not resemble any longer the sensorial and sentimental phenomenal reality in which man's everyday life is spent. But this medium of our lives appears in a new dimension, structured, refined, more *subtle;* and in acting back upon sensory life, the medium that was its matrix, the scientific system reshapes life itself.

The foundation of the science of value, thus, meant the breaking down, the resolution of the value medium within which we live, our value life, and its reconstruction in axiomatic form. Our axiom was based on a simple postulate: the identification of value with one-to-one correspondence of intensions. We hooked on, as it were, the phenomenon of value to a central notion of a formal system, in this case, logic, and utilized the logical relations to reflect the phenomenal value relations. The phenomenal realm thus received structure and precision, and the science of value was, if not born, at least conceived. The resulting scientific object, axiological value, does not resemble any longer the sensorial and sentimental phenomenal reality in which we spend our lives, but this medium of our lives, value reality, appears in a new dimension, structured, refined, more subtle; and in acting back upon value reality, the medium that was its matrix, the scientific system is bound to reshape moral life itself.

The definition of valuation in terms of intensional correspondence brought with it certain logical refinements of which the most fundamental were the following: sets of properties in intension were regarded as analogous to sets of instances in extension, or classes. Quantification and qualification were applied to both sets, so that extensional propositions based on the notion "ε" have analogous intensional propositions based on the notion "ω". Value properties qualify sets of descriptive properties, which means that they are second-order properties, and cannot possibly themselves be descriptive properties. It also means that value properties can be arithmetically expressed in terms of descriptive properties. The set of descriptive properties, the intension, thus serves as the common denominator for the whole value realm. This leads to an inversion of the relation between fact and value: the factual set of descriptive properties is normative for the value field.

The logical and arithmetical relations between descriptive and value properties explained Moore's paradox of *goodness:* that good is *not* a descriptive property but depends *only* on the descriptive properties of the thing that is good. Logically, Moore's paradox means that good is *not* a first-order, or predicative, function of individuals *x*, but that it depends *only* on such functions. It is a "reduced" second-order function, one that appears in predicative form. Arithmetically, it means that a *set* of natural predicates, *n*, is equivalent to *one* value or nonnatural predicate, Goodness, for which the set serves as a standard of measurement. This implies that the natural properties *describe* in their function as primary value properties.

These definitions of *goodness* are synthetic rather than analytic: "good" is defined as a network of formal relations rather than as a nest of analytic concepts, such as "pleasure," "purpose," "preference," and the like. It is defined constructively rather than abstractively. Once "good" is defined this way the system takes over and leads on from definition to definition, from one value theorem to the next. From the logical definitions of the value terms follow the definitions of value relations and thence of value propositions; from their arithmetical definitions follow the definitions of value addition and subtraction, multiplication and division. From the axiological structure of analytic concepts follow, in elaboration of the intensional relation ω, the structures of synthetic and singular concepts, and from the fulfillment of these three types of concepts the three forms of value, extrinsic, systemic and intrinsic. The three kinds of concept — analytic, synthetic, and singular — thus exhibit three kinds of axiometric structure, and these in turn determine three dimensions of value. A thing's intension, axiometrically structured, thus becomes the measure of the thing's value properties — just as certain extensional structures are the measures of a thing's descriptive properties. The complex system of *formal axiology* isomorphically reflects the value realm, which latter becomes structured by the system, isomorphically transfigured itself, and emerges as the *practical counterpart of formal axiology.*

With the practice of axiology begins the new science. The creation of the system is only the structure which enables us to climb up to the high plateau. Before us now stretches the new horizon. Formal axiology, to be a genuine science, must be applicable to the whole vast panorama of the realm of values. Compared with this task, our applications were only samplings, testing the new soil, cautious explorations of the new domain. We were samplers, but our samples were all demonstrated to be consistent elements of the system. Thus, a wide range of value phenomena was accounted for by one and the same method, from the ontological proof of God to the value puzzle of a lady's red nose.

The task before us is to include more and more ethical as well as moral phenomena and to map out systematically the new realm. The axiological system itself, outlined only in so far as to enable us to reach and recognize the plateau, must be elaborated and expanded, to become the exact network of coordinates, of latitudes and longitudes, of the new territory. The territory itself must be systematically defined, explored, and cultivated. The borderlines between the various value realms — the moral, aesthetic, religious, economic, etc. — must be drawn with precision. Within each realm the explorer must determine the last details of the geological, topological, ecological, hydrological layout. All this must be done systematically and consistently, employing the new instrument of formal axiology.

Although this task transcends the present foundations, we have gathered enough information to permit ourselves a glance ahead. Let us look at the moral realm. Since goodness is conceptual fulfillment, moral value will appear as the fulfillment by a person of his own concept of himself. This concept is a singular concept, "I", whose intension, axiometrically structured according to the logic of singularity, will appear as the axiological measure of a person's moral worth. The person will be the more moral the more he fulfills his concept of his Self. Hence, moral terms such as "honest," "sincere," "genuine" will receive an exact axiological meaning, as will their opposites, "dishonest," "insincere," "not genuine." The structure of a person will become the axiometric structure of its self-concept applied to its situational actuality. The person's own choice of his self-concept will be analyzed in terms of the axiological definition of choice, based on that of "ought," and various self-concepts chosen, singular, analytic, or synthetic, will be compared in terms of the hierarchy of values. The "knight of faith" as well as the "organization man" will become a priori types of formal axiology. The unfamiliarity of such a procedure will be outweighed by the exactness and precision with which familiar moral phenomena will be defined within a consistent framework, and the ease with which new and unsuspected phenomena of this kind will make their appearance.

Not only the moral, but also the aesthetic, the economic, and other value realms could be mapped out in detail with the help of the new tool. In general, the applications of the axiological value types — intrinsic, extrinsic, and systemic value as fulfillments, respectively, of singular, analytic, and synthetic concepts — to the various fields of human activity will originate the various social and moral sciences. We shall limit ourselves to an outline of six kinds of applications, namely, to persons, to groups of persons, to things, to groups of things, to concepts, and finally to words.

a) APPLICATIONS OF EXTRINSIC VALUE

The application of *extrinsic* value to individual persons shows each person as a class of functions. This yields the science of *Psychology* — not as a naturalistic but as a value science, showing how the person fulfills or does not fulfill the various functions which it presents. Within the general field defined as psychology — the extrinsic value dimension applied to individual persons — all three value dimensions may again be used. The applications of these dimensions to the various functions yield the various psychologies, with the intrinsic dimension leading to Gestalt psychology, the extrinsic to Behaviorism, the systemic to formal systems of psychology, such as Clark L. Hull's or Thurstone's. Through being inserted in a formal system all these sciences should lose their present arbitrary character.

Extrinsic value applied to groups of persons shows these groups as classes of functions performed by persons as functions within social groups. This yields the science of *Sociology*. Again, within this general field — the extrinsic value dimension applied to groups of persons — all three dimensions may be applied. The results are the various sociologies, the application of intrinsic value leading to *Gemeinschaft*-sociologies in the sense of Tönnies, the extrinsic to *Gesellschaft*-sociologies and others that are purely functional, the systemic to formal sociological systems such as Talcott Parsons' or Dodd's. Again, the system of formal axiology validates these disciplines.[1]

Extrinsic value applied to individual things shows each thing as a class of empirical functions. This valuation of things as functions leads to the science of *Economics;* the things thus regarded are "goods." Their economic value depends upon their availability, and this is governed by the inverse proportion between the intension and the extension of analytic concepts.[2] The universal concept serving as the measure of the value of each thing is called "money" and the thing's specific extrinsic value in question "price." Again, within the general field defined as economics the three value dimensions can be applied, with intrinsic applications leading to organism theories like Spann's, extrinsic to utility theories, and systemic to formal theories of economics.

Extrinsic value applied to groups of things shows things as functions within groups of similar things covered by one concept, each thing fulfilling part of the connotation of the concept. This gives us the science of *Ecology*. The application of the three value dimensions within the general field thus defined gives us organismic, functional, and formal ecological theories.

Extrinsic value applied to concepts shows concepts as functions, i.e. with specific references and meanings. This yields the science of *Epistemology*. The application of the three value dimensions within the general field thus defined gives organismic epistemologies — knower and known as one unit, as in Bergson's "intuition," Kierkegaard's "subjective truth," the Zen's "satori"; functional epistemologies, such as Dewey's pragmatism; and formal epistemological systems, such as that of Kant or the semantical one of Charles Morris.

Finally, extrinsic value applied to words shows words as functions in speech, resulting in the sciences of *Rhetoric, Semantics* or, more generally, of *Value Linguistics*, again in its organismic (aesthetic), functional (contextual, forensic) and systemic ("speech science") forms. Here belong the various theories of performatories, ceremonials, etc. which have been proposed as prototypes of value theories but are only very specific *applications of* value theory.

b) APPLICATIONS OF INTRINSIC VALUE

Intrinsic value applied to individual persons shows the uniqueness of each person and its fulfilling or failing to fulfill its own self. This is the science of *Ethics*. Here appear the philosophical anthropologies (in the continental sense of that word) and aspects of existentialism, certain kinds of psychologies concerned with the self, such as Karen Horney's, Abraham Maslow's and Erich Fromm's, and an ethics such as Kant's. The majority of traditional ethics do not belong here but into the fields of value psychology, sociology, and metaphysics. Again, the application of the three value dimensions within the general field defined as ethics brings about intrinsic, functional and formal ethics, respectively, represented by the theories mentioned — anthropologico-existential, psychological, philosophical — in the order mentioned.

Intrinsic value applied to groups of persons shows the uniqueness of such groups, the involvement in their symbols and institutions, and yields the science of *Politics* and *Social Ethics*. Again, application of the three value dimensions within the general field defined gives us organismic and personalistic, functional, and formal political theories, as we have them in Recaséns-Siches, Marx, and Karl W. Deutsch and Rousseau, respectively, to mention only four.

Intrinsic value applied to individual things shows things as unique values and leads to *Aesthetics*, and a science of *Symbolism*. Again, the various value applications give us three kinds of aesthetic theories, and three kinds of symbolic theory with their respective subjects, from sacraments to signals to semicolons.

Intrinsic value applied to groups of things shows the unique value of such groups of configurations and yields the science of *Civilization* or *Culture*. The application of the three value dimensions gives us three kinds of cultural theories, represented, respectively, by Toynbee, Marx, and Zipf,[3] to mention three representatives of intrinsic, functional, and systemic cultural valuations, respectively.

Intrinsic value applied to concepts shows concepts as unique, involved in, and with reference to, the essence of human life. Here we have the science of *Metaphysics*. Again, within this general field, the three value dimensions can be distinguished. The struggle between the realists and the nominalists is a classic example of intrinsic versus systemic metaphysical valuation of concepts.

Intrinsic value applied to words shows words as unique values, each with universal reference and meaning, that is, as metaphors. Here we have the yet unborn science of private and other intrinsic languages, belonging to *Value Linguistics*, and the science of *Poetry* or *Literary Criticism*. The three value dimensions within this general field yield three different activities, from traditional to "realistic" to "abstract" poetry.

c) APPLICATIONS OF SYSTEMIC VALUE

Systemic value applied to individual persons shows the individual as a system and gives us the science of *Physiology*, in all its aspects, from the study of the human body to that of robots. Applied to groups of persons it originates a system of interpersonal relations, such as the *Law* of persons and of institutions, military regulations, and the like. Again, the three value dimensions yield three kinds of legal theories, from Gierke to Pound to Kelsen and García Máynez.

Systemic value applied to individual things shows each thing as a system, that is, as a machine and the like, and yields the science of *Technology*, and *Engineering* in particular. The application of the three value dimensions within this general field gives us technological theories and practices from the organismic view of some German philosophers and some inventors, such as Edison, to the everyday calculations and techniques carried out at the Massachusetts Institute of Technology or the next factory. In one respect, the value science of technology may be said to begin where that of physiology leaves off, the cybernetic robots.

Applied to groups of things systemic value produces the sciences of *Industrial Technology* and *Civil Engineering*, which may be combined under the name *Technological Ecology* or *Ecological Technology*. The application of the three value dimensions within this general field gives rise to other sciences of systemic things in groups and interrelations, such as *Ritual*, *The Rules of Games* and *Operational Research*, and the *Law of Things* (*Law of Property*), which arise from the application of the intrinsic, the extrinsic, and the systemic dimensions, respectively.

Applied to concepts, systemic value shows us concepts in systematic interrelation, that is, as subject matter of *Logic*. Again, the application of the three value dimensions to this general field gives us the various logics, from the "organic" or "integrative" logic of Vasconcelos, Gabriel and Spann, through the pragmatic logic of Dewey to the symbolic logic of Russell and Quine.

Finally, systemic value applied to words gives us words in systems, or the science of *Grammar*, again as part of *Value Linguistics*, with the various theories of grammar, from the "symbolic forms" of Cassirer and the speculations of Plato's *Cratylus*, to the word-families of Wittgenstein, and the strict systems of the grammar texts, theories of communication, word-counts and the like, as intrinsic, extrinsic, and systemic applications, respectively.

The applications of the value dimensions mentioned are summarized in the following Table:

19. *Axiological Sciences*

APPLICATION TO	INTRINSIC VALUE	EXTRINSIC VALUE	SYSTEMIC VALUE
Individual Persons	Ethics	Psychology	Physiology, Jurisprudence of "Person"
Groups of Persons	Political Science, Social Ethics	Sociology	Law of Persons and Institutions
Individual Things	Aesthetics	Economics	Technology
Groups of Things	Science of Civilization	Ecology	Industrial Technology, Civil Engineering. Games, Law of Property, Ritual
Concepts	Metaphysics	Epistemology	Logic
Words	Poetry, Literary Criticism	Rhetoric, Semantic, Linguistic Analysis	Grammar, Theory of Communication

The application of axiology to actual situations through the applied axiological sciences is a task for new generations of pure and applied axiologists, pure and applied social and moral scientists, and finally, the mechanics and craftsmen of social and moral situations. As the achievements of the natural scientists analyzing natural situations in terms of mathematics, have led to the building of factories turning out new and undreamed of things, so the achievements of the moral scientists of the future, analyzing moral situations in terms of formal axiology, will lead to the building of a new society with new people, living on higher levels of awareness and possessing undreamed of insights into the subtleties and depths of moral reality.

NOTES

INDEX

NOTES

Introduction

1. Fustel de Coulanges, *Monarchie Franque*, p. 31, Theodor Lessing, *Studien zur Wertaxiomatik*, Leipzig, 1914, pp. 104–5. The mottoes of the book are from Hermann Lotze, *Kleine Schriften*, 1847, II, p. 207, and III, p. 568; Alfred North Whitehead, *Science and the Modern World* (Cambridge, 1933), pp. 41, 44; Susanne K. Langer, *The Practice of Philosophy* (New York, 1933), pp. 206–7.

2. Robert S. Hartman, "The Definition of Good: Moore's Axiomatic of the Science of Ethics," *Proceedings, Aristotelian Society*, LXV (1964/65), 235–56.

3. Albert Einstein, "On the Generalized Theory of Gravitation," *Scientific American*, CLXXXII (April, 1950), 13–17. The history of physics shows clearly the pattern of growth, of resistance to, and acceptance of, new theories. "Thirty years commonly pass between the recognition of a puzzling phenomenon and the birth of the idea and the working out of its major consequences. The first 30 years are a time of struggle and searching for a solution, the second 30 a time of readjustment and assimilation of strange conceptions. From Michael Faraday's discovery of electromagnetic induction to Maxwell's theory was 30 years. From Maxwell's theory to Heinrich Hertz's demonstration of electromagnetic waves, or to Pupin's transmission lines, was another 30. It took 30 years to go from Rutherford's nucleus to a rough understanding of nuclear structure and nuclear reactions. It took 30 years from Heisenberg's quantum mechanics to John Bardeen's new theory of superconductivity." Freeman J. Dyson, "Innovation in Physics," *Scientific American*, CXCIX (September, 1958), 81. In value theory, the first 30 years were Moore's discovery of, and struggle to explain, the puzzling phenomenon of nonnatural properties. This conception the present book seeks to develop systematically.

4. Ernst von Aster, *Philosophie der Gegenwart* (Leiden, 1935), p. 147.

5. "A land of rigorous abstraction, empty of all familiar land-

marks, is certainly not easy to get around in. But it offers compensations in the form of a new freedom of movement and fresh vistas. The intensified formalization of mathematics emancipated man's mind from the restrictions that the customary interpretation of expressions placed on the construction of novel systems of postulates." "With the enlargement of the deductive and theoretical sphere grows the free vitality of theoretical research, grows the wealth and fecundity of methods." (E. Nagel and J. R. Newman, *Gödel's Proof* [New York, 1958], p. 13; E. Husserl, *Logische Untersuchungen*, I [Halle, 1928], p. 248).

6. *Principia Ethica* (Cambridge, 1903), p. ix.

7. Ernest Nagel, "The Debt We Owe to G. E. Moore," *Journal of Philosophy*, LVII (Dec. 22, 1960), 815.

8. Kant, *Prolegomena to Any Future Metaphysics*, ¶ 3.

9. By the end of an academic year "Moore would have got through about two-and-a-half of the possible kinds of analysis" of propositions of the form "This is a pencil." Yet, Moore did want to answer questions. The reason that he did not was, he thought, that probably he had "not gone about the business of trying to solve them the right way." R. B. Braithwaite, "George Edward Moore, 1873–1958," *Proceedings of the British Academy* (1959), 360 f.

10. R. F. Harrod, *The Life of John Maynard Keynes* (London, 1951), p. 76. Italics supplied. Also John Maynard Keynes, A *Treatise on Probability* (London, 1952), p. 19.

11. Morton White, "Memories of G. E. Moore," *Journal of Philosophy, op. cit.*, 806.

12. J. M. Keynes, *Two Memoirs* (London, 1949), p. 85.

1. *Philosophical Ethics and Scientific Ethics*

1. Cassirer, "Erkenntnistheorie nebst den Grenzfragen der Logik," *Jahrbücher der Philosophie*, I (1913), p. 13; Leibniz, *On Universal Synthesis and Analysis*, or *The Art of Discovery and Judgment* (1679).

2. Sometimes philosophy is "difficult" out of all proportion to its content. John Dewey, for example, has a very simple message, but delivers it in an extremely ponderous manner. On the other hand, physics has to say extremely complex things, but says them in the simplest possible manner. Thus, if the relation between form and content – the *economy* of the discipline – is considered most of philosophy is unnecessarily difficult and most of science fairly easy. That in spite of this philosophy appears easy and science difficult demonstrates how infinitely more complex is what science has to say and how infinitely more it has differentiated its subject matter than

philosophy; in a word, what an infinitely more advanced kind of knowledge it is.

3. As when John Dewey speaks of the "forces" that disintegrate old customs. Dewey-Tufts, *Ethics* (New York, 1942), p. 113. Ethics is for Dewey "but a more conscious and systematic raising of the question which occupies the mind of anyone . . . in the face of moral conflict." "Moral theory is but an extension of what is involved in all reflective morality." *Op. cit.*, pp. 173–74.

4. To be "in accord with ordinary usage" (cf. E. M. Adams, "'Ought' Again," *Philosophical Studies*, VIII [Dec., 1957], 88) is in today's ethical discussion regarded as a sign of the adequacy rather than the inadequacy of an ethical theory – the very opposite situation from that in science. Cf. Philipp Frank, *Philosophy of Science: The Link Between Science and Philosophy* (Englewood Cliffs, N.J., 1957), Ch. 2; *Modern Science and Its Philosophy* (New York, 1955), Ch. 7. Also Wilhelm Dilthey, *Das Wesen der Philosophie* (1907), *Gesammelte Schriften* (V. Band, Leipzig und Berlin, 1924), p. 414.

5. We use the term *moral philosophy* in the traditional sense of that which is not natural philosophy, that is, today's humanistic and social disciplines. In all other contexts, the term *moral* refers exclusively to the subject matter of ethics, morality, and the term *ethical* exclusively to ethics, the theory of morality.

6. This is the terminology of Kant. The Kantian doctrine of analytic and synthetic *concepts* must be distinguished from that of the analytic and synthetic *judgments.* The former is *logical* and is dealt with in Kant's *Logik.* The latter is *epistemological* and is dealt with in the *Critique of Pure Reason.* Analytic concepts have also been called substantial, material or abstractive, and synthetic concepts functional, formal, or constructive. The difference has been elaborated in the writings of Ernst Cassirer, especially *Substance and Function* (Chicago, 1923). It is not usually discussed in logic textbooks. An exception is David García Bacca, *Introducción a la lógica moderna* (Barcelona, 1936). Dilthey has systematically used it, in the Kantian terminology, for his foundation of *Geisteswissenschaften,* especially in *Ideen über eine beschreibende und zergliedernde Psychologie* (1894), *Gesammelte Schriften* (V. Band, Leipzig und Berlin, 1924), pp. 139–240. For a detailed discussion of the Kantian distinction see Robert S. Hartman, "The Analytic and the Synthetic as Categories of Inquiry," *Perspectives in Philosophy* (Columbus, Ohio, Ohio State University Press, 1953), pp. 55–78. Prior to Kant, the terms "analytic" and "synthetic," introduced by Galileo and Newton, were used for concepts only. Kant extended the usage to judgments. We return to the pre-Kantian, originally scientific, usage. (On Newton's use see P. Wiener and A. Noland, *Roots of Scientific Thought* [New York, 1957], pp. 417–18. Analysis was for Newton, as for Galileo *Induc-*

tion, Synthesis for Newton, as for Galileo, *Hypothesis* and *Construction*.)

7. Daniel Christoff, "Le fondement logique des valeurs," *Proc. X Int. Cong. Philosophy* (Amsterdam, 1949), p. 454.

8. The reason for this is the foundation, the *Fundierung*, of all analytic thought in sense experience. See Husserl, *Formale und transzendentale Logik* (Halle, 1929), p. 130 and *Logische Untersuchungen*, Vol. II (Halle, 1922), p. 276. Also Kant, *De mundi sensibilis atque intelligibilis forma et principiis* (1770), ¶ 5; Friedrich Eduard Beneke, *System der Logik*, Vol. I (Berlin, 1842), pp. 38 ff.

9. Etymologically, "implication" means "being contained within," "being woven (or folded) within." The idea of "class" or "order" is connected with that of "series" in many ways. The Sanskrit root is srtā, meaning knot, tie, join, string together. Hence Greek *artao* (string together, join) *artyo* (arrange, think out), Latin *sero* (string together, join), *series, ordiri* (arrange in series, begin a weaving), *artus* (joint), *ars* (art), German *Art* (kind, class), English *articulation*. Analytic discursiveness thus means stringing together implied conceptual meanings. This implies both the production and the cutting off of series. Hence Latin *sero* (sow, bring forth) as well as *serro* (cut off, lock), *sera* (bolt), and *serra* (saw). Another Latin form for "cut off" is *caedere*, hence the various prepositional combinations *decide, incisive*, etc. The Sanskrit root for "cut off" is *rikhati*, hence English *right*, German *richtig, richten*, and *hinrichten* (right, to decide, and to decapitate), Russian *rjeshit* (to decide). Greek for "lock," "shut off" is *kleio*, hence *klesis* (division), *kalesis* (class), and Latin *classis* (a division named or called up), Greek *kaleo*, Latin *calo* (call, name). Late Latin for "cut off," "clip off" is *taliare*, connected with *talea* (bulbous root) as well as with *talus* (knuckle, heel, joint) and *talis* (such). Hence Spanish *talón* (heel, coupon), English *tally*, and *entailment* – serial origination or germinal succession. An entailment is a sequence of propositions made up of implied, or "contained within," concepts. On the philosophical significance of chains of implications as obscuring rather than elucidating meaning see the founder of the first scientific philosophy, Descartes, *Principia Philosophiae* I, LXXIV; *Regulae*, XII, XIII. Cf. above Introduction, note 9.

10. Mathematically $2^{50,000}$ is only an infinitesimal part of infinity. From the point of view of the infinite there is no difference between 2, $2^{10,000}$, $2^{50,000}$, $2^{1,000,000}$ or any other finite number. Mathematical infinity, $\aleph_0$, only begins "after" the last countable number. According to Leibniz, the number of all possible statements is $10^{7,300,000,000,000}$ (L. Couturat, *Opuscules et fragments inédits de Leibniz*, Vol. II [Paris, 1903], p. 96). For other calculations see Max Bense, *Programmierung des Schönen*, (Baden-Baden, 1960), pp. 120–

22; Stuart C. Dodd, "An Alphabet of Meanings for the Oncoming Revolution in Man's Thinking," *Educational Theory*, Vol. IX (July, 1959), 174–92.

11. It does, however, have a significant interrelationship, emphasized by Donald Williams. Given any properties ϕ and ψ present in a given proportion among the members of a class – e.g. ⅔ of it being white, ⅓ black – more subsets of the class will have ϕ and ψ in this proportion than will not have it. Inversely, if in an observed part of a class the properties ϕ and ψ are present in a certain proportion, then there is a calculable probability for the presence of the properties in the same proportion among unobserved members of the class. (Donald Williams, *The Ground of Induction* [Cambridge, Mass., 1947].)

12. Especially by Plato and Aristotle, and by Kant. Cf. *Philebus*, 16C–17A; Aristotle, *Metaphysics* Z xii, H vi; Kant, *Logik* and *Critique of Judgment*, Introd. and Sec. V. Central works on this phase of classical philosophy are those by Julius Stenzel, especially *Gestalt und Zahl bei Platon und Aristoteles* (Leipzig, 1924), and Nicolai Hartmann, "Aristoteles und das Problem des Begriffs," und "Zur Lehre vom Eidos bei Platon und Aristoteles," *Kleinere Schriften*, II (Berlin, 1957). A detailed discussion of the problem in Kant is Robert S. Hartman, "The Analytic and the Synthetic as Categories of Inquiry," *Perspectives in Philosophy* (Columbus, Ohio, Ohio State University Press, 1953), pp. 55–78. Cf. Lewis White Beck, "Can Kant's Synthetic Judgments Be Made Analytic?" *Kant-Studien*, XLVII (1955/56), 168–81. Also see David García Bacca, *Introducción a la lógica moderna* (Barcelona, 1936), Introduction and Part V. For an application of combinatorial analysis to the structure of intensions see Robert S. Hartman, "The Logic of Value," *The Review of Metaphysics*, Vol. XIV (March, 1961), 398–405. For an application of transfinite arithmetic see Benno Erdmann, *Logik* (Halle, 1907), Chs. 21, 24. For an application of matrix calculus see Henry Lanz, *In Quest of Morals* (Stanford, Cal., 1941), pp. 57–59.

13. García Bacca, *op. cit.*, pp. 229–40, Hermann Lotze, *Logik* (Leipzig, 1912), pp. 46–53. See however the ingenious attempt at structuring analytic intension made by Lambert, 1782; García Bacca, *op. cit.*, pp. 28–30, C. I. Lewis, *A Survey of Symbolic Logic* (Berkeley, 1918), pp. 19–35.

14. See Giorgio di Santillana, note 10, p. 17 of Galileo Galilei, *Dialogue on the Great World System* (Chicago, 1953); Descartes, *Regulae*, XII.

15. The seemingly abstruse reasoning of the alchemists falls into a consistent pattern as soon as it is recognized as constituting implications of analytic concepts.

16. "Phlogiston" was a typically analytic concept, which Lavoisier transformed into a synthetic one by introducing quantitative relations in the intension of the fundamental chemical concept "combustion." He thus did in chemistry what Galileo did in physics. In his famous "Table of Simple Substances Belonging to All the Kingdoms of Nature, which May be Considered as the Elements of Bodies" he contrasted the new and the old concepts:

New Names	*Old Names*
Light	Light
Caloric	Heat Principle or element of heat. Fire. Igneous fluid Matter of fire and of heat
Oxygen	Dephlogisticated air Empyreal air Vital air, or base of vital air
Azote (Nitrogen)	Phlogisticated air or gas Mephitis, or its base
Hydrogen	Inflammable air or gas, or the base of inflammable air

(Lavoisier, *Elements of Chemistry*, Introduction to Part II.) But Lavoisier was less clear-sighted concerning his method than Galileo. He did not see that what was fundamentally new in his procedure was the quantitative method – the law of the constancy of weight throughout a chemical change – and regarded as his fundamental achievement his classification of the chemical elements – a procedure of extensional analytic logic. Hence he called this kind of logic, mistakenly, "the logic of all the sciences" (Lavoisier, *op. cit.*, Preface, ¶ 19).

17. J. J. Fahie, *Galileo, His Life and Work* (London, 1903. Reprint, Dubuque, Iowa, 1963), p. 103. The Aristotelian model for this kind of argument may be found, for example, in *De Caelo*, Book II, Ch. 4. Sizzi's argument appears in his *Dianoia astronomica* (1610).

18. Ockham and others, who had anticipated Galileo in the dehypostatization of motion, had done so philosophically and not scientifically, hence less efficiently than Galileo. They had neither aroused the Church against nor, substantially, the public for the new doctrine. Cf. H. Shapiro, *Motion, Time and Place according to Ockham*, Franciscan Institute Publications, Philosophy Series, No. 13.

Also *Franciscan Studies,* XVI (Sept., 1956), 213–303, especially pp. 248–49.

19. Cf. the famous Mr. Tompkins books by George Gamow, *Mr. Tompkins in Wonderland* ("Dedicated to Lewis Carroll and Niels Bohr") (New York, 1940); *Mr. Tompkins Explores the Atom* (New York, 1944).

20. Some of the wonder of Galileo's reduction is captured, in modern dress, in Gerald Holton, *Introduction to Concepts and Theories in Physical Science* (Cambridge, Mass., 1952), pp. 1–6. Also see Ch. 2, pp. 17–34, and Ch. 3, pp. 50–51.

21. Cf. Descartes, *Regulae,* XII. Cf. L. G. Beck. *The Method of Descartes* (Oxford, 1952), pp. 172–76, and E. von Aster, *Die Philosophie der Gegenwart* (Leiden, 1935), p. 205.

22. Pierre Duhem, *The Aim and Structure of Physical Theory* (Princeton, 1954, Philip P. Wiener, transl.), pp. 187 f., 167.

23. *Op. cit.,* p. 210.

24. García Bacca, *op. cit.,* p. 239. García Bacca's example is the concept of complex numbers. This concept is defined "by a larger number of notes than the real numbers; and therefore the extension, the number of complex numbers, is larger than that of real numbers." What we call "analytic concept" García Bacca calls with the scholastic name, "concept of total abstraction," and what we call "synthetic concept" he calls "functional concept." Cassirer calls the former "substantial," the latter, as García Bacca, "functional." L. S. Stebbing calls the former method that of "common-sense analysis" and the latter that of "functional analysis" (L. S. Stebbing, *A Modern Introduction to Logic* [London, 1948], p. 352). The rule of inverse variation of analytic extension and intension has some qualifications which, however, are irrelevant in our context.

25. Beck, *op. cit.,* p. 281.

26. Norman Kemp Smith, *New Studies in the Philosophy of Descartes* (London, 1952), p. 92.

27. Ernst Cassirer, *Substance and Function* (Chicago, 1923), pp. 6, 19–20.

28. Edward V. Huntington, "The Duplicity of Logic," *Scripta Mathematica* (July–October, 1938), 11.

29. Another striking illustration of this relation is the difference in complexity, on the one hand, and applicability, on the other, between Euclid's and Descartes' geometry. (See Paul R. Halmos, "Innovation in Mathematics," *Scientific American,* CXCIX [Sept., 1958], 66–73.) Halmos' example is the relation between the Euclidian and the Cartesian proofs, of the proposition that equal chords of a circle are equidistant from the center.

30. Nicolai Hartmann calls this *stigmatic intuition* (*Grundzüge einer Metaphysik der Erkenntnis* [Berlin, 1949], Chs. 67–70),

Plato calls it *noesis*. Noesis is followed by *dianoia*, the development of the formal relations between the elements or terms of a system: the unfolding of the axiom that was seen in noesis. For the deduction of the singular nature of the axiom see Robert S. Hartman, "The Logic of Value," *Review of Metaphysics*, XIV, no. 3 (March, 1961), p. 401.

31. For a demonstration of this see Alfred North Whitehead, *The Concept of Nature* (Cambridge, 1955), Chs. III–V. Whitehead's method, applied by him only to space and time, is universally applicable. See below, Ch. 2, note 50.

32. Kant, *Logik*, Introduction, Sec. I.

33. On "ideological or organizational man" as the end product of history – and the end of history – see Roderick Seidenberg, *Post-Historic Man* (Chapel Hill, 1950); Max Lerner, *America as a Civilization* (New York, 1957), p. 946. On the end of history as God's possible choice for man and God's indifference to human suffering see the Archbishop of Canterbury in Philip Toynbee, *The Fearful Choice* (London, 1958). Political quackery is often in practice based on the same confusions that wreak havoc in philosophical theory, especially the confusion of words and actions (the fallacy of method), as in the Soviet conviction of non-conforming writers or the American insistence on the "aggressiveness" of China. On the lure and danger of ideologies see Senator J. William Fulbright, "The University and American Foreign Policy," *Center Diary: 12* (May–June, 1966), Center for the Study of Democratic Institutions, Santa Barbara, Cal.

34. See on this point Nicolai Hartmann, "Aristoteles und das Problem des Begriffs," *Kleinere Schriften*, Vol. II (Berlin, 1957), pp. 100–129.

35. Whenever, says Susanne Langer, philosophers characterize a subject as "peculiar," "subtle," and the like, one may be sure they are not clear about it. She herself uses these words with respect to the class-membership relation: "The relation 'ϵ' or 'membership' is a peculiar and subtle one." (*An Introduction to Symbolic Logic* [New York, 1937], p. 117. Cf. Bertrand Russell, *The Principles of Mathematics* [Cambridge, 1903], pp. 26, 77–78). Husserl regards modern extensional logic as a "*bedenkliche Umdeutung*" (risky misinterpretation) of syllogistic. (*Formale und transzendentale Logik* [Halle, 1929], pp. 65, 73.)

36. What is called "intensional logic" today – "modal" logic and the like – is not intensional in our sense. It does not structure and elaborate meanings; it qualifies copulas. It is a logic not of the concept but of the proposition.

37. The problem of the *scientific* nature of logic goes back to its origins with Aristotle and appears in the transition from *Topica* and *De Sophisticis Elenchis* to *Analytica Priora*. In the former, the *principle* of the syllogism had not yet been found and what we have

are materials not yet fully organized. In the latter "an entirely new start" was made and we get the *science* of logic, "found as a product of abstract construction rather than as an object of empirical observation." (Ernst Kapp, *Greek Foundations of Traditional Logic* [New York, 1942], pp. 67–68.) No similar transition has ever taken place with respect to the concept "concept." Kapp uses it in the sense of "what we comprehend when we know a definition" (*op. cit.*, p. 30).

38. The *Logik* of Kant is widely ignored today. The most recent compendium of logic, Bocheński's *Formale Logik* (Freiburg/München, 1957), does not mention Kant's *Logik* – as little as Wolff's – in the historical bibliography, and denies it any historical importance for (intensional) formal logic. Neither are the developments of this logic in Lotze and Erdmann mentioned.

39. One modern logician discussing the Kantian notion of analyticity, attacked as "analytic" the kind of concept which for Kant would be synthetic, and based on this confusion a widely accepted "rejection" of the distinction between "the analytic and the synthetic" as "an unempirical dogma of empiricists, a metaphysical article of faith," resting on a "metaphorical" notion of conceptual containment (Willard van Orman Quine, "Two Dogmas of Empiricism," *Philosophical Review*, LX [Jan., 1951], 20–43, and *From A Logical Point of View* [Cambridge, Mass., 1953], pp. 20–46). Actually, Kant's notion of containment is elaborately explained not only in his *Logik* but also in other writings. Quine's confusion is the fundamental one between Kant's logical and his epistemological doctrines of analyticity. This confusion appears most flagrantly in Section 4 of Quine's article, "Semantical Rules." The "analytic statements" of an artificial language L_0 whose analyticity is specified by certain semantical rules, are in Kantian terminology statements *where "analytic" is a synthetic concept.* They are the methodological opposites of "analytic statements" in the Kantian sense as which Quine discusses them. Moreover, his very objection against their supposed analyticity – that the semantical rules presuppose it – is Kant's definition of the *syntheticity* of this kind of concept. See on the relation of Kantian and modern "analyticity" H. J. Paton, *Kant's Metaphysics of Experience*, Vol. I (London, 1936), pp. 156, 201, 213.

40. There are "metaphysical" writings by symbolic logicians but they deal with metaphysical problems in a manner which appears illegitimate as compared with authentic metaphysical writings such as those of Nicolai Hartmann, Paul Tillich, Josiah Royce, Paul Weiss, or Charles Hartshorne. The reason is that the symbolism of extensional logic is not fitted for metaphysical argument. For an example of such "metaphysical" discussion see E. J. Nelson et. al. "Symposium on the Relation of Logic and Metaphysics" in *Philosophical Review*, LVIII (Jan., 1949), 1–34. Cf. Bertrand Russell, *My Philosophical Development* (London, 1959), pp. 234–38.

41. "We must end with my first love – Symbolic Logic: When in the distant future the subject has expanded, so as to examine patterns depending on connections other than those of space, number and quantity – when this expansion has occurred, I suggest that Symbolic Logic, that is to say the symbolic examination of patterns with the use of real variables, will become the foundation of aesthetics. From that state it will proceed to conquer ethics and theology." ("Remarks," *Philosophical Review*, XLVI [1937], 186.)

42. "The only way that mankind can develop an ethics and a philosophy commensurate with its achievement in building the atomic bomb is to make full use of symbolic logic in criticizing and correcting the past system of ethics and philosophy and in constructing new and better ones. To do anything less than this is very much like trying to do research in modern physics while using the old Roman arithmetic that lacked even the number zero." ("Attribute and Class," *Philosophic Thought in France and the United States*, Marvin Farber, ed. [Buffalo, 1950], p. 545.)

43. "The Elements of Ethics," *Readings in Ethical Theory*, W. Sellars and J. Hospers, ed. (New York, 1952), p. 1.

44. Galileo's letter to Kepler. F. W. Westaway, *The Endless Quest* (London, 1936), p. 157. The professor of astronomy at Bologna, Magini, promised to have Galileo's new planets "extirpated from the sky."

45. "The intuitionists claim to have at least a dim awareness of a simple unique quality or relation of goodness or rightness which appears in the region which our ethical terms roughly indicate, whereas the [naturalists] claim to have no awareness of any such quality or relation in that region." (W. Frankena, "The Naturalistic Fallacy," *Mind*, XLVIII [Oct., 1939], 474. Also in W. Sellars and J. Hospers, *Readings in Ethical Theory* [New York, 1952], p. 112.) Compare Hasdale's letter to Galileo: "Magini has written three letters, confirmed by 24 men of the profession from Bologna, stating that they had been present when you tried to demonstrate your discoveries, and that you were saying: 'Don't you see that, and that, and that?' and not one of them admitted he did, but all asserted that they saw nothing of what you pretended to show them" (di Santillana in Galileo Galilei, *Dialogue on the Great World Systems* [Chicago, 1953], p. 98).

46. This formulation of "value" is the logical counterpart to the ontological formulation of "good" in Plato, as the class of all classes, in the intensional sense of "idea." (Cf. E. Vernon Arnold, *Roman Stoicism* [London, 1911], p. 57.) A specific instance of this "good" is a "good of its kind" (on the logical nature of this "good" see A. C. Ewing, *The Definition of Good* [New York, 1947], pp. 104–5. On its meaning as the Platonic "virtue" see R. L. Nettleship, *Lectures on the Republic of Plato* [London, 1951], p. 226).

47. The notion of "similarity" or "one-to-one-correspondence" is itself neither extensional nor intensional. It may therefore be used for both extensional and intensional units (both subject variables and predicate variables).

48. *Die Grundlagen der Arithmetik* (Breslau, 1934, Oxford, 1933), p. 34.

49. *Knowledge of the External World* (London, 1962), p. 191.

50. G. E. Moore, *Principia Ethica* (Cambridge, 1903), p. 14.

51. On the difference see Kant, *Logik*, Introduction, Sec. III; *Inquiry Concerning the Evidence of the Principle of Natural Theology and Morals; Critique of Pure Reason*, "The Transcendental Doctrine of Method" (the naturalistic fallacy is found in B866); and Robert S. Hartman, "The Analytic and the Synthetic as Categories of Inquiry," *op. cit.* For Moore, "definition" means "analytic definition," *Principia Ethica*, p. 10. This use contradicts the purpose of *Principia Ethica* as prolegomena to a *science* of Ethics built upon the notion of good. See Robert S. Hartman, "The Definition of Good: Moore's Axiomatic of the Science of Ethics," *Proceedings, Aristotelian Society*, LXV (1964/65), 235–56.

52. See Werner Heisenberg, *Physics and Philosophy: The Revolution in Modern Science* (New York, 1957).

53. Actually, "Aha" is of profound, perhaps *the* most profound scientific significance. See on the "aha experience" in science, Norman L. Munn, *Psychology* (Boston, 1946), pp. 109, 187, Jacques Hadamard, *The Psychology of Invention in the Mathematical Field* (Princeton, 1945), and especially Michael Polanyi, *Personal Knowledge* (London, 1958), p. 122. Among the most famous Aha-experiences are St. Anselm's, Kepler's, and Archimedes'.

54. One of the founders of quantum mechanics once characterized to the present author a world-famous book on science, including quantum mechanics, by one of the founders of the positivistic school, as "pure nonsense."

55. "Natural philosophy, as a speculative science, I imagine, we have none; and perhaps I may think I have reason to say, we never shall be able to make a science of it. The works of nature are contrived by a wisdom and operated by ways far surpassing our faculties to discover, or capacities to conceive, for us ever to be able to reduce them into a science." (*Some Thoughts Concerning Education. Locke, Selections,* Sterling P. Lamprecht, ed. [New York, 1928], p. 11.)

56. See in particular *Philebus*, which in turn was the model for the ethics of Aristotle and hence of that of all the Middle Ages, and the lectures "On the Good." For details see Hans Joachim Krämer, *Arete bei Plato und Aristoteles* (Heidelberg, 1959), Ch. III.

57. *Dialogue Concerning the Two Chief World Systems,*

Stillman Drake, transl. (Berkeley and Los Angeles, 1953), p. 113. There was, of course, also a long-standing scholastic tradition sympathetic to Galileo, dating back to St. Ambrose, Anselm of Canterbury, and Nicolaus Cusanus. They all, in one way or another, proclaimed their aim to understand God not through the study of authorities but through their own thinking. Cusanus ridiculed the "logical circle" of scholastic theology and in his *Apologia doctae ignorantiae* joined St. Ambrose in the prayer: "From dialecticians, O Lord, deliver us!"

58. Benjamin Ginzburg, *The Adventure of Science* (New York, 1930), p. 109.

59. Roger Garaudy, "De l'empirisme logique a la sémantique," *Revue Philosophique de la France et de l'Etranger*, No. 2 (April–June, 1956), 235.

60. Language had expressed this long before Russell discovered it. To think "about" means to think "near by the outside of" (OE. *by-utan*. Cf. Spanish "acerca de," "sobre" and German "über." "Über etwas nachdenken" means to think above and beyond something). Even logic had at least discovered the principle before Russell. See Ockham, *Summa tot. log.* III, 3, c.38, f. 70 v.: *Nunquam pars potest significare totum cuius est pars* – an almost literal Latin version of Whitehead-Russell's "vicious-circle principle": "Whatever involves *all* of a collection must not be one of the collection." (Carl Prantl, *Geschichte der Logik im Abendlande*, Vol. IV [Darmstadt, 1957], p. 42, note 163. Whitehead-Russell, *Principia Mathematica* [Cambridge, 1935], p. 37. Also compare Whitehead-Russell's and Ockham's examples of fallacies. Cf. Spinoza, *Tractatus de intellectus emendatione*, *Opera I* [The Hague, 1913], p. 11).

61. Cassini in his sermon on the text: "Ye men of Galilee, why stand ye gazing up into heaven?" thundered against Galileo that "geometry is of the devil" and "mathematicians should be banished as the authors of all heresies." To the heresies of using mathematics and sense observations, and the degradation of "movement" which apparently degraded the moral and divine world itself, Galileo added the ultimate heresy of sponsoring the Copernican system which dethroned man from his moral and divine position in the universe. Considering the mortal blows Galileo dealt the medieval world it is a wonder that he fared as well as he did.

62. Some of this literature advocates the procedure of explaining logic by value disciplines rather than value disciplines by logic. Logic then appears as "generalized" economics, sociology, psychology, jurisprudence etc. Methodologically, this corresponds to the Aristotelian as against the Galilean procedure. The corresponding proposal in natural philosophy would be advocating that mathematics and natural science be abolished and mathematics be understood by

the categories of natural philosophy, rather than natural philosophy by the axiomatic relations of mathematics. It would mean, in other words, going back to alchemy and astrology. Strangely enough, this analogy is not only not seen, but such proposals are made in the names of Kepler, Galileo, Newton, and Lavoisier. For an example see Stephen E. Toulmin, *The Uses of Argument* (Cambridge, 1958), especially pp. 257–58.

63. A modern example of the shock produced by the encounter between philosophical and scientific procedures was recently reported from India where this struggle is still in full swing in the field of natural philosophy, especially medicine. There are more practitioners of *ayurveda* ("the science of life") than doctors of medicine (96,000 against 92,000). Students who have to learn both the ayurvedic and Western medical systems feel they "are the world's most confused people." (*Time*, August 18, 1958.)

64. As we are actually doing. The Aristotelian procedure of applying the *category* of value to the *system* of logic is, of course, again that of the logical positivists. They regard extensional logic as the highest philosophical value, but do not analyze value. Thus, their preference is uncritical and, according to their own rules, philosophically illegitimate.

65. A world of the unconscious, as Jung's investigations in the nature of alchemy have shown. Logically, he lived in a world of analytic rather than synthetic concepts (Cf. E. von Aster, *op. cit.*, p. 150). Religiously, from the modern point of view, he lived in a world of superstition; scientifically, he lived in a world of quackery and nonsense.

66. Galileo Galilei, *Dialogue on the Great World Systems*, Giorgio di Santillana, ed. (Chicago, 1953), pp. 16, 221.

2. *The Structure of Science*

1. Sir Isaac Newton, *Opticks* (London, 1704), last two paragraphs; Friedrich Dessauer, "Galileo and Newton: The Turning Point in Western Thought," *Spirit and Nature, Papers from the Eranos Yearbooks*, Vol. I (New York, 1954), pp. 300–301.

2. *Novum Organon*, Aphorism XCV.

3. For an elaboration of this simile see Robert S. Hartman, "Cassirer's Philosophy of Symbolic Forms," *The Philosophy of Ernst Cassirer*, Paul A. Schilpp, ed. (Evanston, Ill., 1949), pp. 289–333.

4. On the logical relation between a system and its field of application see Edmund Husserl, *Formale und transzendentale Logik*, Part I, Ch. III.

5. Edmund Husserl, *Die Krisis der europäischen Wissen-*

schaften und die transzendentale Phänomenologie (The Hague, 1954). The *Lebenswelt* is "die stets in frageloser Selbstverständlichkeit vorgegebene Welt der sinnlichen Erfahrung" (the world of sensorial experience always given in unquestionable obviousness, *ibid.*, p. 77). Husserl posits the problem of a science of the *Lebenswelt* (pp. 126–27) and finds it in the phenomenological method of epoché, the transcendental reduction. This method, of course, is not scientific in our sense. A more elegant and truly scientific method would be to regard *the secondary qualities as the primary qualities of a new science.* This is the procedure of formal axiology; which may then be regarded as the science of the *Lebenswelt.* And it does fulfill many of the requirements Husserl set for such a science (see *Symposium sobre la noción Husserliana de la* Lebenswelt, XIII International Congress of Philosophy, Mexico, 1963).

6. Examples are the fiendish, hour-long tortures inflicted upon Babington in Elizabethan England and Damiens in the France of Louis XV. George Sarton, *The Life of Science* (New York, 1948), pp. 179–80, Stefan Zweig, *Maria Stuart* (Wien, 1935), p. 467. Also see the account of the especially slow burning of Servetus in Calvin's Geneva, Stefan Zweig, *The Right to Heresy: Castellio against Calvin* (Boston, 1951), Ch. V. All human suffering, if spectacular enough, was a source of entertainment. Hogarth shows us fashionable people going in parties to Bedlam to laugh at the lunatics. This insensitivity to extremes was matched by insensitivity in ordinary life. Mademoiselle reports from the court of Louis XIV that the Queen's own gentleman-in-waiting thought nothing of dropping the royal hand for a moment "*pour aller pisser contre la tapisserie.*"

7. Sarton, *ibid.*

8. On the relation of moral sanity and science see Henri Bergson, *The Two Sources of Morality and Religion*, Ch. IV, "Mechanics and Mysticism." On the moral failure of the Enlightenment see Robert S. Hartman, "La Ilustración y su enfoque de la ciencia natural," *Memorias del Primer Coloquio Mexicano de la Historia de la Ciencia*, Tomo II (Mexico, 1964), pp. 7–24.

9. Cf. Thurber's cartoon in *Men, Women and Dogs*, of the dejected male at a party, subject of the excited gossip: "He doesn't know anything except facts." A profound and entertaining vision of a world where facts have been overcome is Franz Werfel's *Star of the Unborn.*

10. Cf. the remarkable prediction made by Tolstoy in 1910 that in the second half of the century ethics would be reconstituted by, and an ethical era would follow, the use of symbolism. (Henry James Forman, *The Story of Prophecy* [New York, 1939], p. 254.) Observe the remarkable awakening of young people today to morality, especially in the United States and the Soviet Union.

11. Actually, it is the application of formal axiology that makes an "axiological value" out of a state of affairs. "Phenomenal value," thus, strictly speaking, depends on this application. Before the application, the datum may be, empirically-analytically, either fact or value. This is important for the onto-axiological status of the referents of "value propositions" within formal axiology.

12. On the structure of fact see Susanne Langer, "*The Practice of Philosophy*, pp. 142–43.

13. See the studies and experiments of Adelbert Ames, Jr., Hadley Cantril, Ross Mooney, Hoyt Sherman. Hadley Cantril, *Understanding Man's Social Behavior* (Princeton, 1947); Hoyt Sherman, *The Visual Demonstration Center* (Columbus, Ohio, Ohio State University, 1951).

14. Landscape painting began at about that same time. Compare, for example, a landscape by Giotto and one by Giorgione. Aldo Mieli, *Panorama general de Historia de la Ciencia*, Vol. V., *La Ciencia del Renacimiento* (Buenos Aires, 1952), Ch. VII; and J. Bronowski, "The Creative Process," *Scientific American*, CXCIX (Sept., 1958), 58–65.

15. See the investigations of C. G. Jung, especially *Psychology and Alchemy* (London, 1953), and *Mysterium Coniunctionis* (London, 1963). Much in the world of the Middle Ages, with its witch trials and bizarre superstitions ruling the actual lives of people, is literally a nightmare lived as reality.

16. Herbert Butterfield, *The Origins of Modern Science* (London, 1950), p. 5.

17. Carl G. Hempel, *Fundamentals of Concept Formation in Empirical Science* (Chicago, 1952), p. 36.

18. *Op. cit.*, pp. 46, 47. Concepts with "merely empirical import" may really have *no* import. They may be concepts which connect irrelevant relations, e.g. the *Hage* of a person, the product of his height in millimeters and his age in years. Relevance comes about only within a system.

19. *Op. cit.*, p. 1.

20. The last two steps are often confused. Some contemporary social scientists, who base their "systems" on concepts such as "action," "situation," and the like, call their theories "formal" and "nonempirical," although they are abstract and empirical. Their concepts are analytic, not synthetic. An intermediate step between the two steps is that of shorthand symbols, which replace words by letters but preserve the relations of ordinary discourse. They constitute no scientific advance. Here belong many of the pseudo-systems of social "sciences" and the various ethical "calculi," such as Bentham's or Hutcheson's. In natural philosophy, we have here the symbols of Aristotle, of alchemists and astrologists. On the logical nature of shorthand sym-

bolism see García Bacca, *op. cit.*, pp. 37–44. On the corresponding discussion in antiquity see Jan Łukasiewicz, *Aristotle's Syllogistic* (Oxford, 1951), p. 13.

21. Hempel, *op. cit.*, p. 1.

22. *Op. cit.*, p. 21.

23. *Op. cit.*, p. 33.

24. *Op. cit.*, p. 34.

25. And in this respect is itself an interpretation. It is logic applied to space. This is the point made so powerfully by Whitehead and Russell. All the arguments brought against their deduction of mathematics from logic concern, in my opinion, details but not the principles of the method. Logic is the supreme formal system. Other formal systems, such as the mathematical, arise through axiomatic interpretations of logic. Once the science of mathematics is created it serves as the (mediately) logical system for nonlogical fields and their elements, such as gravitation, force, energy etc. On the controversy whether logic and mathematics are identical see Bertrand Russell, *Principles of Mathematics*, Introduction to Second Edition (Cambridge, 1943), pp. v–vi; and Jørgen Jørgensen, A *Treatise of Formal Logic*, Vol. III (Copenhagen-London, 1931), pp. 101–41.

26. A. L. Lavoisier, *Elements of Chemistry*, Preface, p. 1. As we have seen, Lavoisier confuses the analytic and the synthetic.

27. *Ibid.* Italics supplied.

28. Hempel, *op. cit.*, p. 37. Cf. W. H. Werkmeister, A *Philosophy of Science* (New York, 1940), pp. 22–23. Scientific "fact" is the result of the whole operation.

29. The whole realm of what we call *systemic value* is new to value theory.

30. See on this subject Jørgen Jørgensen, *loc. cit.* For Kant's logic see Immanuel Kant, *Logik, Ein Handbuch zu Vorlesungen*, G. B. Jaesche, ed., 1800 (Leipzig, 1920). Kant's logic is also found in the section "The Transcendental Doctrine of Method" in the *Critique of Pure Reason* and in various smaller essays such as *The Mistaken Subtlety of the Four Syllogistic Figures* (1792); *Inquiry Concerning the Evidence of the Principles of Natural Theology and Morals* (1764) and others. For details see Robert S. Hartman, "The Analytic and the Synthetic as Categories of Inquiry," *op. cit.* For a modern version of the analytic a posteriori see José A. Benardete, "The Analytic A Posteriori and the Foundations of Metaphysics," *Journal of Philosophy*, LV (June 5, 1958), 503–14. Formal axiology, in the sense of this interesting essay, is the posterior analytics of value.

31. Kant faced this problem in the creation of his science of metaphysics and solved it, in the metaphysical deduction of the *Critique of Pure Reason*, in the transition from the analytic unity of both the forms of judgment and the pure categories to the synthetic unity of the schematized categories.

32. Kant, *Logik*, ¶ 104.

33. Here belong writings in the phenomenology and psychology of creation, such as Jacques Hadamard, *The Psychology of Invention in the Mathematical Field* (Princeton, 1945). The logic of the process has been discussed by Ernst Cassirer and some of his followers, especially Susanne K. Langer.

34. Kant, *Logik*, ¶ 110.

35. *Critique of Pure Reason*, A 730–31. Mathematics, on the other hand, ought to *start with* definition, but this definition is *synthetic*. Analytic definition, thus, is the *end point* of its corresponding argument – the philosophical – while synthetic definition is the *starting point* of the argument that corresponds to it, the mathematical and scientific. On the other hand, analytic definition is the *starting point* of the lexical method and synthetic definition is the *end point* of creative thinking: the axiomatic identification. The lexical method is that of today's analytic philosophers. On the Gulliver-Swift-like launching of a successful such philosopher by the lexical method see John Wilkinson, "The Civilization of the Dialogue," *Center Diary:* 12 (May–June, 1966), Center for the Study of Democratic Institutions, Santa Barbara, Cal.

36. Metaphysics, Z xii, H vi. Cf. the notion of "Foundation" in medieval logic and that of "Fundierung" in Husserl's logic.

37. Nicolai Hartmann, "Die Lehre vom Eidos bei Platon and Aristoteles," *Kleinere Schriften*, II, pp. 136–37, 150–52. Also see p. 106. The "hiatus" between *atomon eidos* and *kath' hekaston* is that between the infinitely specified particular and the singular.

38. Galileo, *Two New Sciences*, Henry Crew and Alfonso del Salvio, transl. (Evanston, 1946), p. 162.

39. Ernst von Aster *Die Philosophie der Gegenwart* (Leiden, 1935), p. 147. Heinrich Rickert, *Zur Lehre von der Definition* (Tübingen, 1929), p. 50; E. Husserl, *Formale und transzendentale Logik*, p. 274. Ortega y Gasset calls "term" what we, with Kant, call "analytic definition" (*La idea de principio en Leibniz* [Madrid, 1958], ¶ 9).

40. The individual is then defined as the extension of an intension with content $\aleph_1$. Such an extension must be singular, for $\aleph_1$ signifies a nondenumerable continuum, that is, a Gestalt. Particularity, on the other hand, or nonsingularity, presupposes abstraction, that is, denumerability of the intensional properties. For, properties common to at least two things cannot be abstracted except one by one, or set by set. Particularity thus may be defined as the extension of an intension on content $\aleph_0$. If the extension of an intension of content $\aleph_1$ is two things, then the two form one organic whole. The definition of an *organic whole* then is: an extensional plurality with intension of content $\aleph_1$. See Robert S. Hartman, "Value Theory as a Formal System," *Kant-Studien*, L(1958/59), 287–315, and "The Logic of

Value," *Review of Metaphysics*, XIV (March, 1961), 389–432. For an application of transfinite number to intension see Benno Erdmann, *Logik* (Halle, 1907), Chs. 21, 24. For an application of transfinite mathematics to value theory see Edwin T. Mitchell, A *System of Ethics* (New York, 1950), pp. 126–29. On the nature of finite and transfinite operations in logic see García Bacca, *op. cit.*, pp. 69–71. The connection between value, on the one hand, and the finite and infinite, on the other, is, of course, at the very basis of Western philosophy. See Aristotle, *Metaphysics*, Book A, Ch. 5, and Plato's *Philebus*. For details see Hans Joachim Krämer, *Arete bei Platon und Aristoteles* (Heidelberg, 1959).

41. Bertrand Russell, *The Principles of Mathematics* (Cambridge, 1903), p. 53. The confusion between analytic concept and its referent, which Russell eradicated in mathematics by defining number synthetically, is still prevalent in axiology, the fallacy of method.

42. Galileo, *Two New Sciences*, p. 265. Italics supplied.

43. The conviction that a logic of value was possible, covering the subjects of moral philosophy as mathematics covers those of natural philosophy, was Leibniz's inspiration for his "General Science" and "Universal Characteristic." See his *Preface to the General Science* and *Towards a Universal Characteristic* (1679).

44. Heinrich Rickert, *Zur Lehre von der Definition* (Tübingen, 1929), p. 50. While terms are "crossings of relations," concepts, says Rickert, are "crossings of judgments."

45. Bertrand Russell, *The Principles of Mathematics*, Introduction to Second Edition (Cambridge, 1943), p. xi.

46. Rudolf Carnap, *Foundations of Logic and Mathematics* (Chicago, 1939), p. 12. Cf. Ludwig Wittgenstein, *Tractatus Logico-Philosophicus*, 4.461. For Carnap, what we call the analytic and the synthetic method, is the "first" and "second" method, respectively. *Op. cit.*, Secs. 24, 25.

47. Since the term is an element of a system, once a *term* is given as making sense the whole system is given and hence the intension of the synthetic concept, while when a *concept* is given as part of an analytic intension the intension is not yet given at all.

48. *Terminus* is the Latin translation of the Greek *horos*, limit, demarcation, definition. *Horoi* were originally the boundary stones of a piece of land.

49. Michael Polanyi, *Personal Knowledge* (London, 1958), pp. 195–202. Cf. John Laird, "Synthesis and Discovery," *Proceedings, Aristotelian Society*, Vol. XIX (1918/19), 46–85.

50. See below Ch. 6, Sec. 2. Alfred North Whitehead, *The Concept of Nature* (Cambridge, 1955), Chs. III–V; C. D. Broad, *Scientific Thought* (London, 1952), pp. 36–52. The Chinese-box

simile is found in *The Concept of Nature*, p. 61, *Scientific Thought*, pp. 42, 44. On a similar procedure of Leibniz – the ordering of the unknown by ordering the known with the help of infinite approaches – see "Letter of Mr. Leibniz on a General Principle Useful in Explaining the Laws of Nature" (1687), Loemker, *op. cit.*, pp. 538–43. Leibniz's *petites perceptions*, of course, were infinitesimals of consciousness, and considered by Leibniz in the mathematical sense. This sense was made even more explicit by Maimon, for whom the thing-in-itself is the limiting conception for the infinite decreasing series from complete consciousness to complete unconsciousness which is an *irrational* quantity, and where consciousness merges with the given. Nothing can be thought outside of consciousness, hence the given can be defined only as the lowest grade of the completeness of consciousness. Consciousness thus must be thought of as a diminishing series of infinite stages down to nothing, and *the idea of the limit of this infinite series (comparable to* $\sqrt{2}$*) is that of the thing-in-itself*, the merely-given. Things-in-themselves are, for Maimon, in direct reference to Leibniz's *petites perceptions, differentials of consciousness* (Salomon Maimon, *Versuch über die Transzendentalphilosophie* (Berlin, 1790), pp. 27–45; Leibniz, *Nouveaux Essais*, Avant-propos, Gottfried Wilhelm Leibniz, *Opera Philosophica*, J. E. Erdmann, ed. (Berlin, 1840, Aalen, 1959), p. 198; Also Liv. II, chap. I, p. 225; Kuno Fischer, *Gottfried Wilhelm Leibniz* (Heidelberg, 1902), pp. 494–99.

51. Nicolai Hartmann, *Grundzüge einer Metaphysik der Erkenntnis* (Berlin, 1949), Chs. 64–70, especially pp. 514–19, 531–32.

52. The richness lies in the combination of the world of philosophy with that of science, that is, the infinite applicability of the latter to the former. It applies only to synthetic systems, not to analytic "system." And it applies only to systems in their totality, not to their elements. (Pierre Duhem, *The Aims and Structure of Physical Theory* [Princeton, 1954], pp. 183–88, Joseph Fourier, *Theory of Heat*, 1822, Preliminary Discourse.) To fulfill their enriching function systems must never degenerate into mere tools; but their intrinsic origin must be kept in mind; nor must their elements be used out of the whole cultural context. Otherwise their functions become impoverishing rather than enriching and *value transpositions rather than value compositions result*. Culture then degenerates into what Ortega y Gasset calls "specialist barbarism" (*The Revolt of the Masses* (New York, 1932), Chs. IX, X, XII), and to "calculations" such as Edward Teller's, who writes that since the annual production of the United States is $500 billion and "the total value that exists in the country . . . is only about $1500 billion . . . survivors of an all-out nuclear attack . . . could rebuild our industrial complex in a very short time" (*The Legacy of Hiroshima* [Garden City, N. Y.,

1962], pp. 254–55. On the total neglect of cultural development and context in this "calculation" see Ralph A. Lapp, *Kill and Overkill* [New York, 1962], p. 101).

53. One result of our argument is that phenomenology *is itself a formal axiology*, and that special formal axiologies, such as Theodor Lessing's or Husserl's own are *multiplicationes praeter necessitatem*.

54. *Philosophie der symbolischen Formen*, Vol. III. (Berlin, 1929), pp. 119, 466–67, 462–63. See Robert S. Hartman, "Cassirer's Philosophy of Symbolic Forms," *The Philosophy of Ernst Cassirer*, Paul A. Schilpp, ed. (Evanston, 1949), pp. 291–333.

55. Cf. Kant, *Critique of Pure Reason*, A 834 = B 862. Nicolai Hartmann is not clear about the symbolic nature of the ideal object.

56. G. E. Moore, *Principia Ethica*, pp. 7–10, 20.

57. G. E. Moore, "The Conception of Intrinsic Value," *Philosophical Studies* (London, 1948), p. 275; "Reply to My Critics," *The Philosophy of G. E. Moore*, Paul A. Schilpp, ed. (Evanston, Ill., 1949), pp. 590–91.

3. *The Concept of Axiological Science*

1. Nicolai Hartmann, *Kleinere Schriften*, I (Berlin, 1955), pp. 36–37; G. E. Moore, *Principia Ethica* (Cambridge, 1903), p. ix.

2. These *logical* levels of fact are not identical with the previously examined *methodological* levels of fact, namely "fact," fact, and the combination of both, scientific fact. Both particular and singular fact are *facts* in the second methodological sense. The first logical and the first methodological sense, "fact," are identical.

3. These *logical* levels of value are not identical with the previously discussed *methodological* levels, "value," value, and the combination of both, axiological value. Both particular and singular value are *values* in the second methodological sense. The first logical and the first methodological sense, "value," are identical.

4. J. O. Urmson, "On Grading," *Mind*, LIX (1950), 145–69. Also in *Logic and Language*, II, A. G. N. Flew, ed. (Oxford, 1953); Paul Edwards, *The Logic of Moral Discourse* (Glencoe, Ill., 1955).

5. Philosophers have found a variety of reasons, and rationalizations, to avoid this conclusion. Blanshard recoils from it because Moore's "theory makes goodness too abstract" (*Reason and Goodness* [London, 1961], p. 269). Frankena misinterprets the naturalistic fallacy as the "definist fallacy" and hence as absurd (*Mind*, XLVIII [1939], 473). Actually, for Moore, it is *not* fallacious to define good (*Principia Ethica*, p. 20). It is, *in case one defines it*, to define it falsely, namely, by confusing logical types. To deny that good is in-

definable involves a fallacy *only because it involves contradictions* (*Principia Ethica*, p. 77).

6. *Principia Ethica*, p. 20.

7. What Moore calls "Ethica" should really be "Axiologica," for Moore endeavored to find not what is moral good but what is "good in general" (*Principia Ethica*, p. 4). The distinction Moore makes between ethics and casuistry corresponds to that between a theoretical and an applied science. For details see Robert S. Hartman, "The Definition of Good: Moore's Axiomatic of the Science of Ethics," *Proceedings, Aristotelian Society*, LXV (1964/65), 235–56.

8. *Philosophical Studies* (London, 1922), p. 273.

9. *Philosophical Studies*, p. 274; "Reply to My Critics," *The Philosophy of G. E. Moore*, P. A. Schilpp, ed. (Evanston, Ill., 1942), pp. 590–92.

10. This axiom is the formulation in *logical* terms of a principle common to all classical value theory. It has been expressed in ontological, teleological, epistemological and other terms. Ontologically, a thing has been called good in the degree of its *perfection*, teleologically, in the degree of fulfilling its *purpose*, epistemologically, in the degree of possessing its essential *properties*. It has been called good in the degree that its actuality corresponded to its ideality, or its ideality was fulfilled in its actuality (Weiss), and in the degree that there was "fulfillment of its essential nature" (Tillich), or that it had "the special complex of characters which justify us in calling it good" (Moore), etc. Surveying the history of axiology we have here a general consensus which may be called *axiologia perennis*. It becomes *theologia perennis* when the notion of specific perfection is generalized into the perfection of an absolute being which lacks nothing and has the abundance of all properties. On the axiomatic character of Moore's own determination of good see Robert S. Hartman, "The Definition of Good: Moore's Axiom of the Science of Ethics," *Proceedings, Aristotelian Society*, LXV (1964/65), 235–56.

11. Jerrold J. Katz, "Semantic Theory and the Meaning of 'Good,'" *Journal of Philosophy*, LXI (December 10, 1964), 739–66. Katz's result concerning the meaning of 'good' almost coincides with ours, even though derived at in an entirely different manner: "The meaning of 'good' is a function that operates on other meanings, not an independent attribute. Apart from combination with the conceptual content of other words and expressions, the meaning of 'good' does not make sense. Since the meaning of 'good' cannot stand alone as a complete concept, we shall say that the meaning of good is *syncategorematic*." We elaborate this syncategorematic meaning of 'good': 'good' is the *universal axiological quantifier*. Katz proposes, as do we, "that Moore's intuition was an intuition of the syncategorematicity of the meaning of 'good.'"

12. Cf. Bertrand Russell, *Knowledge of the External World*, pp. 212–13.

13. For details see Robert S. Hartman, "Group Membership and Class Membership," *Philosophy and Phenomenological Research*, XIII (March, 1953), 353–69.

14. *Collected Papers of Charles Sanders Peirce*, Vol. VI, Charles Hartshorne and Paul Weiss, ed. (Cambridge, Mass., 1935), Sect. 484.

15. A. L. Hilliard, *The Forms of Value* (New York, 1950), pp. 278–79.

16. See the Preface to Lawrence Durrell's *The Alexandria Quartet*: "A suitable descriptive subtitle might be 'a word continuum.' . . . The whole was intended as a challenge to the serial form of the conventional novel: the time-saturated novel of the day." The work, says Durrell, could be continued in any direction without losing its character as a continuum.

17. Whose axioms are metaphors. For details see Robert S. Hartman, "The Logic of Value," *Review of Metaphysics*, XIV (March, 1961), 389–432.

18. For details see Robert S. Hartman, "Prolegomena to a Meta-Anselmian Axiomatic," *Review of Metaphysics*, XIV (June, 1961), 637–75. Since existence is *defined* as "having $\aleph_0$ properties," Hartshorne's objection against the axiological proof does not seem to apply, nor does Findlay's paradox. When I posit something as having this predicate I posit it as existing; and "existence is only one of the possibilities of varying the given in imagination." See below motto to Ch. 6 and Charles Hartshorne, *Anselm's Discovery* (La Salle, 1965), p. 261.

19. For details see Robert S. Hartman, "Four Axiological Proofs of the Infinite Value of Man," *Kant-Studien*, LV (1964/65), pp. 428–38.

20. E. V. Huntington, *The Continuum* (Cambridge, Mass., 1942), p. 32.

21. "On the General Characteristic," Leibniz, *Philosophical Papers and Letters*, Vol. I, Leroy E. Loemker, tr. and ed. (Chicago, 1956), pp. 344–46. For the application of transfinite mathematics to intensional structure see Benno Erdmann, *Logik* (Berlin, 1923), Chs. 21–24.

22. *Was sind und was sollen die Zahlen?* (1887), ¶ 66.

23. *The World and the Individual*, Vol. I (New York, 1923), pp. 501–38, especially p. 534, 578–88. Cf. Richard Hocking, "The Influence of Mathematics on Royce's Metaphysics," *Journal of Philosophy*, LIII (Feb. 2, 1956), 77–91.

24. See Robert S. Hartman, "Prolegomena to a Meta-Anselmian Axiomatic," *Review of Metaphysics*, XIV (June, 1961), 669. The consistency of all the predicates thinkable in Being (God)

has been demonstrated by Anselm and by Leibniz. The proportionate richness of extension and intension of Being shows that Being cannot be a category. It must be determined axiomatically, as was done by Anselm. Cf. Royce, *loc. cit.*

25. Theologically, the secondary predicate of Being may be called God. God then is the value predicate of the world. Any particular goodness is then indeed a hierarchical subdivision of God's perfection as the scholastics held. For details see Robert S. Hartman, "The Good as a Non-Natural Property and the Good as a Transcendental," *Review of Metaphysics,* XVI (Sept., 1962), 149–55.

26. *Principia Ethica,* p. 15.

4. *The Axiological Reinterpretation of Moore's Ethical Theory*

1. Bertrand Russell, *Our Knowledge of the External World* (London, 1952), p. 209; Plato, *Republic* VII, 534.

2. "Philosophie als strenge Wissenschaft," in *Logos* I (1911) p. 339. Cf. above Ch. 2, note 38.

3. Note the quotation marks. The fallacy lies in using Darwin's *theory* for this conclusion, not in interpreting the *fact* in question axiologically.

4. These philosophers, in their endeavor to make ethics scientific, thought they could apply the formal framework that made natural philosophy scientific, to moral philosophy. Descartes' goal was not only to reformulate a natural science, but also a "mathematical morality." For Leibniz the differential calculus was only part of a large calculus of universal logic applicable to all the sciences and humanities, so that "two philosophers who disagreed about a particular point instead of arguing fruitlessly would take out their pencils and calculate." Spinoza applied the geometrical method to ethics in an *Ethica ordine geometrico demonstrata.* Locke wrote his *Essay* as prolegomena to "a subject very remote from this," namely morality and revealed religion, and showed "that moral knowledge is as capable of real certainty as mathematics." (Locke, *Essay Concerning Human Understanding,* Bk. III, Ch. XI, Sec. 15–18; Bk. IV, Chaps. III, Sec. 18–20; IV, Sec. 5–10, XII, Sec. 7–8. See also Epistle to the Reader, ¶ 15). The full title of Hume's Treatise is *A Treatise on Human Nature. Being an Attempt to Introduce the Experimental Method of Reasoning into Moral Subjects.* Even Berkeley used epistemology only as a tool for theological ethics, the rules of which "have the same immutable universal truth with the propositions of Geometry." (*Passive Obedience,* Sec. 53. The older Berkeley lost this vision because of what we call the fallacy of method.)

5. In his Biblical studies, especially on the Book of Daniel.

6. Cf. Planck's famous *Gedankenexperimente,* thought experiments. Max Planck, *Die Physik im Kampf um die Weltanschauung* (Leipzig, 1935), pp. 20–22.

7. See on this point most emphatically, William James, *The Principles of Psychology,* Vol. I (New York, 1905), pp. 196–98, 240. James calls this fallacy "The Psychologist's Fallacy."

8. On the speciousness of the Kantian distinction see Theodor Lessing, *Der Bruch in der Ethik Kants* (Bern, 1908). See below Ch. 7, note 48.

9. Nicolai Hartmann, "Vom Wesen sittlicher Forderungen" ["On the Nature of Moral Obligations"], *Kleinere Schriften,* I (Berlin, 1955), pp. 302–3. Cf. the same author's *Grundzüge der Metaphysik der Erkenntnis* (Berlin, 1949), Ch. 72, and *Ethik* (Berlin, 1926), Ch. XVI.

10. Alfred North Whitehead, *Science and the Modern World* (Cambridge, 1933), pp. 41–48.

11. For details see Robert S. Hartman, *El conocimiento del Bien: Crítica de la razón axiológica* (Mexico-Buenos Aires, 1965).

12. E.g. Charles L. Stevenson, *Ethics and Language* (New Haven, 1944), pp. 271–73. Brand Blanshard, *Reason and Goodness* (London, 1961), pp. 319–21. Cf. above Ch. 3, note 5.

13. A good point may be made for holding that Moore's "end of the matter" is also the beginning of the matter: it is an analytic end but a synthetic beginning, a categorial end but an axiomatic beginning, a philosophical end but a scientific beginning. Good is analytically indefinable, but not synthetically. It cannot be defined as an empirically abstracted concept, but it can be defined as a logical construct. It is not a philosophical category but may well be a scientific axiom. Hence its "indefinability" is that of axioms: it is that of the self-evident, incapable and in no need of demonstration (*Principia Ethica,* pp. x, 143–45), it opens up a new science (pp. ix, 3–5), it is *sui generis* (pp. 6–7, 21), all propositions with it are synthetic (pp. 7, 58, 143), it is unknown at present but not unknowable (p. 20), and may be subject to formal structurization as is number (pp. 111, 117, 125, 145). *Principia Ethica,* in other words, is a treatise on the *unknown axiomatic nature of good.* See Robert S. Hartman, "The Definition of Good: Moore's Axiomatic of the Science of Ethics," *Proceedings, Aristotelian Society,* LXV (1964/65), 235–56.

14. *The Philosophy of G. E. Moore,* Paul A. Schilpp, ed. (Evanston, Ill., 1942), p. 590.

15. *Ibid.*

16. "Reply to My Critics," *op. cit.,* p. 588.

17. *Principia Ethica,* pp. 41, 206.

18. "Reply," *op. cit.,* p. 584.

19. *Ibid.*, p. 585.
20. *Philosophical Studies,* p. 274.
21. *Principia Mathematica,* Vol. I (Cambridge, 1935), p. 37.
22. "Reply," *op. cit.*, p. 591.
23. "Reply," *op. cit.*, pp. 591–92. Mr. Stevenson held that certain ethical predicates had primarily emotive rather than cognitive meaning.
24. *Philosophical Studies,* pp. 274–75.
25. *Op. cit.*, p. 275.
26. *Op. cit.*, pp. 259–60.
27. *Op. cit.*, p. 260.
28. *Reply,* p. 554: "I am *inclined* to think that this is so, but I am also inclined to think that this is not so [that value predicates have only emotive and no cognitive meaning]; and I do not know which way I am inclined most strongly. If these words, in their ethical uses, have only emotive meaning, or if Mr. Stevenson's view about them is true, then it would seem that all else I am going to say about them must be either nonsense or false (I don't know which). But it does not seem to me that what I am going to say is either nonsense or false; and this, I think, is an additional reason (though, of course, not a conclusive one) for supposing both that they have a 'cognitive' meaning, and that Mr. Stevenson's view as to the nature of this cognitive meaning is false." In later years Moore was so completely inclined to think that "this was not so" that he forgot he had ever been inclined to think that this *was* so. See below note 46.
29. *Philosophical Studies,* p. 265.
30. Edel, who discusses this problem in *The Philosophy of G. E. Moore,* p. 145, is not free to use it, and therefore his "solution" is no solution. The replacements he makes are precisely those which must not be made as long as the relation between the thing's nature and value is unknown. Edel does not clarify this relation. However, Edel's essay is the only one I know explaining Moore's theory of value from a *logical* point of view. It is significant that with no author is Moore as exasperated in his "Reply" as with Edel.
31. *Philosophical Studies,* p. 273.
32. H. J. Paton, *The Categorical Imperative* (Chicago, 1948), p. 123; G. E. Moore, *Principia Ethica,* pp. 143, 5; C. D. Broad, "Is Goodness a Name of a Simple Non-Natural Quality?" *Proceedings, Aristotelian Society,* XXXIV (1933/34), 249–68; A. C. Ewing, "G. E. Moore," *Mind,* LXXI (1962), 251. See below note 46.
33. *Philosophical Studies,* p. 274.
34. C. H. Waddington, *Science and Ethics* (London, 1942), p. 7.
35. Such as the notion of "oddness." See P. H. Nowell-Smith,

Ethics (London, 1954); Paul Ziff, *Semantic Analysis* (Ithaca, N. Y., 1960). Ziff operates with what he calls the ordering of adjectives, e.g. "That is a red heavy good table," "That is a heavy red good table," "That is a red good heavy table," etc. All such expressions are "odd." The only expression "not the least bit odd" is "That is a good heavy red table." Ziff feels that "some principle other than simple privilege of occurrence must be at work here. Semantically speaking, it appears to be one having something to do with natural kinds but I can provide no satisfactory syntactic characterization" (pp. 204 and 206). Our axiom provides this characterization; Good is a second-order predicate and thus should have the rank of position before first-order predicates. If it does not, the expression sounds odd. The reason is that in this case a first-order predicate is used to characterize a second-order predicate, and thus is elevated to the third order. This no descriptive predicate can stand without strain. The strain would be relieved by a comma after it. Thus, while it sounds odd to say "That is a red good heavy table" it does not sound odd to say: "That is a red, good heavy table." Here "red" is restored to its first-order position. The only predicates that may characterize value predicates are value predicates themselves. Thus, while it sounds odd to say: "That is a heavy good red table" it does not sound odd to say: "That is a lovely good red table."

36. A. C. Ewing, *The Definition of Good* (New York, 1947), p. 105.

37. *Science and the Modern World* (Cambridge, 1933), p. 41.

38. Moore's "intuition" is not a psychological but a logical notion. It means any kind of cognition that cannot be proved. An axiomatic identification would be an intuition in Moore's sense (*Principia Ethica*, pp. x, 144–45).

39. The controversies between the Aristotelians and Galileo are well known. As to Kepler, see the instructive controversy between him and Fludd, in W. Pauli, "The Influence of Archetypal Ideas on the Scientific Theories of Kepler," in C. G. Jung and W. Pauli, *The Interpretation of Nature and the Psyche* (New York, 1955), pp. 147–212, especially 198–200.

40. Paul Kecskemeti, *Meaning, Communication and Value* (Chicago, 1952), p. 217.

41. Here belong, for example, the value theories of Everett W. Hall, *What Is Value?* (New York, 1952), and A. C. Ewing, *The Definition of Good* (New York, 1947), and *Second Thoughts in Moral Philosophy* (London, 1959). For details see Robert S. Hartman, *El conocimento del Bien.*

42. In this sense, Galileo's *Dialogue on the Two Great World Systems* may be compared to Moore's *Principia Ethica.* Both are sharp and irrefutable analyses of their adversaries' position. Both un-

cover a fallacy, Moore the naturalistic fallacy, Galileo the metaphysical fallacy.

43. *Mysterium Cosmographicum* (1596), where the distances of the planets are derived from the relations between the five regular solids; *Astronomia nova* (1609), which contains the laws of elliptical orbits and of equal areas; and *De Harmonice Mundi* (1619), containing the famous "third law" connecting planetary periods and distances. Kepler, like Moore, never understood the importance of his discovery. To him, its importance "was that it furthered his chimerical quest—and nothing else." Arthur Koestler, *The Sleepwalkers* (New York, 1959), p. 395.

44. *Principia Ethica* (1903); "The Conception of Intrinsic Value" (1922), which states the "two different propositions [which] are both true of *goodness*," Moore's "first and second law," as we may say; "Reply to My Critics" (1942), which clearly separates natural and nonnatural intrinsic properties, states the logical entailment relation between the two – Moore's "third law" – and projects the logical solution of the value problem through a possible analysis of "description." There is a truly marvellous parallelism between Kepler's and Moore's discoveries, with the difference only that Kepler consummated his own discovery while we must consummate Moore's. Kepler saw himself that the opposition of librations and ellipse dissolved into their identity. Moore saw the opposition between the set of descriptive properties and the value property but did not *quite* dissolve it into identity; as we are doing in the value axiom. (On Kepler's discovery see the spirited account in N. R. Hanson, *Patterns of Discovery* [Cambridge, 1958], pp. 82–83.)

45. Except the stimulating essay by Edel. In the case of Kepler, it was Newton who was the only one to grasp the fundamental importance of the three laws.

46. From a letter by Professor Brand Blanshard to the author, of March, 1957: "You may be interested that I spent last Christmas with Moore. I asked him whether he still felt equally drawn to the objectivism of his *Ethics* and to the emotive theory that was tempting him in the early forties. He has gone back to his earlier position so completely that he couldn't even remember the statement he had put in print about being 'equally drawn' to the two." Quoted by permission of Professor Blanshard. Also Blanshard, *Reason and Goodness*, p. 269n and A. C. Ewing, "The Work of G. E. Moore," *The Indian Journal of Philosophy, I* (Dec., 1959), and *Mind*, LXXI (1962), 251. In the same conversation, Ewing reports, Moore "said he thought that true judgments of intrinsic value were all 'logically necessary.' (They would of course have to be synthetic a *priori* on his view.)" Cf. Robert S. Hartman, "The Definition of Good: Moore's Axiomatic of the Science of Ethics," *loc. cit.*

47. Frankena, "The Naturalistic Fallacy," *Mind*, XLVIII

(1939), 474: "The definists are in all honesty claiming to find but one characteristic [namely the one in terms of which they define 'good'] where the intuitionists claim to find two," namely both that characteristic and the property of being good.

48. This is true without qualification only for the ideal hierarchy of systemic, extrinsic and intrinsic value (with finite, denumerably infinite and nondenumerably infinite intensional cardinalities, respectively). Within extrinsic value, the degree of goodness of a thing within the totality of all things (the scholastic "perfection") must be distinguished from its goodness as a thing of its kind. In the latter case, the one-to-one correspondence of intensions only applies to the intensions had by the things of the class. See Robert S. Hartman, on "Good as a Non-Natural Quality and Good as a Transcendental," in *Review of Metaphysics,* XVI (Sept. 1962), 149–55. But see below Ch. 5, Sec. 2.

5. *The Elements of the Axiological System*

1. Edmund Husserl, *Formale und tranzendentale Logik* (Halle, 1929), p. 121.

2. *Modern Science and Human Values* (New York, 1956), pp. 105–6. Hall calls Galileo's formula for velocity "a miniature geometry" devised to account for motion.

3. G. E. Moore, *Principia Ethica,* pp. 7–8. On the relation between Moore's "complex" and the Kantian "containment," see Robert S. Hartman, "The Analytic, the Synthetic, and the Good: Kant and the Paradoxes of G. E. Moore," *Kant-Studien,* XLV (1953/54), 67–82, XLVI (1954/55), 1–18. It must be noted that containment is different from and related to entailment. A subject-predicate proposition *S-P entails* another such proposition *S-Q* if *P contains Q.*

4. For simplicity's sake, we shall in the following use the term "proposition," even if the expression is a (logical or axiological) propositional function. Cf. Felix S. Cohen, *Ethical Systems and Legal Ideals* (New York, 1933), pp. 170–71.

5. Considering not only "good" but all value terms, we may say that the difference between "*x* is a C factually" and "*x* is a C valuationally" is represented by the difference between "ϕx" and "$\phi ! x$", the latter being the matrix for the second-order functions $(\phi)\phi x$, $(\exists\phi)\phi x$, etc., which define one aspect of the value terms.

6. *Principia Mathematica,* Vol. I (Cambridge, 1935), pp. 58–59. That there is a relation between the theory of types and value theory has been mentioned by recent writers in ethics and aesthetics. Cf. Rosamond Kent Sprague, "Negation and Evil," *Philos-*

ophy and Phenomenological Research, XI (1951), 566 and Morris Weitz, *Philosophy of the Arts* (Cambridge, Mass., 1950), p. 142. There are writers in axiology who come close to the solution without seeing the logical connection (e.g. W. D. Ross, *The Right and the Good* [Oxford, 1930], pg. 121–22) and writers in logic who come close to it without seeing the axiological connection (e.g. Bertrand Russell, *Principia Mathematica,* p. 56; *Inquiry into Meaning and Truth* [New York, 1940], pp. 250–51, *Introduction to Mathematical Philosophy* [London, 1938], p. 189; *My Philosophical Development* [London, 1959], p. 115). Although the properties which Russell discusses in connection with the problems of logical orders, in particular the axiom of reducibility, are all value properties – Napoleon's greatness or viciousness, the typicalness of a Frenchman or an Englishman, the virtues or vices of Elizabeth I – he makes no use of this in his axiological writings. Actually, to do so would open a Pandora's box. The axiom of reducibility, the quantification of predicates, and other devices of present-day logic we are forced to employ have a precarious position in this logic. Their use is tentative; they will have to be redefined in a developed intensional logic in our sense, that is, a logic of the relation ω. The parallelism between this relation and the relation ε is qualified; between the terms of the relation ω, for example, there are all kinds of type differences.–On the identity and difference of concept-terms and value-terms see Daniel Christoff, "La valeur en général et les valeurs spécifiques," *Symposium sobre valor in genere y valores específicos,* XIII International Congress of Philosophy (Mexico, 1963), p. 39.

7. W. J. Rees, "Continuous States," *Proceedings, Aristotelian Society,* LVIII, 1957/58, 223–44.

8. *Op. cit.*, p. 241. Also cf. Ryle's "second order" inclinations, motives, etc. e.g. Gilbert Ryle, *The Concept of Mind* (London, 1949), p. 112.

9. See Maria Ossowska, "Qu'est ce qu'un jugement de valeur?" *Proc. X Int. Cong. Phil.* (Amsterdam, 1949), 443 f. In Spanish, the difference in question may be expressed by two different verbs, *ser* and *estar.* Ossowska's example of the redness of poppy and the lady's nose, respectively, is solved in formal axiology by the doctrine of composition and transposition, respectively.

10. *Time,* September 1, 1958, "Headline of the Week."

11. Headline in *The News,* Mexico City.

12. *Time,* June 22, 1962.

13. Disagreements based on confusions of logical orders are almost certainly value disagreements. One such confusion gave rise to an international incident during World War II. On November 24, 1944, British Field Marshall Lord Alanbrooke recorded in his diaries, published after the war, "a very unsatisfactory state of affairs in

France, with no one running the land battle." Then he continued: "Eisenhower, though supposed to be doing so, is on the golf links at Reims – entirely detached and taking practically no part in running the war." President Eisenhower's Press Secretary Jim Hagerty retorted angrily: "From the time the Allies landed in Europe until the victory was won, the President didn't have a golf club in his hands – much less play at a golf course." At this, Alanbrooke hastened to explain: he had not meant that Ike was actually playing golf just before the critical Battle of the Bulge; he was merely referring to the fact that Ike's headquarters at Reims was on a golf links (*Time* [Nov. 16, 1959]). – Such disagreements are subject to analysis by formal axiology. Whether the higher-order function is hidden or overt, it must probably always be regarded as valuation. To do so would, for example, solve the value questions connected with adverbs. In *He drove slowly, He ate with his hands,* "slowly" and "with his hands" are valuations, *qualifications* of "he drove" and "he ate." A value theory ought to account for this value character of adverbial expressions. The expansion of our theory in regarding all higher-order functions as valuations, would take care of this. Moreover, there seems to be no example of higher-order functions that is not charged with value content. Cf. the suggestive example of Franz Crahay and its explanation:

> "Y., a mediocre poet, to be sure, but, for a poet, a very skilfull politician, and, among those most skilfull, the least obnoxious."
>
> To *type* t_0 (individuals) belongs: "Y."
>
> To *type* t_1 (predicates of individuals or classes) belongs: "poet," "politician."
>
> To *type* t_2 (predicates of predicates or classes of classes) belong, classified in orders:
>
> *of order* O_0: "mediocre," which qualifies "poet" and does not explicitly refer to the totality of poet-individuals.
>
> *of order* O_1: "skilfull," which qualifies "politicians" and explicitly refers to the totality of politician-individuals.
>
> *of order* O_2: "The least obnoxious," which explicitly refers to the totality of "most skilfull politicians."

(Franz Crahay, *Le formalisme logico-mathématique et le problème du non-sens* [Paris, 1957], p. 48). Yet, we are not prepared to identify all nonpredicative functions, or even all "reduced" predicative functions, with valuation. To do so may constitute the next, "Newtonian," step in formal axiology.

14. If the genus of the definiens is said to be of the same type as the definiendum, then the differentia is of a higher type than the definiendum. If x is human, to be human means to be rationally animalic, and to be animalic means to possess a certain set of prop-

erties, then *x is human, x is animalic,* and *to be human is to be rational* are predicative functions; but *x is rational* is a "reduced" predicative function. Cf. García Bacca, *Introducción a la lógica moderna,* pp. 59–60; Henry Lanz, *In Quest of Morals* (Stanford, Cal., 1941), pp. 57–59.

15. For our present purpose we do not have to make a distinction between definitional and expositional analyticity, that is, between propositions concerning the properties contained in the *definition* of C – say α and β – and propositions concerning the properties contained in the *exposition* of C – say, γ, δ, ζ. Both kinds of properties are "contained in" C, and the propositions in question are therefore analytic. The value pattern, therefore, is a pattern of "analytic entailments." About the difference between expositional and definitional analyticity and its significance for the value pattern see Robert S. Hartman, "The Analytic, the Synthetic, and the Good: Kant and the Paradoxes of G. E. Moore," *Kant-Studien,* XLV (1953/54), 67–82, and XLVI (1954/55), 3–18. See also K. Marc-Wogau, "Kants Lehre vom analytischen Urteil," *Theoria,* XVII (1951), 140.

16. For a similar "square of oppositions" see William Kneale, "Objectivity in Morals," *Philosophy,* XXV (1950), 149–66, reprinted in W. Sellars and J. Hospers eds., *Readings in Ethical Theory* (New York, 1952), pp. 681–97. Kneale's terms are "obligatory" and "wrong" as contraries, and "right" and "non-obligatory" as subcontraries. For a discussion of the distinctions between *x is good,* on the one hand, and *x is no(t) good* and *x is bad,* on the other, see Francisco Romero-Eugenio Pucciarelli, *Lógica* (Mexico, 1958), pp. 31–32.

17. See Edmund Husserl, *Formale und transzendentale Logik* (Halle, 1929), especially Part II; Kant, *Critique of Pure Reason,* B114, where *bonum* is categorially interpreted as *qualitative completeness* (totality) of a concept. Cf. below note 30.

18. Cf. the English uses *x better be* C and *x better had be* C, which are strong forms of "ought."

19. However, if we let the proposition *It is better for x to fulfill than not to fulfill its intension* not only have the form *It is better that* x*R*y *than that* x*S*y but also the form *It is better that* x*R*y *than that* x*R*z, then a thing may fulfill something other than its conceptual intension. In this case, we would get different axiologies, whose pattern would not be logic but, say, semiotics, jurisprudence or the law, ritual, etc. According to what was said in Ch. 1, sec. 7, such a procedure would not be legitimate.

20. This form is equivalent to Ewing's definition of good as the object of a pro-attitude (*The Definition of Good,* pp. 148–49). This definition thus appears as a theorem of formal axiology. If for "choose" we put the more general term "to have a pro-attitude toward" then "*x* ought to choose what is good" is equivalent to "*x*

ought to have a pro-attitude toward what is good," which in turn is equivalent to " '*z* is good' means '*x* ought to have a pro-attitude toward it' "; from which follows "*z* is good" means "*z* ought to be an object of a pro-attitude." QED. A logic of preference may be based on formal axiology.

21. See Ch. 6, note 3, and Edwin T. Mitchell, *A System of Ethics* (New York, 1950), pp. 126–27. Also E. F. Carritt, *An Ambiguity of the Word 'Good'*, Brit. Acad., Vol. XXIII (London, 1937). Note that *x is good* is a logically synthetic but *x is a good* C a logically analytic proposition. See below note 32.

22. From German "GuT" and English "BaD." The "u" in "gut" reminds us of *u*niversal, the "a" in "bad" of p*a*rticular. The "T" reminds of nega*t*e. The "D" is the weaker phonetic form of "T."

23. Whether B-propositions are positive or negative depends on whether partial fulfillment of the intension ("fair") is meant positively or negatively. We take it positively, as indicating fulfillment.

24. Even in the analysis of the validity of *logical* propositions the analysis of the horizontal propositional structure alone is insufficient and "depth analyses" have to be made, as in the assumption of existence of the subject's referent in particular propositions and its nonexistence in universal propositions, or the assumption of existence of the predicate's contradictory in the partial inverse, and the like. Here also belong the so-called "laws of thought," Husserl's "formal ontology" of logic, and the like.

25. Kant, Logik, ¶ 30; *Critique of Pure Reason*, B 100–101.

26. Such propositions, although "normative" materially – though not formally – are factual and not valuational. The dichotomy "declarative"-"normative" is not identical with the dichotomy "factual"-"valuational." The identification is due to what we call the normative fallacy. Its philosophical status is due to Kant's mistaken assignment of the moral to the nontheoretical, the noumenally metaphysical. This gave "ought" a metaphysical status which it does not have and is pure hypostatization. "Ought" is the axiological copula, applicable, within the framework of formal axiology, to mixed axiological (materially factual) and pure axiological (materially valuational) states of affairs. These states of affairs, as applications of formal axiology, are what we call "axiological values" (as against formal and phenomenal values).

27. Where also belong performatories, ceremonials, and similar statements. Thus, "I promise" as p_1 means I do promise, i.e. I do mean to keep my promise; as p_2 that I am not sure I shall, and as p_3 that I am sure I shall not. "I am lying" as p_1 means that I *am* lying (hence what I say is the truth); as p_2 it means that I may or may not be lying (hence what I say may or may not be true); as p_3 that I am not lying (hence what I say is a lie). If we designate the

proposition "I am lying" by "p" and the underlying judgments by "p_1" "p_2" "p_3", respectively, then p does not contradict p_1 nor p_2 but p_3. In all cases of this kind of statements, the question is one of fulfillment or nonfulfillment of certain concepts, hence, according to our definition of value, a question of value. These statements, then, are valuations, that is, *subject matters of* value theory, and cannot be used as prototypes of value *theories*. To do so is to commit the fallacy of method.

28. See Robert S. Hartman, "The Analytic and the Synthetic as Categories of Inquiry," *Perspectives in Philosophy* (Columbus, Ohio, Ohio State University), pp. 55–78.

29. For the pure axiological forms of these relations see Robert S. Hartman, "A Logical Definition of Value," *The Journal of Philosophy*, XLVIII (1951), 413–20.

30. Eventually a different symbol may have to be used here; the more so as what we are dealing with here is the logical form of intrinsic rather than extrinsic value. See *ibid.* Cf. Kant, *Critique of Judgment*, ¶ 4: "In order to find anything good I must always know what sort of a thing the object ought to be, i.e., I must have a concept of it." This may be called Kant's statement of axiological – rather than moral – "good" and "ought."

31. *Logic* (Cambridge, England, 1921), I, p. 199 and (1924), III, Ch. VII. This points to the determinable rather than the class character of the "good" here in question.

32. The analyticity is, of course, axiological and not logical. Logically, such propositions are, as Moore held, synthetic. The predicate "good" can never be contained in the concept of the subject. The continuity of the subject throughout the variations of its expositional properties is a problem that not only troubled the ancients and Kant, but also the Korzybskians. See above, note 21.

33. For details see "A Logical Definition of Value," *loc. cit.* and "The Logic of Value," *Review of Metaphysics*, XIV (March, 1961), 408–23.

34. A question arises in the case of propositions with qualified subjects, such as *All my children are good*, which, according to this rule, must be axiologically false. That this actually is the case is easily seen. "Good" refers back to the subject, which is "my children." Thus, as the proposition stands, it means "All my children are good (my children)" – my children fulfill the exposition of "my children." This exposition states merely the fact of my parenthood or the "mineness" of my children. "Good my children" then merely means that my children *are* really my children – and mineness allows of no degrees, hence the proposition is systemic. In other words, the qualification of the subject excludes the subject's exposition, and hence the intended meaning of "good," which of course is intended to refer to "children" and not to their mineness. The proposition, then, is

elliptic for "Some children are good, among them my own" or "My own children are the kind of children that are good" or "My own children are good children." These are compound axiological propositions, and they may be true. Cf. above note 21.

35. To give a hint at applications of this rule, it applies to the class of angels and the class of God, as well as the class of the world, insofar as the latter necessarily proceeds from God. Therefore, Spinoza's world is a systemic class. From this follows, in accordance with our rules, that value predication can only be applied by a mind which is in error about the nature of the world (Spinoza, *Ethics*, Part IV, Preface). A mind which knows the world as what it is, is beyond using the value predicates, "good," "bad," and the like, in their usual meaning. For the world as logical system is exempt from this kind of valuation (Spinoza, Letter to Blyenbergh, January 28, 1665). A mind that understands the world in this way becomes itself systemic. Cf. Kant's "perfectly good" divine will which is beyond the ought. *Foundations of the Metaphysics of Morals*, Akademie edition, p. 414.

36. This kind of constitutive "ought" most closely resembles the "ought" of hypothesis. Some historical such propositions are: *The earth ought to revolve around the sun, There ought to be a sealane to India, There ought to be the ruins of Troy, There ought to be a law to Balmer's numbers, The atom ought to be split(able).* This constitutive "ought" refers to the spatially or the timelessly (cf. the law of Balmer's numbers) existing but to the human mind as yet unknown, except by intuition (cf. Gauss's remark: "The result I have, if only I knew how to get there"). It does not refer to the potential which has yet to be created, such as *The Thirteen Colonies ought to be free, Europe ought to be united.* This is a different kind of constitutive "ought." For the epistemological idealist there is not much difference between the two kinds.

37. But *You* (being the janitor) *ought to close the door* is an analytic mixed axiological proposition and hence a-true(A→C). The difference appears well when the equivalent is used: *It is better for you to close the door.* In the case of the janitor this means *You better . . . (or else)*; in the case of anybody else it means, it is better for your health, and the like. There is no reason why imperatives should not be interpreted as "ought" -propositions and thus subsumed under our rules.

6. *The Systematic Import of Formal Axiology*

1. G. E. Moore, *Principia Ethica* (London, 1903), p. 206. Edmund Husserl, *Erfahrung und Urteil* (Hamburg, 1948), p. 423.

2. For a discussion of the structure of the intensional norm for *intrinsic* value see Robert S. Hartman, "The Logic of Value," *Review of Metaphysics,* XIV (March, 1961), 408–23.

3. This also holds for terms of the axiological system. Hence "Good is good" makes as much or as little axiological sense as "Circles are good". What is *defined as* good *is* not good, whether it be defined by nominal or by real definition. In the first case, what is defined as "good" is a synthetic concept, Ψ, such that whatever is said to be Ψ is good and whatever is said to be good is Ψ; and what is said to be good may be systemically, extrinsically, or intrinsically good. In the second case, what is defined as "good" is the definitional part of an analytic intension and what *is* good is extrinsically good only. Both kinds of "good" are, as definitions, systemic goods or perfections, the former as a construct, the latter as a schema in the sense to be defined. The former is an axiom, the latter a category, the former is *applicable to* all three kinds of value, the latter *refers to* extrinsic value. As systemic values, neither of the two kinds of "good" can be bad; but what they define, except systemic value, may be both good or bad.

4. Aristotle, *Metaphysics,* Bk. Δ, Ch. 27.

5. See Robert S. Hartman, "A Logical Definition of Value," *Journal of Philosophy,* LXVIII (June 21, 1951), 413–20; and "The Logic of Value."

6. The "fulfillment" of such an "intension" is a transposition of values. See "The Logic of Value" and below Ch. 7, note 38.

7. This does not mean, as Kant held, that St. Anselm was wrong. On the contrary, as we have seen above, Ch. 3, Sec. 3,c, he was right on axiological grounds.

8. That there is a difference between "horse" and "Equidae," "dog" and "Canidae," "lilac" and "Syringae" can be seen by simple experiments. Substitute, for example, "Syringa vulgaris" for "lilac" in the famous lines about going down to Kew. The extrinsic and the corresponding systemic terms, thus, are not synonymous; for their intensional structures are not similar. This also goes for the famous example of "brother" and "male sibling." I am my brother's keeper, but not my male sibling's. (Cf. for some aspects of this subject Gilbert Ryle, *Dilemmas* [Cambridge, 1954], Chs. V, VI).

9. J. D. García Bacca, *Introducción a la lógica moderna,* pars. 5, 9, 47, 48.

10. Aristotle's *hyle noete, Metaphysics* Z, Ch. 10 (1036 a 9).

11. This does not mean that, in particular cases, certain sets of properties within a thing may not be *weighted* in a particular manner.

12. This word is here not being used in its mathematical sense.

13. Note that by "intension" is here meant "exposition."

14. See Robert S. Hartman, "Group Membership and Class Membership," *Philosophy and Phenomenological Research*, XIII (March, 1953), 353–70.

15. "The Conception of Intrinsic Value," *Philosophical Studies* (Cambridge, 1922), p. 274.

16. "Reply to My Critics," *The Philosophy of G. E. Moore*, Paul A. Schilpp, ed. (Evanston, 1942), pp. 588–92.

17. Strictly speaking, since we defined "bad" as the subaltern of "no good" we should say that $\frac{n}{2} - m$ is the range of "no good" except the values > 1, which are the range of "bad."

18. Two comments are in order. First, Moore did think in terms of arithmetical operations with (intrinsic) values (see on this point Austin Duncan-Jones, *Philosophy*, XXXIII [July, 1958]), 245–46). Second, the problem is more complex than Moore thought. It is, for example, not at all certain what is the unit or "one" property. Since analytic properties contain, and are contained in, one another, a property containing another is different in its "oneness" from a property contained in another. The exact determination of "one" depends on the complete logical solution of the intensional structure, especially the distinction of denumerable and nondenumerable (discursive and nondiscursive) such structures. Cf. Moore, "Reply to My Critics," *op. cit.*, pp. 586–87

19. This gives a theoretical basis to Hilliard's empirical formulation of the value content of a glass of burgundy which is 3.6×10^{46} (based on 158 properties). A. L. Hilliard, *The Forms of Value* (New York, 1950), pp. 278–79.

20. Closest to the solution here presented is Edmund Husserl in *Die Krisis der europäischen Wissenschaften und die transzendentale Phänomenologie* (The Hague, 1954). Unfortunately, Husserl never connected what he said in this book with what he said on formal axiology; nor did Theodor Lessing, who elaborated Husserl's suggestions on the subject. See Edmund Husserl, *Formale und transzendentale Logik* (Halle, 1929), pp. 121–22; *Ideas* (London, 1931), pp. 330, 405–7; Theodor Lessing, *Studien zur Wertaxiomatik* (Leipzig, 1914). Also cf. Max Scheler, *Der Formalismus in der Ethik und die materiale Wertethik* (Bern, 1954), p. 102.

21. Cf. H. Bergson, *Creative Evolution* (New York, 1937), pp. 11–12.

22. *Logik*, ¶ 105. The Definition was, for Kant, a *measure*, both *abgemessen* (*praecisio*) and (*angemessen* (*adaequatio*). *Logik*, Introduction, Sec. VIII.

23. Actually, what was cut out must be at least two things whose common properties are being determined. For what has been

described is the process of abstraction, but from a purely intensional point of view: a set of properties p is being differentiated into a set of n properties Φ, where $\frac{2^n - 1}{n} = \frac{p}{\log_2(p + 1)}$. Only when the n properties of Φ have thus been determined follows the extensional application: all things which have properties ϕ are numbers of the class in question. Or: there are as many things of the class in question as there are extensions of set Φ. It may well be that abstraction really is not a drawing off of common properties from different things but, on the contrary, an application of a fixed set of properties Φ to certain things which seem to fit it. This would explain not only analogy but also induction. Induction, then, would not be the transference of properties abstracted from a set of known things to an infinite set of unknown things but the same as "abstraction": the application of a set Φ intensionally determined by the process described, to any thing that fits it. Since the set Φ is at the same time the value standard of the things which Φ fits, induction and abstraction, thus defined, would constitute the process of producing value standards. In this case, the analytic concept would be a synthetic or formal concept: that construct of properties whose application to things results a] in classes, b] in extrinsic valuation. The synthetic process by which analytic intensions would be constructed would, precisely, be the process defined by $\frac{2^n - 1}{n} = \frac{p}{\log_2(p + 1)}$. This formula is almost identical with the prime number theorem, $p = \frac{n}{\log_e n}$, and would signify that the proportion of relevant items to a total number of items is equal to the proportion of a number of primes to the total number containing them. This may open up a new deductive approach to induction. (On the general relationship of induction to valuation see J. Bronowski, *Science and Human Values* [New York, 1958]. Also Albert Einstein on the "combinatory play" in creative thinking, in J. Hadamard, *The Philosophy of Invention in the Mathematical Field* [Princeton, 1945], pp. 142–43.)

24. Bertrand Russell, *Introduction to Mathematical Philosophy* (London, 1938), pp. 133–37; Whitehead-Russell, *Principia Mathematica*, Vol. II (Cambridge, 1957), pp. 183–84.

25. It is conceivable that other exponentiations of 2 may give rise to entirely different value hierarchies, e.g. exponentiation with π or with i.

26. The logarithmic power of systems as compared to everyday discourse would have to be examined in detail. Such examination would constitute both the structure of systemic value in formal axiology and an axiological theory of systems.

27. In the Kantian sense. In the axiological sense, description

or depiction is intrinsic intension. The Kantian description is a higher value type of extrinsic definition, and the intrinsic description, or depiction, a higher value type of Kantian or extrinsic description.

28. According to what was said in Ch. 2, sec. 2, the same would be the relation between philosophy and science. Philosophy spells out in longhand what in science results from shorthand (in the sense of scientific symbolism, not in García Bacca's sense). Yet—and this was made clear by Husserl as against the presumption of Galileo and his scientific followers who saw in primary properties what they called reality—this longhand view of the world is of greater value than the shorthand view. As we see now, it is to the shorthand view as value is to fact. On the other hand, as was said in Ch. 2, note 52, there is an infinite power of applicability in a system. Thus, although the system has the "number of properties" $P = \log_2 n$, this same system, as the intension of an axiom, has the number of properties $\aleph_1$. The reason is that most values for $\log_2 n$ are transcendental numbers and belong to cardinality $\aleph_1$. In this sense, a system is a Gestalt; and as such it is *more* valuable than an analytic intension ($\aleph_0$). And in this sense the term is to the concept a limit, an intensional Dedekind cut. See above Ch. 2, Sec. 2,b.

29. On this matrimonial Gestalt see Nicolai Hartmann, *Ethics*, Vol. II (London, 1932), Chaps. XXXIII, XXXVII.

30. On the mystic experience see for example, Aldous Huxley, *The Perennial Philosophy* (New York, 1945); Evelyn Underhill, *Mysticism* (New York, 1911), especially Ch. VIII; William James, *Varieties of Religious Experience*. On other experiences of expanded awareness see Aldous Huxley, *The Doors of Perception* (London, 1954); *Heaven and Hell* (London, 1956); Robert S. de Ropp, *Drugs and the Mind* (New York, 1960); Constance A. Newland, *My Self and I* (New York, 1962). Transfinite experiences used to be possible only by life-long dedication. Today, LSD makes possible transfinite "trips" at cut rate. These could match in intensity and fulfillment the experiences of the mystic elite of the past if part of a dedicated life.

31. The significance of intensional sets may well be found, eventually, in their characteristic ordinal rather than cardinal number.

32. Gaylord M. Merriman, *To Discover Mathematics* (New York, 1942), p. 354, quoting from Edgar Allan Poe, "The Mystery of Marie Roget."

33. Particularly striking illustrations of this are the Parables of Jesus. The parables are situations in infinity, expressed metaphorically in finite language. Their moral is: "For whosoever hath, to him shall be given, and he shall have more abundance; and whosoever hath not, from him shall be taken away even that he hath." (Matt. 13:12). He who has infinite spirit can only add to it; he who

has not has only the finite which he will lose in his death. For a particularly clear illustration of the principle of transfinite cardinality, that the part equals the whole, see the parable of the Laborers in the Vineyard (Cf. Maurice Nicoll, *The New Man: An Interpretation of Some Parables and Miracles of Christ* [London, 1955], pp. 52–56.)

34. "Latin *norma* was a carpenter's square. Hence pattern, rule . . . With the sense of rule or standard already acquired in Latin, it came into English as norm" (Joseph T. Shipley, *Dictionary of Word Origins* [Ames, Iowa, 1955]). The Greek root is *gnomon*, which originally meant *the means of knowing, the one who knows, the expert*. Thus, the meaning of "norm" is the opposite of "normative" in the contemporary axiological sense; it is "cognitive" – as indeed all classic philosophers held, including Aristotle.

35. H. Butterfield, *The Origins of Modern Science* (London, 1950), Ch. I; Galileo, *Dialogue on the Great World Systems*, First Day. See especially the edition by Giorgio di Santillana (Chicago, 1953), p. 17; Everett W. Hall, *Modern Science and Human Values* (New York, 1956), pp. 99–105.

36. This equivalence appears in the logical – rather than arithmetical – solution of the problem as the reducibility of the value property. In the logical solution, the value property appears as a second-order descriptive property, in the arithmetical solution the descriptive property appears as a primary value property. In the first case, the value property appears as a quantification of the descriptive properties, in the second case, the descriptive properties appear as a specification and differentiation of the value property. Both solutions, thus, are complementary, one the inverse of the other.

37. Henry James Forman, *The Story of Prophecy* (New York, 1939), p. 254.

38. Cf. Abraham Edel, "The Logical Structure of G. E. Moore's Ethical Theory," *The Philosophy of G. E. Moore*, Paul A. Schilpp, ed. (Evanston, 1942), p. 168. Also Kant, *Critique of Practical Reason*, last paragraph and "Critical Examination of the Analytic of Pure Practical Reason," ¶ 6 (Akademie edition V, pp. 93, 163).

39. On the "fluidification" of things when looked at valuationally see Ortega y Gasset, *Ensayo de estética a manera de prólogo*, Sec. V, "La metáfora," *Obras Completas*, Vol. VI (Madrid, 1955), p. 259. On the "flight from reality" in art – the aesthetic counterpart to Moore's "nonnatural quality"–see Ortega y Gasset, *The Dehumanization of Art* (Princeton, 1948). The classic of this view of art is Friedrich Schiller's *Über die aesthetische Erziehung des Menschen* (1794). Schiller's "Schein" is Ortega's "irreality," Susanne Langer's "semblance" (Susanne K. Langer, *Feeling and Form* [New York, 1953], Ch. IV), and our intrinsic Gestalt.

40. The responsibility of a supervisor of several workers "is

not different from those below him, but is rather the sum of them all." E. F. L. Brech, *Organisation: The Framework of Management* (London, 1958), p. 290.

41. The prohibition above p. 233 may mean the *M*'s here must be different.

42. Perhaps here a modulus symbolization could be introduced: "$G \equiv B$ (ought D)." This would bring out the etymological root of *ought*, as that which is owed, the *Soll*, *Debe* or *Tang:* that which must be delivered to complete the given and fulfill the norm. In formal axiology the norm is the name; the ought thus is what a thing owes its name–a universal *noblesse oblige.*

43. For the corresponding case in arithmetic see N. R. Campbell, *Physics: The Elements* (Cambridge, 1920), p. 306. Arithmetical commutativity hides logical noncommutativity. Logically, $2 \times 3 \neq 3 \times 2$, even though both are 6, for "2×3" means two sets of three elements and "3×2" means three sets of two elements. In the first case, the set of 6 consists of 2 sets ${}_6C_3$, whereas in the second case it consists of 3 sets ${}_6C_2$.

44. For the *mathematical* meaning of this see N. R. Campbell, *loc. cit.*

45. This does not seem to apply, however, to the forms which result in pure numbers > 1, such as Xb, unless these numbers, e.g. "2," are given intensional meanings, similar to the value meaning given "1," above. "2" may mean a set of two properties, such as genus and differentia. In this case, $M = 1$, and $2M = 2(2 - (2 - M) = G = 2(2 - (2 - 1) = 2$. In this case it makes sense to say that "M ought to be 2" namely, to fulfill the two properties which constitute G. For any value of $M > 1$, one would have to say "*M* ought to be double"; and "2" would have to mean "double."

46. See Edwin T. Mitchell, *A System of Ethics* (New York, 1950), pp. 126 ff. Toward each other, these values have the same relationship as fact and value have within the finite series, n, 2^n, 2^{2^n}, etc.

47. Cf. The universal distribution of both subject and predicate in E-propositions. I have to know more of a predicate when it denies than when it affirms a subject. Similarly, there is more value in the disapproval of good than in good (although not than in the approval of good).

48. The formula $n - x = o$ ("*x* ought to be *n* through *o*") is equivalent to $-n + x = -o$. Only *x* is positively there, *n* and *o* are both lacking: *x* ought to be *n*, which is not yet, through *o*, which is not yet either.

7. *The Empirical Import of Formal Axiology*

1. Kant, *Critique of Pure Reason*, B 19; Aristotle, *Fragment from Statesman* (Werner Jaeger, *Aristotle* [Oxford, 1948], pp. 87, 261; Rudolf Stark, *Aristotelesstudien* [Munich, 1954], p. 27); Edwin R. Mitchell, A *System of Ethics* (New York, 1950), p. 129.

2. "The Logic of Values," *Review of Metaphysics*, XIV (March, 1961), 408–23. Since the axiom is an intrinsic value, it is itself a kind of metaphor. See William Kent, "Scientific Naming," *Philosophy of Science*, XXV (July, 1958), 185–93, and J. Bronowski, "The Creative Process," *Scientific American*, CXCIX (Sept., 1958), 58–65. For a related subject see Edward Rosen, *The Naming of the Telescope* (New York, 1947).

3. Risieri Frondizi, in *What is Value?* (La Salle, Ill., 1963), p. 26 says: "There is no yardstick to measure the yardstick" in value philosophy. But there is in value science. Formal axiology is the measure that measures the value measures used in life situations. See the author's discussion of Frondizi's book (it's Spanish original, Mexico-Buenos Aires, 1958) in *Philosophy and Phenomenological Research*, XXII (Dec., 1961), 223–32.

4. See Robert S. Hartman, "The Logic of Value," *op. cit.*, 418–19, and Susanne K. Langer, "Abstraction in Science and Abstraction in Art," in *Structure, Method and Meaning, Essays in Honor of Henry M. Sheffer* (New York, 1951), pp. 171–82. What Langer calls "abstraction in art," namely Gestalt differentiation, is precisely intrinsic value measurement.

5. Richard L. Nettleship, "Plato's Conception of Goodness and the Good," *Philosophical Remains* (London, 1901), p. 312. Cf. Ernst Kapp, *Greek Foundations of Traditional Logic* (New York, 1942), p. 35.

6. E.g. Dante's use of the intensions of the concepts "God" and "man" as common standards for the value measurements of Pope and emperor. *De monarchia*, III, 12, 62. Cf. Ernst H. Kantorowicz, *The King's Two Bodies* (Princeton, 1957), pp. 459–60.

7. Crane Brinton, *Ideas and Men* (New York, 1950), p. 153.

8. Cf. Nowell-Smith's suggestive distinction between "I ought" and "you ought," in *Ethics* (London, 1954), pp. 193–97, 261–62, which is an analytic account of a distinction we make synthetically. The distinction, for us, rests not merely on the forms "I" and "you" but on the entire axiologically defined content of the proposition.

9. Cf. George Bernard Shaw's Devil in *Man and Superman* on "Man, the inventor of the rack, the stake, the gallows, the electric chair, of sword and gun and poison gas: above all of justice, duty,

patriotism, and all the other isms by which even those who are clever enough to be humanely disposed are persuaded to become the most destructive of all destroyers." Also the speech of Senator J. William Fulbright on *Pacem in Terris* at the New York Conference of the Center for the Study of Democratic Institutions, February 20, 1965. See above Ch. 1, note 33.

10. Cf. Jacques Maritain, *Man and the State* (London, 1954); William H. Whyte, Jr., *The Organization Man* (New York, 1956); Chris Argyris, *Personality and Organization* (New York, 1957); Erich Kahler, *The Tower and the Abyss* (New York, 1957); Paul Goodman, *Person and Personnel* (New York, 1965).

11. *Foundations of the Metaphysics of Morals*, Sec. II.

12. *Social Contract*, Bk. I, Ch. VII.

13. *Op cit.*, Bk. III, Ch. XVI; Aristotle, *Metaphysics*, Bk. Δ, Ch. 27.

14. *Op. cit.*, Bk. II, Ch. III.

15. *Op. cit.*, Bk. II, Ch. XII.

16. *Op. cit.*, Bk. II, Ch. IV.

17. *Op. cit.*, Bk. III, Ch. XV. Cf. Charles de Gaulle's characterization of politicians as "the system," "the trade union of placeholders."

18. *Op. cit.*, Bk. III, Ch. IV.

19. Jacques Maritain, *Man and the State*, pp. 15–16.

20. *Op. cit.*, p. 47.

21. Classes of individuals Kierkegaard called by the technical term "the crowd" [*The Point of View* (London, 1939), p. 114]. The low valuation of the system, both analytic and formal, does not extend to the axiom which is a noetic individual, a *hyle noete*, nor to the formal system in its axiomatic aspect.

22. Paul Weiss, "God and the World," in *Science, Philosophy and Religion: A Symposium* (New York, 1941), pp. 422–23. Also see Paul Weiss, *The World of Art* (Carbondale, Ill., Southern Illinois University Press, 1961), pp. 68–73, and *Review of Metaphysics*, XIII (March, 1960), 370–72.

23. Cf. Rilke's last letter, written on the eve of his death: "Madame: Miserably, horribly ill, and painfully so, to a degree which I had never dared imagine. It is this nameless suffering which is named by doctors, but which is, itself, satisfied, if it but teaches us two or three cries, in which our voice is not to be recognized. That voice, which had been educated to nuances!" The terror in this outcry is so stark, so inconceivable to others, that even to cite it seems sacrilege.

24. Cf. Wittgenstein, *Philosophical Investigations* (Oxford, 1953), pp. 89–111. It is quite difficult to know what really Wittgenstein meant by a private language owing to the fragmentary nature of

his own account and the differences, and often contradictions in the accounts of others (see for a good presentation N. Malcolm, "Wittgenstein's Philosophical Investigations," *Philosophical Review*, LXIII [1954], 530–39. Also L. Linsky, "Wittgenstein on Language and Some Problems of Philosophy," *Journal of Philosophy*, LIV [May 9, 1957], 285–92, and David Pole, *The Later Philosophy of Wittgenstein* [London, 1958], Ch. III). In formal axiology, the language of intrinsic value *is* a private language, which does not mean that it is *no* language, but that it is a language in a definite logical sense: that of using singular concepts whose intensional structure is infinite in a precisely defined way. For a clear distinction of systemic and intrinsic language see G. E. Hutchinson, *Scientific Language and Religious Faith* (New York, The National Council, n.d.). Formal axiology supplies, in its theory of intensional structure, the logical patterns of private languages. See "The Logic of Value," *loc. cit.*

25. Music is the articulation of feeling, and so, especially in its intrinsic dimension, is formal axiology. It may well be that the most appropriate measure of the intrinsic intensional Gestalten may be found in musical harmony. Cf. Susanne K. Langer, *Philosophy in a New Key* (Cambridge, Mass., 1942), pp. 101, 153; *Feeling and Form* (New York, 1953), Ch. VIII. On the relation of music and metaphor see *Philosophy in a New Key*, pp. 101, 139–41, 240, 281. There is possible a musical notation of the value calculus.

26. Ortega y Gasset, *The Dehumanization of Art and Notes on the Novel* (Princeton, 1948), pp. 14–19.

27. $\aleph_1^{\aleph_1} = \aleph_2$; $\aleph_1^n = \aleph_1$; $\aleph_0^n = \aleph_0$; n^n is of the same value dimension as n. The painter's attitude, as here described by Ortega, is not typical. It would be involvement if he were actually painting the scene.

28. *The Nature of the Physical World* (Cambridge, 1933), pp. 316–21.

29. *The Quest for Certainty* (New York, 1929), p. 240.

30. For details see Robert S. Hartman, "The Logic of Value," 408–18.

31. Francis Hutcheson, *An Inquiry Concerning the Original of our Ideas of Virtue or Moral Good* (London, 1726), Sec. III, xi.

32. Above Ch. 2, note 20. For details of the illegitimate mixing of analytic and synthetic procedures see Robert S. Hartman, *El conocimiento del Bien: Crítica de la razón axiológica* (Mexico-Buenos Aires, 1965), Ch. 5. When no such confusion exists, e.g. arithmetical operations do not intervene in the combinations of letters standing for philosophical concepts, such combinations are legitimate. Thus, Paul Weiss combines the modes of being in an increasingly complex and logical pattern, from the symbol of one mode, e.g. A (Actuality) applied to any of the other modes *G*, *E*, *I* (God, Existence, Ideality),

to secondary forms *AG, AE, AI,* to tertiary forms, the application of *AG, AE, AI,* to A, *E, I,* in *AGE, AGI, AEG, AEI, AIG, AIE* etc., to quatenary forms, *AGAE, AGAI, AEAI* etc. "AGAE for example is the explication of individualized meaning imposed on Existence as individualized." (Paul Weiss, *Philosophy in Process,* Vol. I [Carbondale, Ill., Southern Illinois University Press, 1965], p. 22, also see pp. 27–28, on the construction of various disciplines). Our procedure in the following is the same, only with the addition of the arithmetical basis of each concept symbolized, and hence the use of arithmetical operators. Both procedures realize a method suggested by Leibniz, of intensional symbols and their combinations (*ars combinatoria*).

33. Even in the literal sense that the base of the exponentiation is lowered.

34. Inversely, any bad thing can turn into the constituent of some good, that is, a composition. Thus, evil can be overcome by good. The constituents of a transposition, having no overlapping intensions, are "bad for" one another, those of a composition are "good for" one another. The technique of overcoming evil by good thus consists in finding common denominators of meaning.

35. It is not, however, an organic whole. This notion we reserve for intrinsic value compositions, such as I^I, E^I, etc.

36. Elizabeth Sewell, *The Field of Nonsense* (London, 1952); A. Liede, *Dichtung als Spiel: Studien zur Unsinnpoesie an den Grenzen der Sprache* (Berlin, 1963).

37. E. Husserl, *Logische Untersuchungen* (Halle, 1921–1928); *Formale und transzendentale Logik* (Halle, 1929); J. D. García Bacca, *Introducción a la lógica moderna* (Barcelona, 1936), pp. 47–49; Alejandro Rossi, "Sentido y Sinsentido en las *Investigaciones Lógicas,*" *Dianoia: Anuario de Filosofía* (Mexico, 1960), 91–114.

38. This contradiction was "fulfilled" by the car collision mentioned above

39. Cf. *anathema,* dedicated to evil, *anatheteon* to be cancelled, withdrawn (Plato, *Laws,* 935 e). The Kantian logic of judgment applied to the singular proposition leads to the notion of *synlytic judgment.* In this proposition the subject has a content so infinite as not only to "contain" but as a Gestalt, to embrace and enclose the predicate, forming one organic whole with it. The proposition thus becomes closed, in the Gestalt sense of "closure," and itself a Gestalt. Such a proposition is more than analytic; its analytic nature is "closed," fixed into a tighter order, a *synlytic judgment,* a judgment, that is, where the analytic itself is condensed (*syn-* = *con-*. The *closer density* is the mark of $\aleph_1$ as against $\aleph_0$).

40. The inverse of the transfinite has clear axiological meaning ($1/E$, $1/I$), although the arithmetical meaning is undefined.

41. Since "n" merely stands for "finite" it makes no difference here whether we write "n" or "n^2". We use the value n rather than $\log_2 n$ for S, because in everyday valuation the schematic rather than the axiomatic aspect of systemic value is used.

42. See, for example, Matt. 5:27–48; 12:35–37; 25:28–29; Luke 19:24–25; also I Cor. 11:27, where taking the communion meal in a spirit of eating is equated with killing Jesus.

43. An electronic computer is at present being programmed to turn out the value forms of the first eleven levels, in all 1,306,125,-378 different value forms. This will constitute a *Table of Axiological Forms.* The Table of Secondary Value Combinations serves as basis for an axiological test, the *Hartman Value Inventory* (Miller Associates, Boston, 1966), with 10^{15} theoretically possible value profiles.

44. The quotes are from Thomas Mann, *The Magic Mountain* (New York, 1939), pp. 272, 279 f., 327–30, 353, 363, 433, 552.

45. For a different, and classical, treatment of the same theme, the relationship of intrinsic and systemic value, see St. Paul's letter to the Galatians. Also the books mentioned in note 10.

46. Which makes the novelist himself more conscious of his own meaning. From a letter by Thomas Mann to the author, concerning a manuscript on the same subject: "On the passages concerning me I have naturally dwelt with particular curiosity; they . . . have made me thoughtfully conscious of my voice as a counterpoint in the vocal movement of the age." See *Thomas Manns Briefe, 1937–1947* (Frankfurt, 1963), p. 304.

47. John Burnet, *Plato's Euthyphro* (Oxford, 1954), pp. 4–8.

48. These norms abound in the law, where the formal rules state what the officers of the law are to do in given cases. See Eduardo García Máynez, *Introducción a la lógica jurídica* (Mexico, 1951); *Lógica del juicio jurídico* (Mexico 1955); *Lógica del concepto jurídico* (Mexico 1959); *Lógica del raciocinio jurídico* (Mexico, 1964). But these kinds of norms appear also in the natural sciences, as in the geometrical directions for constructions. On these kinds of norms is based Kant's a priori syntheticity of judgments. See in particular preface to the Second Edition of the *Critique of Pure Reason.*

49. For details see Robert S. Hartman, "Is a Science of Ethics Possible?" *Philosophy of Science,* XVII (July, 1950); "The Moral Situation: A Field Theory of Ethics," *Journal of Philosophy,* XLV (May 20, 1948). The axiological basis of teleology is value exponentiation. On the "translation" of axiology into teleology see Robert S. Hartman, "Die Wissenschaft vom Entscheiden" (The Science of Decision Making), *Wissenschaft und Weltbild* (Vienna, 1966).

50. *Prolegomena to Any Future Metaphysics,* Introduction. Such complaints are justified against philosophical critics of the kind

described, S_I, but not against those of the kind S^I nor against those of the kinds S^E, S_E, S^S, or S_S.

51. Bertrand Russell, "L'importance philosophique de la logistique," *Revue de la métaphysique et de morale* (1911), 281.

Summary and Outlook

1. For details see Robert S. Hartman, *El conocimiento del Bien: Crítica de la razón axiológica* (Mexico-Buenos Aires, 1965), Ch. 6.

2. According to the axiom of formal axiology, a thing is the more valuable the richer it is in intensional properties. Since economic value is extrinsic, economic subjects must have analytic concepts, which means that the law of inverse proportion between intension and extension applies. Hence, the more valuable, economically, a thing is, the rarer it is. Thus, the axiom of scarcity, and the law of supply and demand, follow from the axiological definition of economic value.

3. George Kingsley Zipf, *Human Behavior and the Principle of Least Effort* (Cambridge, Mass., 1949).

INDEX

Absoluteness of value, 107, 109, 296
Abstraction: 83, 113, 198; application of fixed set of properties, 351; arithmetical determination, 350–51; in art, 355; characteristic number of, 113, 118; extensive, 89–90; from intensional point of view, 351; intensive, 90, 92; levels of, 201, 202; process of, 118, 202, 351; a value combination, 272. *See also* Construction
Abundance value, 245
Action, 77. *See also* Method; Practice
Addition of values, 228–29, 231–33, 242
Ad hoc linguistic constructions, 145
Ad hoc logics, 11
Adverbs, 344
Advertisers, 113
Aesthetic "good," 180
Aesthetics, 48, 108, 114–15, 180, 227, 309, 311, 324
A-falsity, 186, 187, 192
"A good," 120, 156, 157, 160, 171, 181, 218, 295, 296, 342, 346
Agreement, 108, 110
"Aha," 9, 325
A-indeterminacy, 186, 187, 192
Alanbrooke, Lord, 297, 343
Alchemy, 15, 47, 124, 126, 228, 296, 319, 327
Aleph. *See* Cardinality
Alexandria Quartet, 336
Alice, 35
Alienation, 273
"All," 154, 169, 347. *See also* Quantification
Allergy, 278
"All right," 211
Ambrose, 326
Ames, Adelbert, Jr., 329
Analogy between natural science and moral science: 146, 149; between mathematics and axiology, 19–20, 51–52
Analysis: 25, 317; categorial, 46; empirical, 73; induction, 317; linguistic, 308, 311
"Analytic," 317
Analytic: a-falsity, 187; a posteriori, 80; a priori, 80; opposition of subject and predicate, 271; "ought," 184, 192, 247; philosophers, 331; properties, 222; synthetic method, 216–17; value propositions, 174, 175, 179. *See also* Concept; Containment relation; Definition; Intension; Judgment; "Ought"; Oxford School; Proposition; System; Value Proposition
Analyticity: axiological, 171, 180, 182; definitional, 345; expositional, 345; Kantian and modern, 323; logical, 171, 180
Anathema, 358
Anathetic, 271, 358
"A new concept of," 162
Anselm of Canterbury. *See* St. Anselm
Anthropology, 11, 47, 48
Apodictic modality, 171
Apophantic truth-value, 164
Applicability, universal and precise, 36
Application: 65; of extrinsic value, 307–8, 311; of formal axiology, 254, 306–11; of formal axiology to concept of value, 254; of formal system, 108–9, 273; of intrinsic value, 308–9, 311; and reference, 66, 92; subjectivity of, 110; of systemic value, 309–10, 311; value of, 292. *See also* Method
Applied: axiological propositions, 180; axiological sciences, 114–15; axiological terms, 180, 182, 306; formal axiology, 107; mathematics, 107; pure axiology, 107; value sciences,

Applied (*continued*)
114–15; value terms, 180, 182, 306
Approval, 11, 245–48
A priori syntheticity, 359
A-proposition, 169
Arbitrariness of empirical value determination, 78
Archbishop of Canterbury, 322
Archimedes, 325
Argyris, Chris, 356
Arguments, axiological, 115–20, 296
Aristotle: first and second sense of actuality, ix; analytic concept, 31, 34; ethics and physics, 27, 61; good, homonym, 18, 99; Good, measure, 249; ordinary language, 27; logic, 84, 322
Aristotelian method: 326; motion and rest, 225; movement, 31; naturalistic fallacy, 124; number, 194, 253; teleological principle, 36, 61. *See also* Galileo; Method
Aristotelians, 146. *See also* Galileo
Arithmetical pattern of value terms, 208–15
Arithmetical value operations, 350
Arnold, E. Vernon, 324
Ars combinatoria, 358
Articulation of feeling, 129, 357. *See also* Sensitivity
Artists, 113
Assertion, 178
Assertory modality, 171, 172, 179
Assumption of existence, 346
Aster, Ernst von, 315, 327, 331
Astrology, 15, 47
Atomon eidos, 331
A-truth, 186, 187, 192
Attitude, 173
Augustine, 119
Automobile, 103, 268
Average, 211, 212, 215, 232, 234
Aversion against formalization, 13, 62
Awareness, 224
Axiologia perennis, 335
Axiologic, 105, 107, 300
Axiological: agreement and disagreement, 108, 110; analyticity, 172–74; arguments, 115–20, 296; calculus, *see* Calculus; class, *see* Class; copula, 165–67; dimensions, 112–14, 251, *see also* Dimensions; ethics, 115, 306, 308, *see also* Ethics; fallacy, 122–31; formalization, 110; harmony, 258; hypotheticity, 172–74; interpretation, 110, 292–93; inversion of values, 61, 217–20; logic, 300, *see also* Logic; modes, 172–74, *see also* Ought; pattern, 155, 171; quantification, 156, 161–62, 169; positivism, 54–58; propositions, 107, 168–92; relations, 162–68; science, 114–15, 128, 311; specificity, 106–20; syntheticity, 172–74; system, as standard, 109, 256, *see also* System; terms, 154–62; truth values, 107, 178–92; validity, 186; value, *see* Value; variations, 258, 261
Axiologies, 345
Axiologist, 6, 7, 77
Axiologistic, 107
Axiology: applied, 107; axiom of, 101–6; for ethicists, 78; formal, 30, 60, 107, 131, 153; framework of moral philosophy, 53; and metaphysics, viii, 58–63, 309; like music, 8, 10, 128, 258, 357; practice of, 65–66, 305; pure, 107, 109; scientific, 10, 105, 303; structure of, 153; and teleology, 36, 61, 62, 243, 297, 298, 359; applied to valuation, 203
Axiom: axiological and mathematical, 51–52, 105; of classes, 160; and category, 16, 46, 91; definition of, 76, 97, 101; extension of, 97, *see also* Science; of formal axiology, 101–6, 122, 132, 149; a formula, 105; ideal individual, 202, 356, *see also Hyle noeté;* infinity of categories, 91; intension of, 91, 97, *see also* System; an intrinsic value, 91, 355; a limit, 90–91; and logic, 102; a metaphor, 336; objectivity of, 110; reducibility, 157, 160, 343; singular concept, 41, 113; specifics of, 106; synthetic a priori, 80; and synthetic method, 83; of value, x, 99, 144, 153, 209
axiomatic: deduction, 80; identification, 73, 74, 75, 90, 102, 301, 303, 304, 331; interpretation, 73, 74, 75, 78–79, 330; synthesis, 46
Axiometric, 193, 202, 249–52, 254, 258. *See also* Intension; Measurement
Ayurveda, 327

Babington, A., 328
Baby, 272

Bacca, David García. *See* García Bacca
Backhaus, W., 127
Bacon, F., 65
"Bad": arithmetical definition of, 211; basic value term, 162, 178, 199, 345; combinatorial definition of, 215; and descriptive properties, 210; formal axiological feature, 295; formal specific of value, 107; logical definition of, 160; measure, 250; nonfulfillment, 161; and "ought," 234; problematical modality of, 179; quantifier, 169; scoring, 232; in Spinoza, 348; subaltern of "no good," 350; transposition, 161, 268, *see also* Transposition; topological pattern of, 207; truth-value pattern of, 207; typical, 206, 207. *See also* D-proposition; Value terms
"Bad for," 107, 163, 166, 203, 295, 358
Badness: and badness for, 203; extrinsic value, 195; transposition of concepts, 112; of world, 111, 296
Balmer's number, 348
Barbarism, specialist, 333
Bardeen, John, 315
"Beautiful," 125, 135, 180. *See also* Aesthetics
Beck, L. W., 319
Behaviorism, 307
Behrens, Hofrat, 281–86
Being: no category, 337; consistency of, 336; and Fact, 96; and good, viii, 119; Goodness of, 119; modes of, 358; intensional-extensional scale, 97; and science, 44; transcendental, 101, 119; of value, 108
Benardete, J. A., 330
Beneke, F. E., 318
Bense, M., 318
Bentham, J., 267, 329
Bergson, H., 119, 255, 328, 350
Berkeley, G., 337
"Best," 108, 166
"Better": definition, 114, 163; formal axiological feature, 295; formal specific of value, 107, 108; and "ought," 165, 184, 345; relation, 162, 166; applied to value, 254, 257. *See also* Value relations
"*Better be*," 345
"Better for," 107, 163, 166, 184. *See also* Value Relations
"Better had be," 345
Better value, 254, 257
Blanshard, B., 334, 338, 341
Bocheński, I. M., 323
Bohr, N., 321
"*Bon*," 110
Bonum, 164, 345
Botanist, 5, 6, 7
B-proposition, 169–70, 228–48, 346
Brahe, Tycho, 146
Braithwaite, R. B., 316
Brech, E. F. L., 354
Bridgman, P. W., 126
Brinton, C., 355
Broad, C. D., 332, 339
Bronowski, J., 329, 351, 355
"Brother," 255, 349
Buber, M., 115
Buick, 268
Burnet, J., 286
Butterfield, H., 329, 353

Calculus of matrices, 178
Calculus of value, xv, 107, 265–93
Calvin, J., 297
Campbell, N. R., 354
Canidae, 199
Cantril, J., 329
Cantor, G., viii, 224
Captain MacWhirr, 263, 264
Cardinalities of value, 112, 117–19, 193, 195, 202–3, 221–24, 265, 332. *See also* Gestalt
Carnap, R., 332
Carritt, E. F., 346
Carroll, Lewis, 321
Cassini, 326
Cassirer: 46, 92; formal and abstract thinking, 25; substance and function, 39, 82, 321; on symbol, 91; symbolic forms, 310, 331
Castrop, Hans, 281–86
Casuistry, 78, 335
Categorial analysis, 46
Category, 16, 41, 72, 77, 99, 101, 113. *See also* Concept, analytic; *Katagoreuein*; Philosophy
"Causes," 297
Ceremonials, 11, 346
"Chair," 31–32, 154, 205, 207
Change, 31, 34. *See also* Movement
Characteristic number, 113, 117, 118, 193, 221, 267, 274, 352. *See also* Axiometric; Intension

Chauchat, Clawdia, 281–86
Chefs de cuisine, 113
Chess, 33, 252
Chinese boxes, 84, 89, 332
Choice, 166
Christmas tree, 269
Christoff, Daniel, 343
Citizen, 253, 254
Circle, 81, 112, 194. *See also* Systemic value
Civil engineering, 310, 311
Civilization, 309, 311
"Class," 318
Class: 33; axiological, 161, 197; empirical, 187, 189, 197; existential, 197; expositional, 178, 187; extension, 155; intension, 155; logical, 161, 178, 197; null-, 198; schematic, 197; systemic, 187, 188, 197, 198, 348
Classes, kinds of, 197
Class membership: logical, 178; and valuation, 179. *See also* ϵ
Classless society, 125
Closure, 358
Cognitive meaning, 339
Cohen, Felix S., 342
Collision, 268, 358
Combinations: of properties, 213; of values, 213, 228–31, 237, 272–76, 280. *See also* Calculus.
Combinatorial: calculus, 19, 112; relationship between fact and value, 224; variations, 19, 265. *See also* Intension; Mobility
Combinatorial play, 351
Common denominator of values, 260
Common sense, 66, 278
Communication theory, 310, 311
Composition, method of, 303
Composition of values, 245, 246, 248, 268, 272–80, 333, 343, 358. *See also* Transposition
Comprendre, c'est pardonner, 275
Concept: 154; abstract, 113; analytic, 76, 77, 87, 252; analytic and synthetic, xi, 31–43, 64, 79–92, 222, 317; Aristotelian, 34; arithmetical difference between analytic and synthetic, 222; "concept," 48, 323; empirical, 80; formal, 80, 112; functional, 321; impossibility of, 272; irrational, 269, 271; isomorphous, 46; logic of, 43–54; as measure, 252; "a new—of," 162; parts of, 155, *see also* Containment relation, ω; philosophical, 43–54, 81; precision of, 51, 81–83, *see also* Precision; and reality, 88; Rickert on, 332; scientific, 43–54, 81; self-, 306; singular, 41, 113, 252; substantial, 321; synthetic, 73, 76, 77, 79, 82, 112, 194; and term, 84–89; transpositional, 271; value of, 164; world of analytic and synthetic, 36, 64–69. *See also* Axiometric; Category; Extension; Intension
Conception, 218
Conceptual enclosure objects, 90
Concrete: fact, 96; knowledge of, 145; value, 98
Condillac, E. B., 74
Conrad, J., 263, 264
Constancy of weight, 320
Constitutive "ought," 191
Construct, 194, 199, 210
Construction, xiv, 39, 53, 73, 83, 112, 318. *See also* Synthesis
Constructive "ought," 190, 191
Consul, The, 273, 276
Containment relation, 80, 131, 155, 318, 323, 342, 345. *See also* ω
Continuant, 181
Continuous states, 158–59
Continuum, intensional, 113, 195, 201, 265, 331. *See also* Gestalt
Contradicton: 198, 269; of approval and disapproval, 248
Contrary-to-fact conditional, 180
Convergence to simplicity, 89–90, 222
Copernican: inversion of fact and value, 220; theory, 298
Copernicus, 66, 147
Copula: "is," 168, 173; "ought," 165–68, 173. *See also* Propositions, axiological; logical
Correspondence of properties and predicates, 104
Counter-sense, 269, 270
Couturat, L., 318
Crahay, F., 344
Creativity, 331
Criticism, literary, 309, 311
Crossings: of judgments, 332; of relations, 332
"Crowd," 356
Culture, 309
Cusanus, N., 326
Cybernetics, 310

Damiens, R. F., 328
Dante, 297, 355
Darwin, C., 9
Datum, 122
Death of great man, 259–61
Debe, 239, 354
"Declarative"—"normative," 177, 346. *See also* Normative fallacy
Dedekind, 117, 119
Deduction, axiomatic, 80
"Deficient," 162
Definiendum, 199, 201
"Definist," 342
"Definist fallacy," 334
Definition: 71, 78, 81, 82, 121, 187, 199; analytic, xiv, 81, 84, 195, 199, 331; arithmetical definition of, 218–19; characteristic number of, 112; of ethics, 100; and exposition, 179, 188, 195, 196; extensional and intensional, 25, 31–32; as measure, 350, *see also* Axiometric; structure of, 199–202, 344; synthetic, 85, 331, *see also* System; of value, 92, 101–6, *see also* Value, definition of
Definitional analyticity, 345
Definitional class, 187
Definitional equality, 199
Definitional properties, 178, 182
Definitional types, 199–202
De Gaulle, C., 244, 356
Degree: 232; of differentiation, 200–201; of goodness, 342; of specification, 201, 204
Demand, 360
Democracy, 254
Denumerable intensional structure. *See* Cardinalities
Dependence of intrinsic value on intrinsic nature, 104, 137, 141, 257
Depiction: 199, 352; characteristic number of, 112
De Ropp, R. S., 352
Descartes: analysis-synthesis, 39, 82, 83, 321; book of world, 56; impli cations, 318; metaphysical fallacy, 124; method, 39, 82, 83, 321; scientific ethics, 99, 337
Description: 78, 81, 82, 85, 92, 135–37, 199, 208, 305, 341; arithmetical definition of, 218–19; characteristic number of, 112, 222; complete, 144; differentia between fact and value, 209; differentia between natural and non-natural properties, 137; and goodness, 217, 226; norm for value sets, 217; value meaning of, 137; a value set, 217; and value terms, 210
Descriptive properties: 134–35; combinations of value properties, 237; depend on goodness, 226; differentiations of value properties, 353; measure value, 226; mobility of, 215, 228, *see* Fluidity, Mobility; normative for value, 226, 227; "points," 233; primary value properties, 219, 227; secondary value properties, 227; sets of—equivalent to one value property, 218, 226; specifications of value properties, 353
Dessauer, F., 64
Destructive "ought," 184–85, 190, 191
Deutsch, K. W., 309
Devil, 274
Dewey, J., 13, 27, 28, 262, 308, 310, 316, 317
Dianoia, 322
Dialectic, 326
Dichotomies, x–xi, 305
Difference between things, 138
Differentia, 199
Differentiation, 200, 202, 204
Dilthey, W., 128
Dimensional pattern of value terms, 194–99
Dimensions of value: 19, 112–14, 193, 221–23, 251, 257, 305; axiometric sequence of, 258; and experience, 265; as fact, 221; formal specifics of value, 107; leaps, 223; of value measurements, 250; numbers of properties, 267
Dirac, P. A. M., 28
Disagreement, 108, 110, 296, 343
Di Santillana G., 319, 324, 353
Disapproval, 245–48
Discourse: ordinary, 66; philosophical, 66. *See also* Language
Discursiveness, 195, 266, 318
"Dishonest," 306
Dispositional properties, ix
Distance, emotional, 260
Disvalue, 271, 272
Divine spirit, 224
Division of values, 231, 241–42
Dodd, S. C., 307, 319
Dogmatic thinking, 113

Dogmatism, 57
Doing and thinking, 11
D-proposition, 169–70, 228–48
Drake, S., 326
Duhem, P., 38, 60, 333
Duncan-Jones, A., 350
Duration, 255
Durrell, L., 336
Duty, 190, 191
Dyson, F. J., 315

ϵ, 155, 179, 304, 322, 343
Easy chair, 205
Ecological technology, 310
Ecology, 308, 311
Economics, 11, 108, 114–15, 180, 307, 311
Economic value, 360. *See also* Value
Eddington, A. S., 124, 261, 263, 281
Edel, A., 339, 341, 353
Edison, T. A., 9, 28
Edwards, P., 18, 99, 334
Efficiency, 68
"Egghead," 274
Einstein, A.: on aversion against formalization, 13; on creative thinking, 351; and common sense, 66; historic position, 9; and positivism, 54; on scientific passion, 12; scientific style, 28; system of, 36, 39
Eisenhower, D. D., 297, 344
Elegant test of theory, 294, 298, 328. *See also* Parsimony
Elements of value. *See* Value
Emotional distance, 260
Emotive meaning, 339. *See also* Meaning
Empathic thinking, 114
Emphatic thinking, 114
Emperor, 355
Empirical: analysis, 73; bias, 129; class, 187, 189, *see also* Class; concept, 80; fallacy, 126–27; import, 71–79, 154, 249; interpretation, 73; thing, 196; value, 78
Enclosure objects, 90
"Energy," 69
Engineering, 310
Enlightenment, 68, 328
Enrichment, 98, 223, 224. *See also* Richness
Entailment, 132, 160, 207, 318
Entelechy, 34, 72, 76
Enumeration, 106
Epiaxiological logics, 147
Epistemology, 48, 308, 311
Epithet, 180
E-proposition, 169, 354
Equidae, 199
Equus caballus, 195, 196, 205
Erdmann, B., 97, 319, 323, 332
Essence: of fact, 219; of valuation, 215–28; of value phenomenon, 303
Estar, 343
Eternity, 255
"Ethical," 317
Ethical value, 108, 306. *See also* Intrinsic value
Ethics: 48, 308, 311; application of intrinsic value, 114, 308; and aesthetics, 115; axiological, 306; and casuistry, 335; definition of, 25–27, 114; and economics, 115; and logic, 2, 324; and metaethics, 58–60; philosophy of, 43, 227; science of, 43, 56; and sociology, 114–15; systematic, 30, 58; and symbolic logic, 324; a theoretical value specific, 114–15, 180; traditional, 43, 227
Euphemism, 180
Euthyphro, 286–92
Evil, 358
Ewing, A. C., 339, 340, 341, 345
"Excellent," 162
Existence: 336; assumption of, 346; axiological term, 166; God's, 116; and goodness, 197
Existential: a-falsity, 187; a-truth, 186; class, *see* Class; "ought," 184, 192
Existentialism, 11, 254
Existentialist fallacy, 128
"Expensive," 180
Experience, 224, 225, 265
Experimentation, 127
Expert, 110
Exponentiation of values, 243–48, 265, 272
Exposition: xv; arithmetical definition of, 218–19; characteristic number of, 112, 113; and definition, 179, 188, 195–96; in Kantian logic, 81, 82; of metaphysics, 266; modalities of, 179; norm of extrinsic value, 222; and truth value, 187. *See also* Analytic concept; Extrinsic value; Fulfillment
Expositional: analyticity, 345; class, 187–88; differentiation, 201; modality, 182; properties, 178; value, 182

Extension: 155; analytic, 33, 77; of concept "concept," 49; of formal axiology, 131; and intension, 37, 38; scale of, 96; synthetic, 77
Extensional-intensional scale of richness, 114
Extensional logic, 50
Extensional proposition, 155
Extensional richness, 96
Extensionality, principle of, 49
Extensive abstraction, 89
Extrinsic measurement, 251
Extrinsic value: xiv, 19, 164; application of, 307–8, 311; characteristic number of, 113, 222, 267, 272; as fact, 221, 223; fulfillment of exposition, 199; in hierarchy of values, 252, 342; and intensional structure, 193, 194; and intrinsic value, 91; its norm, 222; primary elements, 154; relations, 162–68, 224; richness of properties, 114; and scholastic perfection, 342; terms, 154–62, 196; types, 221; of a value theory, 298. *Also see* Calculus; Goodness; Value

Φ, 155
Fact; any given set of properties, 223; arithmetical definition of, 219; aspect of given, 193; concrete, 96; definition of, 19, 70, 129; empirical, 96; essence of, 219, 227; as extrinsic value, 221; formal, 16, 69, 96, 334; formal specific of value, 108; generic, 96, 97; as ideal case, 76; logical levels of, 96, 334; material methodological feature, 296; meaning of, 227; measure of value, 220, 227; mere, 227; mere fact of a value, 246; three methodological levels of, 69, 334; norm for value, 19, 219, 220; particular, 96; phenomenal, 16, 69; of primary properties, 16; primary property of value, 219; propositions and value propositions, 175–78; relativity of, 220–25; of secondary properties, 16; set of secondary properties, 220; fact properties secondary, 227; scientific, 16, 69; singular, 96; situation, 16, 69; specific of value, 108; unreality of, 220; and value, viii, 4, 19, 156–57, 209, 217–27; as value, 220, 223, 226; value specific, 108; value of, 108, 273; value of mere fact, 246; world of, 64–69, 220, 225
Factuality: fixation of value set, 219; and intension, 227; norm for value, 227; and valuation, 220
Facultative material norm, 292
Fahie, T. T., 349
"Fair": arithmetical definition of, 211, 212; a basic value term, 162, 178, 199; combinatorial definition of, 215; definition of, 160, 211, 212, 215; and descriptive properties, 210; formal axiological feature, 295; formal specific of value, 107, 108; measure, 251; and "ought," 234; problematical modality of, 179; quantifier, 169; scoring, 232; topological pattern of, 207; truth-value pattern of, 207; typical, 206, 207. *See also* B-proposition; Value terms
Fallacies, axiological, 100, 122–31, 296, 326
Fallacy: "definist," 334; empirical, 126–27; existentialist, 128; of genera, 123; of illicit generalization, 188; metaphysical, 123–24, 273, 277, 341; of method, 5, 6, 7, 12, 13, 16, 17, 59, 61, 123, 126–31, 322, 332, 337, 347; moral, 99, 125–26; naturalistic, 16, 17, 52, 63, 92, 100, 108, 109, 124, 128, 341; normative, 128, 225; positivistic, 128, 129; psychologist's, 338; of species, 124
Falsity, 269, 270
Fanaticism, 276
Faraday, M., 9, 45, 215
Feeling, 129, 357
Fermi, 9
Fetishism, 276
Findlay, J. N., 336
"Fine," 162
Fitch, F. B., 50, 324
Flow of constituting cardinalities, 265
Fludd, R., 340
Fluidity of descriptive properties, 215, 257, 353. *See also* Mobility
"Follows from," 132
Foot, 205
"Force," 29
Ford, 268
Forensic, 308
Formal analogy of science and ethics, 149
Formal axiological specifics, 107

Formal axiology: 107; axiom of, 101–6; extension of, 131; Husserl on, 153; intension of, 131; systematic structure of, 153. *See also* Axiology
Formality and materiality, 88
Formalization: 316; aversion against, 13, 62; of nature, xiii
Formal methodological features, 295
Formal norm, 292
Formal ontology, 346
Formal relations, 65
Formal rules, 292, 359
Formal specificity, 121
Formal specifics: 106; of value, 109–12
Formal structure: of goodness, 131; of value, 145
Formal value. *See* Value
Formal system. *See* System
Formulae. *See* Calculus
Fourier, J., 333
Frame of reference, 122
Frankena, W., 324, 334, 341
Franklin, B., 9
Free will, 124
Frege, G.: axiom of mathematics, 105; axiomatic identification, 102; naturalistic fallacy, 52, 122
Freud, S., 277
Fromm, E., 108, 308
Frondizi, R., 355
Fulbright, J. W., 322, 356
Fulfillment: of axiomatic intension, 199; of concept, 164; of construct, 199; of definition, 199; of depiction, 199; of exposition, 199; of intension, 254; of intention, 19, 164; of proposition, 164; of transposition, 272–73, 349; of value, 243
Function, 39
Functional, 321
Fundierung, 318, 331

Gabriel, L., 310
Galilean revolution, 14–15
Galileo: 59, 146, 320; analytic-synthetic method, 42, 45, 61, 216, 300, 317; and Aristotle, 31, 45, 61, 76, 86, 92; and Aristotelians, 50, 57, 62, 121, 146, 148, 340; and Church, 35, 123, 296, 326; and Copernicus, 147; destroyer, 62; and experience, 87; and "facts," 71; formalization of nature, xiii, 87; and Kepler, 57; changed nature of man, 64; mathematical language, 27; and measurement, 18, 105; mechanics, 77; metaphysical fallacy, 341; and metaphysics, 61; and observation, 87; reduction to primary properties, 226; on rest and motion, 225; secularization of science, 63; and seeing, 70–71; and texts, 56
Game, 272
Games, 310, 311
Gamow, G., 321
Garaudy, R., 326
García Bacca, J. D., 39, 319, 321, 330, 345
García Máynez, 310, 359
Gauss, C. F., 348
Gedankenexperiment, 145, 338
Gedankenwelt, 117, 118
Geisteswissenschaften, 128, 317
Gemeinschaft-sociologies, 307
Generality, 198
Generalization, 84
General specification, 46
"Generation of Waves by Wind," 261–62
Generic and specific value, 95–101
"Genuine," 306
Genus: 199; and species, 96–97
Gesellschaft-sociologies, 307
Gestalt: axiometric unit, 250; characteristic number of, 331; continuum, 331; depiction, 199; enrichment, 224; intrinsic intension, 353, 357; judgment as, 358; matrimonial, 352; metaphor as, 267; private language as, 255; psychology, 307; spiritual, 118; synlytic judgment as, 358; of system, 352. *See also* Continuum
Gierke, O., 310
Ginzburg, B., 326
Giorgione, 329
Giotto, 329
Girl, 232
Glass of Burgundy, 113, 350
Gnomon, 353
Goals, 297
God: and Being, 337; ontological proof, 116; and Soviet cosmonaut, 124; value predicate of world, 337; value of values, 116
God's "indifference," 322
God's will, 124
Goethe, J. W., 9
"Good": 48, 110, 345; aesthetic, 180; analytic concept, 154; arithmetical definition of, 210, 211, 212; asser-

"Good" (*continued*)
tory modality of, 179; axiological, 180, 347; axiom of, 101; axiomatic nature of, 144, 338; basic value term, 162, 178, 199; as philosophical category, 101; combinatorial definition of, 215; definable as axiom, 101; definition of, 103, 153, 156, 349; definition of—, synthetic, 92; definitional part of analytic intension, 349; dependence on intrinsic nature, 104, 137, 141, 153, 257; and descriptive properties, 210; determinable, 347; exemplification, 103, 104; follows from descriptive properties, 209; formal axiological feature, 295; formal definition of, 156; formal specific of value, 107; formula of, 101; homonym, 18, 99, 105; indefinable as category, 101; indefinable and definable, 338; and intension, 51; logical nature of, 145; logical operation, 103; a logical term, 131, 153, 160; used against use of mathematics, 62; measure, 249, 250, 251; moral, 180, 306, 347; nonnatural property, 119; odd use of, 340; pattern of, 155–56; predicate, 157; property of concept, 104; principle of, 95; propositions with—, synthetic a priori, 144; quantifier, 169; "reduced" second-order function, 157, 304–5; scoring, 232; second-order property, ix, 119, 157, 340; self-evidence of, 143; similarity of fulfilled intensions, 17, 51; simple, unanalyzable concept, 101; set of intensions *n*, 17, 51; in Spinoza, 348; syncategorematic term, 335; synthetic concept, 349; as systemic value, 349; teleological, 62; term, x; topological pattern of, 207; transcendental, 119; truth-value pattern of, 207; uses of, 145; a variable, 18, 99, 105. *See also* "A good"; G-proposition; "Is good"; "Is a good;" "It is good that;" "Less than good;" "More than good;" "Perfectly good;" The good
"Good for": 166, 203, 204; definition of, 163; formal axiological feature, 295; formal specific of value, 107; and overcoming evil, 358. *See* Value relations
"Good is good," 131, 349
Good automobile, 103
Goodman, P., 355
Good murderer, 125
"Goodness," 43
Goodness: attribution of—to—, 245; of beauty, 125; and being, viii, 119; classless society as, 125; definition by Moore, 140; definition synthetic, 305; degree of, 342; and description, 217, 226; and existence, 197; extrinsic value, 195, 199; formal structure of, 131; in general, 102; and goodness for, 203; of God, 116; and intension, 226; as measure, 226, 249; Moore's paradox of, 18, 102, 104, 119, 143, 153, 304; moral, 306; non-descriptive, 102–3; norm for thing's value possibilities, 226; principle of, 95; species of value, 214; and specific goodnesses, 243; of truth, 125; unanalyzable, 101; of world, 111, 296. *See also* Ethics; Value
Goodness for, 203
Goods, 307
Good star, 255
Good-value, 164, 204
"Go the whole intension," 250
G-proposition, 169–70, 228–48
Grammar, 310, 311
Gravitation, 72, 76
Great man dying, 259
"*Gut*," 110, 346

Hadamard, J., 325, 331, 351
Hagerty, J., 344
Hall, E. W., 340, 353
Halmos, P. R., 321
Hanson, N. R., 341
Harmony: pre-established, 112; of values, 258
Hartmann, N.: analytic and synthetic concept, 91; *diairesis*, 319; feeling of value, 129; the Good, 95; ideal object, 334; intensive abstraction, 92; matrimonial Gestalt, 352; metaphysics, 323; stigmatic intuition, 321; *Wesensschau*, 90
Hartman Value Inventory, 359
Hartshorne, C., 323, 336
Hasdale, M., 324
Heaven, 255
Hegel, G. W. F.: being and nothing, 44; metaphysical politics, 253; natural and moral science, 47
Heidegger, M., 13, 44

Heisenberg, W., 126, 315, 325
Hempel, C. G., 71, 72, 329
Hertz, H., 315
Heuss, T., 47
Hierarchy: of differentiations, 221; of intensifications, 221; of intension and predicates, 265; of intensions, 119; of secondary predicates, 119; ideal—of values, 342; of intrinsic values, 224; of richness, 114; of values, 107, 114, 223, 249–65, 267, 342
Hilliard, A. L., 350
Hobbes, T., 30
Hochzeit, 255
Hocking, R., 336
Hogarth, W., 328
Holton, G., 321
"Holy," 180
Homoiousion, 251
Homonymity of "good," 18, 99, 105. *See* "Good," variable
"Honest," 180, 306
Honeymoon, 255
"Honor student," 180
Horney, K., 108, 297, 308
Horoi, 332
Horse's tail, 202
Hourglass, 83
Hull, C. L., 307
Humanism, 68, 328
Humanities, 46
Hume, D., 20, 337
Huntington, E. V., 40, 336
Husserl, E.: aspects of reality, 284; on extensional logic, 322; on fact, 193, 259, 336; on formal axiology, 153, 350; fulfillment of proposition, 163; *Fundierung*, 318, 331; *Lebenswelt*, 67, 200; phenomenology as axiology, 19; on profundity, 122; pure logical grammar, 271; system and application, 327
Hutcheson, F., 267, 329, 357
Hutchinson, G. E., 357
Huxley, A., 352
Hyle noeté, 202, 349, 356
Hypostatization: of ideologies, 252; of "ought," 346; of value, 9; a value composition, 272
Hypothesis, 71, 72, 318
Hypothetical "ought," 184, 192
Hypothetical value propositions, 174, 175, 179
Hypotheticity, 172, 180, 183
"I," 306
"I am lying," 347
I and Thou, 115
"I"—"you," 355
Idea, 324
Ideal case, 76
Ideal hierarchy of values, 342
Ideal individual, 202. *See also Hyle noeté*
Ideality of world, 124
Ideal object, 334
Ideals, 297
Identification. *See* Axiomatic identification
Ideology, 47, 322
Imperative, 11, 180, 272, 348
"Implication," 318
Implication: 32, 33, 64; contextual, 13; series of analytics, 200–201
Import of science, 71–77
Impossibility, logical, 272
Improvement, 236
"In a class by itself," 162. *See also* "Unique"
Inclination, 343
Indeterminacy principle, 124, 126, 145
Index, modal, 178
Indifference, 299, 322
"Indifferent," 211
Individual, 331
Individuality, 198
Induction: 317–18, 319; arithmetical determination of, 350–51
Industrial Technology, 310, 311
"Inferior," 162
Infinite value of person, 116–17
Infinity of experience, 224
"Insincere," 306
Inspection, 159
Integrative logic, 310
Intelligenzbestie, 273, 274
Intension: 155; analytic, 32, 195, 222; arithmetical definition of, 219, 222; of an axiom, 97; axiometric structure of, viii, 249–51; central role in axiology, 193; common denominator for value realm, 304; of concept "concept," 49; differentiation of, 203; dissolution of, 217; and extension, 37–38; of formal axiology, 131; fulfillment of, 254; and goodness, 17, 51, 226; hierarchy of —s, 119; and hierarchy of values, 265; implicative relations of —s,

Intension (*continued*)
32; logic of, 60; logic of value, 55; measure, 117, 219; norm of value, 227; one value property, 226; scale of —s, 96; set of predicates, 19, 112, 193, 226; similarity of —s, 17, 51; singular, 195, 250, 266; standard for value measurement, 106; structure of, 193, 194, 199; synthetic, 38, 195, 222–23, 250; systemic, 222; transpositional, 269, 271–72; whole of, 250. *See also* Axiometric; Concept; Gestalt; Logic
Intensional cardinalities, 112, 193. *See also* Cardinalities of value; Characteristic number
Intensional configurations, viii. *See also* Combinations
Intensional continuum, 113. *See also* Continuum; Gestalt
Intensional Gestalt. *See* Gestalt
Intensional logic, 82, 322. *See also* Logic
Intensional measurement, 251. *See also* Measurement
Intensional norm, 193, 218, 349. *See also* Norm
Intensional properties, 210
Intensional proposition, 155
Intensional richness, 96, 98
Intensional sets, 112
Intensive abstraction, 90, 92, 222
Intensive measurement, 232. *See also* Measurement
Intention, 164
Intentionality, 19
Interest, 78
Interpretation: 71–74, 180; axiomatic, 73–76, 78–79, 330; empirical, 73; modal, 177
Interrogative, 180
Intrinsic measurement: 251, 355; and metaphor, 267
Intrinsic nature, 135, 142–43
Intrinsicness of value, 140–44
Intrinsic properties. *See* Properties
Intrinsic value: xv, 19, 113, 164, 193, 347; applications of, 308–9, 311; characteristic number of, 113, 267; compositions, 358; definition of, 113, 140–44, 199, 254, 267; dimensions of, 224; as fact, 221; fulfillment of singular intension, 199, 254; hierarchies of, 224; limit of extrinsic values, 91; maximum plenitude, 252; intrinsic nature of, 139; Moore's paradox of, 131–49; relationships, 224; structure of its norm, 349; transfinite, 248; uniqueness, 195, 199, 248; value theory as, 298–99. *See also* Cardinalities; Gestalt; Involvement
Intuition: 41; in sense of Moore, 145, 340; stigmatic, 321
Intuitionists, 342
Inversion of relation between fact and value, 19, 61, 225–26, 254, 304
Involvement, 7, 251
I-proposition, 169
Irony, 173, 178
Irrationality of value, 107, 109, 269, 296
Irreality of value, 353
"Is": 168, 173; and "ought," 175, 225
"Is a," 181, 295, 342
"Is a good," 181, 346
"Is good," 181, 295, 345, 346
Isness, 225
Isomorphism, 9, 46, 65, 97, 99, 104, 131, 249, 260, 267, 305
"It is bad that," 166, 295
"It is better for," 164
"It is better that," 164, 166
"It is good that," 107, 120, 163, 166, 247, 295
"It is worse that," 164, 166, 295

Jabberwock, 269, 272
Jacob's ladder, 224
Jaeger, W., 355
James, W., 338, 352
Janitor, 348
Jeans, J., 124, 296
Jesus, 273, 274, 352, 359
"Jo" (Hungarian "good"), 110
Job: evaluation, 216; fulfillment, 233
Johnson, W. E., 181
Jones, 161
Jørgensen, J., 330
Judgment: analytic, xi, 79–80, 271, 317; analytic a posteriori, 80, 143; analytic a priori, 80; anathetic, 271; parathetic, 271; and proposition, 171, 179; synlytic, 271, 358; synthetic, xi, 79–80, 271, 317; synthetic a posteriori, 80; synthetic a priori, 80; *teratic*, 271; underlying, 171, 175, 176. *See also* Modality
Jukes, 263, 264
Jung, C. J., 327, 329

Juridical value, 274, 277, 310, 311
Jurisprudence, 11, 277, 310, 311

Kahler, E., 356
Kant: on analytic and synthetic, 79–82, 84, 113, 317; on axiological "good," 347; on *bonum*, 345; containment relation, 342; definition, 81–82, 218; description, 81–82, 218; dialectic, 44; epistemology, 308; ethics, 28, 253, 308; existence, 198; exposition, 81–82, 218; formula of a single problem, 249; fulfillment of proposition, 163; and Hume, 20; logic, 39, 80, 81, 82, 323, 330; logical scale of being, 97; logical scale of value, 98; metaphysical deduction, 330; modality, 171; normative fallacy, 225; perfectly good will, 348; philosophy as history, 299; schema, 197; science of ethics, 99; science of metaphysics, 330; style, 29; on syllogism, 160
Kantorowicz, E. H., 355
Kapp, E., 323, 355
Katagoreuein, 42
Kath' hekaston, 331
Katz, J. J., 335
Kecskemeti, P., 57, 146, 340
Keller, Helen, 113
Kelsen, H., 11, 310
Kent, W., 355
Kepler, J.: 10; aha-experience, 9, 325; and Fludd, 340; and Moore, 146–48; moral fallacy, 125; naturalistic fallacy, 124; three laws, 341
Keynes, J. M., 20, 21
"Khoroshii" (Russian "good"), 110
Kierkegaard: "the crowd," 356; *Either/Or*, 115; *Point of View*, 115; *The Sickness Unto Death*, 115; subjective truth, 308; system and faith, 274; value inversion, 254
Kingdom of God, 255
Kneale, W., 345
"Knight of faith," 306
Knowledge: paradox of, 2, 145, 301; of value, 8
Koestler, A., 341
Korzybskians, 347
Krämer, H. J., 325

Laborers in the Vineyard, 353
Laird, J., 332
Lamb, W. E., 264
Lambert, J. H., 97
Landscape painting, 329
Langer, Susanne K.: abstraction in art, 355; on "ϵ," 322; logic of creativity, 331; on logic of ethics, 2; on music, 357; semblance, 353
Language: of intrinsic value, 113–14, 256; of metaphor, 113, 266–67; ordinary, 73; refinement of, 74; private, 356–57; systemic, 199; technical, 73, 199; theoretical or systematic, 73; of value, 76, 78, 107, 255; vagueness of ordinary, 34
Lanz, H., 319, 345
Laplace, P. S., 9, 63
Lapp, R. A., 334
Lavoisier, A. L.: 9, 59; analytic and synthetic, 42; and "facts," 71; language of science, 74, 75; method, 320
Law: of persons, 311; of property, 311; of supply and demand, 360; a systemic value, 310; of things, 310
Laws: 251
Laws of thought, 346
Leap: between analytic and synthetic, 82; between value dimensions: 223
Lebenswelt, 67, 200, 328, 352
Leibniz, G. W.: analytic and synthetic method, 25, 113; *ars combinatoria*, 358; characteristic number, 117; impossibility of concept, 272; logic of value, 332; logical scale of being, 97; metaphysical fallacy, 124; metaphysics, 108, 296; number of possible statements, 318; *petites perceptions*, 333; predicates of world, 337; preestablished harmony, 112; scientific ethics, 87, 99, 337
Length, 250
"Less than good," 51, 52
Lessing, T., 334, 338, 350
Levels of abstraction, 200, 202
Lewis, C. I., 319
Lexical method, 84, 331
Liede, A., 359
Light, 73, 75
Lighthouse, 82
"Lilac," 255, 349
Limit, 89, 222
Limits of experience, 225
Limitation, logical, 112
Linguistic: analysis, 308, 311; quantifiers, 162; *ad hoc* constructions, 145

Linguistics, 308, 309, 310
Linsky, L., 357
Literary criticism, 309, 311
Locke, J., 56, 325, 337
Logarithmic structure of systemic intension, 222
Logic: 48; Aristotelian, 61; and axiom, 102; and cardinalities, 332; of concept, 48–50, 322; of creativity, 331; of discovery, 300; and ethics, 2, 58–60, 324; extensional, 50, 55, 322; of "good," 145; "integrative," 310; intensional, viii, 19, 49, 50, 55, 60, 82, 322; modal, 322; pragmatic, 310; of precision, 51, 80–84; of preference, 346; science of, 49, 323; generalized social science, 326; supreme formal system, 330; as systemic value, 310, 311; and teleology, 61; as value philosophy, 327; of value, 30, 60, 62, 91, 107, 108. *See also* Class; Concept; Proposition
Logical a-falsity, 187
Logical a-truth, 187
Logical grammar, 271
Logical oddness, 339
Logical "ought." *See* Ought
Logical pattern of the value terms, 199–208
Logical positivism. *See* Positivism
Logical proposition. *See* Proposition
Logical relation between philosophy and science, 31
Logics, *ad hoc*, 11
Logistic, 106
Lotze, H., 2, 97, 98, 323
"Lousy with," 162
Lovers, 113
Loyalties, 252
LSD, 352
Łukasiewicz, J., 330
Lying, 178, 347

Mach, E., 4
MacWhirr, 263, 264
Mademoiselle, 328
Magicians, 113
Magic Mountain, 280–86
Magini, 324
Maimon, S., 333
Maius, 116
Malcolm, N., 357
"Male sibling," 255, 349
Man, infinite value, 117
Mann, T., 280–86, 359
Marconi, G., 9
Marc-Wogau, K., 345
Maritain, J., 356
Marx, K., 47, 309
Maslow, A., 108, 297, 308
"Mass," 29
Material features of value theory, 296–98
Materiality and formality, 88
Material methodological features, 295–96
Material norms, 292
Material specifics: 107–8; of value, 115–20
Mathematics: applied, 107; axiom of, 19–20, 51–52; cumulative, 29; and natural science, 53; for physicists, 78
Mathesis universalis, 99
Matrimonial Gestalt, 352
Matrix calculus, 178
Maxim, 272, 293
Maxwell, J. C., 9, 28, 315
Máynez. *See* García Máynez
Meaning: 84, 273; cognitive, 339; emotive, 339; of "good," 103; measure of value, 18; of system, 86; as value, viii, 49
Measure: definition as, 350; the Good as, 249; intension as, 117; as norm, 225; of value, 219, 226, 249–51; of value phenomena, 257
Measurement: 219, 250–51; by degree, 232; of intrinsic value, 355; of value, 18, 105–6, 112, 216, 232, 250–51
Median, typical, 206–7
Medieval world, 36, 67, 327
Melius, 116
Menotti, G. C., 273, 276
Mere fact, 227, 246
Metaaesthetics, 59, 60
Metaepistemology, 59, 60
Metaethics, 30, 58–63
Metametaphysics, 59, 60
Meta-philosophia-moralis, 59
Meta-philosophia-naturalis, 59
Metaphilosophy, 59
Metaphor, 44, 113, 180, 250, 266–67, 273, 336. *See also* Gestalt; Intrinsic value
Metaphysical fallacy, 123–24, 274. *See also* Fallacy
Metaphysics, viii, 48, 60–63, 108, 309, 323

Meter rod, 250
Method: analytic-synthetic, 216–17; Aristotelian, see Aristotelian method; "first" and "second," 332; Galilean, 300; illegitimate, 357; lexical, 84, 331; philosophical, 46, 84; of resolution and composition, 216–17, 219, 303, 304; result of science, 77; scientific, 21, 36, 46, 84–87; synthetic, 79. *See also* Action; Practice
Methodological features of value theory, 295–96
Middle Ages, 36, 67, 327
Mieli, A., 329
Millennium, 255
Miniature geometry, 300, 342
Miniature logic, 155
"Miserable," 162
Missouri (battleship), 70
Mitchell, E. T., 249, 332, 346, 354
Mobility of descriptive properties, 215, 228, 351, 353. *See also* Fluidity
Mockmurder theory, 159
Modal index, 178
Modal interpretation of value proposition, 175
Modal logic, 322
Modes of being, 357
Modes of exposition, 179, 182
Modes of "ought," 184
Modes of value propositions, 171–78
Modulus symbolization, 354
Moment, 90
Moments of thought, 171
Money, 307
Mooney, R., 329
Moore, G. E.: arithmetical operations, 350; Cartesian reduction to "simple," 144; complexity of concept, 342; construction of "value," 42; Copernicus, 147, 149; critique of value philosophy, 56; defining goodness, 140; on "definition," 325; on definition of ethics, 100; on definition of intrinsic natural properties, 136–40; on definition of intrinsic value, 140–44; use of "description," 92; on differentia between fact and value, 209; "end of the matter," 338; ethical theory, 121–49; ethics and casuistry, 335; formula for "good," 102; and Galileo, 50–51, 146–49; on intrinsic properties, 133–36; intuition, 145, 340; judgment of intrinsic value, 341; and Kepler, 10, 146–49, 341; moral fallacy, 125; on nature of thing, 132; on naturalistic fallacy, 52, 53, 341; and Newton, 149; open question test, 119; paradox of goodness, 18, 102, 104, 119, 143, 153, 304; paradox of intrinsic value, 131–49; *Philosophiae Moralis Principia Ethica,* 101; precision of language, 20; primary properties of value, 226; "Prolegomena," 19, 42, 92, 95, 101, 146, 147; questioner, 19–21; and Rees, 159; science of ethics, 325; science of value, 10, 100; secondary qualities, 193; solution of paradox of goodness, 158, 226; solution of paradox of intrinsic value, 144; syncategorematicity of "good," 335; "three laws," 147–48, 341; on value of thing, 132
"Moral," 317
Moral awakening, 328
Moral fallacy, 125–26. *See also* Fallacy
Moral goodness, 306
Moral law, 296
Moral philosophy, 15, 63
Moral sanity, 328
Moral science, 15, 63, 99, 130, 146, 337
Moral sensitivity, 60. *See also* Sensitivity
Moral terms, 306
Moral world, 68–69
"More than good," 162
Morphology, 271
Morris, C., 308
Motion, 31, 35, 37, 62, 225. *See also* Movement
Mouth, 205
Movement, 31, 34, 35, 61, 83. *See also* Motion
M-proposition, 228–48
Mr. Tompkins, 220, 321
Mrs. Jones, 161
Multiplication of values, 230, 238–41, 242
Multiplicative value complement, 242
Multiplier of a value, 240
Munn, N. L., 325
Murderer, 125
Musical notation of value calculus, 357
Music, 8, 10, 30, 129, 357
Mystic, 113

Mystic experience, 224, 272
Mysticism, 129

Nagel, E., 316
Name: 117, 154; equals norm, 109
Naturalistic fallacy, 53, 102, 122, 123, 124, 130–31. *See also* Fallacy
Naturalists, 8, 146
Natural properties, 305. *See also* Properties
Natural philosophy, 45, 130
Natural science, 107, 130, 146, 227. *See also* Science
Nature, intrinsic, 139–40
Nazi, 68, 273
Negative exponentiation, 272. *See also* Transposition
Negative value, 268, 271. *See also* Value
Negatory modality, 171, 172, 179, 184
Nelson, E. J., 323
Nettleship, R. L., 324, 355
"New concept of," 162
Newland, Constance A., 352
Newton: 9, 36, 59; analytic and synthetic method, 42, 64, 113, 317; and Kepler, 341; and Locke, 56; moral fallacy, 125; naturalistic fallacy, 124; *Philosophiae Naturalis Principia Mathematica*, 101, 146; system, 37, 39
Nicoll, M., 353
"*N'importe*," 162
"No," 169
Noblesse oblige, 354
Noesis, 322
"No good": arithmetical definition of, 211, 212; a basic value term, 162, 178, 199; combinatorial definition of, 215; definition of, 160; and descriptive properties, 210; formal axiological feature, 295; formal specific of value, 107; negatory modality, 179; and "ought," 234; quantifier, 169; symbolization of, 244; topological pattern of, 207; truth-value pattern of, 207; typical, 206, 207. *See also* Value terms
Nominalism, 309
Non-denumerable intensional structure. *See* Cardinality
Nondiscursive, 195, 266
Nonnatural quality, 353. *See also* Properties
Nonnaturalistic science of value, 10
Nonnaturalists, 146
Nonvalue: systemic, 112; extrinsic, 244
"Non-obligatory," 345
Nonsense, 54, 55, 180, 248, 269, 325
Norm: 193, 218, 349; any given intensional set, 223; description as, 217; equals name, 109; of extrinsic value, 222; fact for value, 19; factor of value operation, 242; factuality of, 227; facultative material, 293; formal, 292; and goodness, 226; intension seen axiologically, 227; material, 292; measure, 225; number of descriptive properties, 218; obligatory material, 293; predicate as, 184; of value of property, 205; structure of, 194; of systemic value, 223; of value, 220, 227. *See also* Ought-value; Value-ought
Normalcy, 217
"Normative," 177, 346
Normative fallacy, 128, 346
Normative ought-value, 240–44
Normativity of fact, 217, 219
Nose, 159, 306
Nota notae rei ipsius nota, 160
"Not bad," 162
"Not good," 162, 345
"Not good for," 163
Nowell-Smith, P. H., 27, 339, 355
Null class, 198
Number: 17–18, 106, 107, 194; extensional analogue of value, 17–18, 51–52
Nurse on duty, 158, 159

ω (containment relation), 155, 194, 208, 304, 305, 343
ω (ordinal number), 204
Objection against axiological science, 128
Objectivity of value: 110, 296; formal specific of value, 107; of intention of judgment, 164; of truth-value, 164
Obligation, 234, 235
"Obligatory," 345
Obligatory material norm, 292
Observation, 72, 130
Observer in relativity theory, 126
Ockham, W., 4, 320, 326
Oddness, logical, 339
"O.K.," 162
One property, 272, 350

One-to-one-correspondence, 325
Ontological proof, 116, 296, 306
Ontological relation between genus and species, 96
Ontology, 97
Open question test, 119–20
Operational philosophy, 126
Operational research, 310
Operation, axiological, 228–48, 305
O-proposition, 169
Optimism, 111, 296
"Order," 318
Order of value, 221–23
Ordinal number, 204, 352
Ordinary discourse, 66
Ordinary language. *See* Language
Ordinary understanding, 278
Ordinary usage, 317
Organic logic, 310
Organic whole, 193, 331, 358
Organization man, 306
Orozco, 114
Ortega y Gasset, J.: 297; on aspects of reality, 261, 284, 357; "drops of phenomenology," 259–61; on flight from reality, 353; on specialist barbarism, 333; on term, 331
Ossowska, M., 343
"Ought": 108, 114, 165–67, 173–74, 233–48; analytic, 173, 175, 235, 253, 254; applied, 253; and approval, 246; axiological, 347; constitutive, 348; copula, 165–67, 173–74; definition of analytic—, 247; definition of synthetic—, 247; derivation from "is good," 175; destructive, 184–85; dimensions of, 252; existential, 184; formal axiological feature, 295; formal specific of value, 107; forms of, 190–91; hypostatized, 346; of hypothesis, 348; hypothetical, 175; illegitimate use of, 235; and "is," 175, 225; and "it is better that," 165; and "it is good that," 247; legitimate use of, 236; modes of, 184, 192; moral, 306, 347; obstructive, 184–85, 190, 191; "owed," 234; and "perfect," 253–54; senses of, 167, 189, 190, 268; statistical, 184; symbolization of, 354; synthetic, 175, 234; and truth values, 187. *See also* Value relations
"Oughtness," 225
"Ought not," 166, 184
Ought-propositions, 174, 178
"Ought to be," 165
"Ought to be C," 183
"Ought to be good," 165, 182
Ought-value, 233–48
Overcoming evil by good, 358
Oxford analysts, 13, 145, 251, 252

Pain, 255–57
Painting, 329
Parables of Jesus, 353
Parabola, 87, 129
Paradox, 272
Paradox of intrinsic value, 131–49
Paradox of knowledge, 2, 145, 301, 302. *See also* Whitehead
Paradox of types, 104
Parsimony principle, 298, 316
Parsons, T., 307
Partial inverse, 346
Particularity: 198, 331; of value, 109, 296
Party, 254
Pascal, B., 119
Pascal's triangle, 200–201
"Passable," 211
Paton, H. J., 323, 339
Pattern: arithmetical of value terms, 208–15; of axiological propositions, 176; dimensional of value terms, 194–99; of "good," 155–56; logical, of value terms, 199–208; of typical value terms, 206; value expressions, 155–56, 172
"Peach," 266, 272
Pedant, 273
Peirce, C. S., 111
Pejoratives, 180
Perception, 218
"Perfect": formal axiological feature, 295; formal specific of value, 107, 108; and "ought," 253–54
Perfection: applied value term, 253–54, 349; and "ought," 253–54; scholastic, 335, 342; systemic value, 112, 194, 195, 199, 349; of a value, 244, 245
"Perfectly good," 253, 348
Performance measure, 216
Performatories, 346
Person, 116–17
Persuasion, 11
Pessimism, 111, 296
Petites perceptions, 333
Phenomena: of science, 76, 97; structure of, 79

Phenomenal specifics, 107, 108
Phenomenal value, 4, 6, 7, 257, 303, 304, 329
Phenomenological reduction, 90
Phenomenology: 13; a formal axiology, 334; "drops of," 259–60. *See also* Intention
Phenomenon: logical structure of, 79
Phenomenon of value: *See* Phenomenal value
Philebus, 251, 319, 325, 332
Philosophical critics, 359
Philosophical discourse, 66
Philosophical method, 36, 77, 84
Philosophical "system," 27–30, 31, 84
Philosophy: analytic method of, 36, 77, 84; definition of, 77; and metaphor, 44; natural, 71, 130; noncumulative, 29; poetry, 44, 66; and science, xi, 14, 16, 27–30, 31, 46, 77, 92, 352; no synthetic method, 36. *See also* Alchemy
Phlogiston, 34, 320
"Physicist," 45
Physics, 60–63
Physiology, 309, 311
Physiometric, 250
Planck, M., 145, 338
Plato: 12; astronomy and theology in, 108, 296; and logical scale of Being, 97; *Cratylus*, 310; *dianoia*, 322; *Euthyphro*, 286–92; and Good, 121, 324, 325; ideal of science, 4; *Laws*, 251; and mathematics, ix; measurement in, 251; and metaphor, 44; method of, 12; and moral fallacy, 125; *noesis*, 322; naturalistic fallacy in, 124; *Philebus*, 251, 319, 325, 332; scientific ethics, 87, 99; scale of value intensity, 98
Play, 351. *See also* Fluidity; Mobility
Pleasure, 78, 124, 227, 305
Poe, E. A., 352
Poetry, 309, 311
Poets, 113
Point, 90
"Points," 233
Polanyi, M., 325, 332
Pole, D., 357
Political quackery, 322
Political science, 47, 48, 309, 311
Political system, 115, 253–54, 356
Politicians, 113, 356
Polygon, 38
Polyguity, 105. *See also* Homonymity
"Poor," 162, 179
Pope, 355
Positive value, 268
Positivism, ix, 11, 13, 54–58, 129, 325, 327
Positivistic fallacy, 127
Postulate of formal axiology, 53. *See also* Axiom
Pound, R., 310
Practical reason, 66
Practice, 21, 69, 77, 305
Pragmatic logic, 310
Pragmatic thinking, 113
Pragmatism, 11, 13, 308
Prantl, C., 326
Precision, 36, 51, 58, 80–84
Predicates: 34; axiological, 162, 167, 168; in Being, 336; continuum of, 265; contradictory, 198; descriptive 209, 210; of intrinsic value, 138; logical, 167, 168; norms for subject, 184; quantification of, 156, 343; sets of, 251; and subject, 169, 184; of value, 144, 218
Prediction, 127, 130
Pre-established harmony, 112
Preference, 78, 124, 166, 227, 305, 346
Prejudice, 112, 299
Price, 308
Primary elements of extrinsic value, 154
Primary qualities. *See* Properties
Prime number theorem, 351
Principia Ethica, 338. *See also* Moore
Principle, 77
Principle of indeterminacy. *See* Indeterminacy
Private language, 356–57
Privative prefixes, 244
"Prize bull," 180
Pro-attitude, 246–47, 345
Problematical modality, 171, 172, 178
"Proconsul of France," 198
Procrustes, 55
Product of values, 214
Profundity, 85, 122
Proletariat, 124
Promise, 190, 191, 234, 235, 346
Proof: God, 116; of value of person, 116–17
Properties: analytic, 222; chaos of, 218; definitional, 178, 182, 197; and description, 137; descriptive,

Properties (*continued*)
xi, 134, 135, 209, 210, 217; dispositional, ix; expositional, 178, 197, 201; of fact, 227; infinity of, 203; "irreal," 145; intrinsic, 131, 133, 136–40, 137, 139; intrinsic natural, 134, 135, 142; "which are intrinsic," 133, 137; judgment, 206; mobility of, 215; natural, 104, 135, 142; of natural science, 227; natural—, norm for non-natural—, 226; natural—, and value—, 104, 134–36, 215–28; natural—, as primary value—, 226; non-definitional, 188; non-natural, 42, 134–35; nonsensory, 42; normal, 217; "one," 272, 350; place of—, in intension, 202; primary, 16, 105, 226; primary—, of value, xi, 19, 225, 227, 305; secondary, 16, 105; secondary—, of fact, 19; secondary—, as primary, 18, 105, 217, 328; sets of, 19, 217, 225, 273; synthetic, 222; form a priori system of value, 144; tertiary, 227; value complex, 193; value—, 42, 134, 209; value of, 203, 204; and value terms, 210; and unity of intension, viii; weighted, 349. *See also* Description; "Good"; Mobility; Moore; ω
Proposition: 169; "analytic," 79; analytic, xi, 271, 346; anathetic, 271; applied axiological, 180; axiological, 107, 156, 168, 176, 177–78, 185, 192; B-, 169, 175; background of axiological, 186; compound axiological, 171, 181; contrasensical, 270; D-, 169, 175; depth analysis of, 170, 346; E-, 169; enthymematic axiological, 180; extensional, 155; false, 270; fulfillment of, 163, 164; G-, 169, 175; I-, 169; intensional, 155; and judgment, 170, 179; logical, 156, 168, 176, 185; M-, 228; mixed axiological, 168, 183–84, 185, 190–91, 192; mixed logical, 168, 181, 185, 187–88, 192; modes of, 172, 178, 179–85; negative axiological, 170; nonsensical, 270; "normative," 177; O-, 167; particular, 346; parathetic, 271; positive axiological, 170; pure axiological, 168, 176, 177, 182–83, 185, 188–90, 192; pure logical, 168, 171, 180–81, 185, 187, 192; synlytic, 271; synthetic a priori, 144; synthetic, xi, 271, 346; T-, 169, 175; teratic, 271; transpositional, 270; universal, 346; value—, 163, 167, 168–74. *See also* Ought-proposition, Quantification
Propositional function, 156
Prostitution, 115
Pseudo-system, 14
Psychiatrist, 127
Psychologist, 5
Psychologist's fallacy, 338
Psychology, 11, 108, 307, 311
Ptolemaic theory, 146, 298
Pucciarelli, E., 345
Pure axiology, 107, 109, 112–15
Pure logical grammar, 271
Purpose, 78, 297, 305
Pythagoras, 108, 124, 296

Quackery, 322
Qualification, 162, 169, 250, 304
Qualities. *See* Properties
Quantification: 155–56, 304, 347–48; axiological, 156, 161, 162, 213, 250, 335; logical, 155, 161, 169; of predicates, 343
Quantifier, linguistic, 162
Quantitative method, 320
Quaternary value combinations, 278, 279
Questioning, 173
Quinary value combinations, 279
Quine, W. van Orman, 310, 323

Rationality: definition of, 116; formal specific of value, 107; of value, 109, 296; of world, 110
Rationalization, 273, 299
Ray of light, 73, 75
Realists, 309
Reality: and concept, 88; factual aspect, 261; philosophical and scientific, 64–69; value—, 304
Realm of value, 306
Reason and reasons of value, 11
Recaséns Siches, L., 309
Red nose, 159, 306
Reducibility, ix, 157–60, 343, 353
Reduction: analytic-synthetic, 222; Moore's, 226; phenomenological, 90; to primary properties, 67, 226, 303. *See also* Resolution; Synthetic method
Rees, W. J., 158, 159, 343

Reference, 66, 92
Refinement, 68–69, 129
Relations: axiological, 166; "crossings of," 332; and senses, 65; of value, 109, 162–68; synthetic, 87–88; between value terms, 228–48
Relativity of fact and value: 220–25; theory, 124; of value, 107, 296
Religion, 180
Renaissance, 68
Resolution and composition, 216–17, 219, 303, 304. *See also* Galileo
Rest and motion, 225
Revolution, scientific, 45
Rhetoric, 11, 308, 311
Richness: extensional, 96, 98; hierarchy, 114; intensional, 96, 98; within intrinsic value, 224; scale of, 96, 98; of system, 333. *See also* Enrichment
Rickert, H., 331, 332
Rickover, Admiral, 9
Rightness, 298. *See also* Action
Rilke, R. M., 356
Ritual, 310, 311
Robots, 310
Romero, F., 345
Rosen, E., 355
Ross, W. D., 343
"Rotten," 162
Royce, J., 119, 323
Rousseau, J.-J., 253–54, 309
"Rules," 297
Rules: of application of axiology, 292; of games, 310; of valuation, 107; of value relativity, 223
Russell, B.: 310; his axiomatic identification, 102; and definition, 121; and formalism, 88; on mathematiical logic, 300; his definition of number, 17, 105; and naturalistic fallacy, 52, 122; axiom of reducibility, 157, 160, 343; vicious-circle principle, 135, 326
Ryle, G., 158, 159, 343, 349

Sagredo, 85
St. Ambrose, 326
St. Anselm, 116, 325, 326, 349
St. Augustine, 119
St. Paul, 359
Salviati, 85
Sarton, G., 328
Satan, 273
Satisfaction, 227
"Satori," 308
Scale: of increasing specificity, 97; logical, of Being, 96–97; logical, of value, 97–98; of richness, 114
Scarcity, axiom of, 360
Schein, 353
Scheler, M., 27, 350
Schema, 197, 198, 199, 222
Schematic class. *See* Class
Schematic thinking, 113
Schiller, F., 353
Schlemihl, 278
Schönberg, A., 245
Science: x; axiological, 3, 4, 8, 9, 105, 107, 114–15, 301, 303, 304, 311; axiom of, 41, 90, 336; creation of, 77, 114; cumulative, 11; definition of, 76, 77, 122; efficiency, 68; empirical and theoretical, 72, 335; extension-intension, 38, 76, 77; formal analogy of moral and natural, 146, 149; network of formal relations, 65; import of, 71–79; language of, 74–75; measurement, 251; as method, 21, 36; method of, 46, 84–87; moral, 4, 56, 130, 146, 337; natural, 71, 107, 130, 146, 220; and naturalistic fallacy, 53; orderly thinking, 14; passion of, 12; political, 47; practical, 67; reduction in, 67; secularization of, 63; and philosophy, xi, 14, 15, 16, 31, 46, 80, 92, 352; social, 47; structure of, 69–92; system, 27–30, 31, 84–92; theoretical and applied, 72, 106, 292–93, 335; transformation of man, 60, 68; universal and concrete, 36; of value, 3, 4, 8, 9, 105, 107, 114–15, 301, 303, 304, 311; vision of, 9, 11. *See also* Axiom; Measurement; System
Scientific detachment, 14
Scientific revolution, 45
"Scientist," 45
Scoring, 233, 242
Secondary qualities, 227, 328. *See also* Properties
Secondary underlying judgment, 175
Secondary value combinations, 273–74
Secondary value product, 215
Secondary value properties, 227
Second-order inclinations, 343
Secret: of valuation, 214; of value, 158

Seeing, 70
Seidenberg, R., 322
Self, 306
Self-concept, 306
Semantics, 308, 311
Semblance, 353
Senary value combinations, 279, 280
Sensitivity, 68, 69, 129, 258, 279, 311
Ser, 343
Servetus, M., 297
Set of properties. *See* Properties
Set theory, 225
Sewell, E., 358
Shapiro, J., 349
Shaw, G. B., 297, 355
Sherman, Hoyt, 329
Shirham, King, 33
Shorthand symbols, 329
Siches. *See* Recaséns Siches
Signals, 309
Similarity: between things, 138; logical, 325; of intensions, 18, 52, 75, 250. *See also* Measurement
Simple, Cartesian, 83
"Sincere," 306
Singular, 201
Singular concept, 252. *See also* Concept
Situation of value, 107, 297
Sizzi, F., 34, 35, 349
Slavery, 115
Social ethics, 309, 311
Social science, 46–47, 329
Sociology, 11, 47, 48, 114, 307, 311
Socrates, viii, 56, 286–92
Soll, 354
"Some," 162, 169
"Some . . . not," 169
"So-so," 162, 179, 211, 232, 250
Sovereignty, 254
Soviet cosmonaut, 124
Space-time universe, 118
Spann, O., 308, 310
Specialist barbarism, 333
Species, fallacy of, 124
Specific: generalization, 46; number, 106; valuation, 107; value, 107, 108
Specification, 83, 155, 200, 201, 204
Specificity, 106–20, 121
Specifics: of axiom, 106; formal, 107; material, 107; material of value, 115–20; phenomenal, 107; theoretical, 107, 108; theoretical of value, 108, 112–15
Speech science, 308
Spinoza: 296; ethics, 28, 337; metaphysical fallacy, 124; metaphysics, 108; scale of value intensity, 98; science of ethics, 99, 337, style, 29; theory of types, 326; on value terms, 348; on world, 348
Spiritual life, 225
Spiritual value, 252
Sprague, R. K., 342
Square circle, 198
Square of oppositions, 169, 345
Spring of chair, 205
Stagnant value, 244
Standard for value, 205, 217
Stark, R., 355
State, 253, 254
Statistical a-truth, 187
Statistical "ought," 184
Stebbing, L. S., 321
Stenzel, J., 319
Stevenson, C. L., 137, 141, 338, 339
Stigmatic intuition, 321
Structure of intension. *See* Intension
Structure of value, 3, 249–51
Subject-predicate relation, 169
Subject of value proposition, 168
Subjective intention of judgment, 164
Subjective truth, 308
Subjective truth-value, 164
Subjectivity of value, 107, 110, 296
"Substance," 81
Substance, 181
Substance and function, 39
Subtraction of values, 229–30, 233–38, 242, 246
"Such a thing," 218
Sum of values, 213, 323–33. *See also* Addition
Summa virtus, 244
Summum bonum, 244
Summum genus, 101, 119
"Superior." *See* "More than good"
Supply, 360
Survival of the fittest, 124
Syllogism, 86–87
Symbol, 45, 185, 309
Symbolic logic, 324
Symbolism, 91, 260
Syncategorematic, 335
Synlytic, 271, 358
Synonymous, 255, 349
Synthesis, 25, 46, 318, 346
Synthetic: 317; a posteriori, 80; a priori, 80, 359; concept, 31–43, 72–92,

Synthetic (*continued*)
252, 271, 317; definition, 85, 331; intension, 222; judgment, xi, 79–80, 271, 317; method, 79, 216–17; "ought," 184, 192, 247; properties, 222; value propositions, 174, 175, 179
Syntheticity, axiological, 172, 180, 183, 184
Syringae, 199
Syringa vulgaris, 255, 349
System: analytic, 32–33, 333; application of, 65, 108–9, 272; axiological theory of, 352; characteristic number of, 351, 352; formal, xiv, 72; Gestalt, 352; intension of axiom, 76, 77, 97; meaning of, 86; political, 115, 253–54; power of, 351; logic, 330; philosophical, 27–30, 84; social, 329; scientific and philosophical, 14, 27–30, 31, 36, 84, 199, 223; as systemic value, 277; norm of systemic value, 222–23; synthetic, 38, 65, 83, 101–2, 333. *See also* Isomorphism
Systematic import, 71–79, 154, 193
Systemic class, 179, 187, 188
Systemic concept, 252. *See also* Concept
Systemic goods, 349
Systemic intension, 222
Systemic intention, 164
Systemic measurement, 250
Systemic things, 194
Systemic truth-value, 164
Systemic value: xv, 19, 164, 193; application of, 253–54, 309–10, 311; calculus, 267; characteristic number of, 112, 222, 267; dimension, 114; as fact, 221, 222, 223; fulfillment of definition or construct, 199; in hierarchy, 252; perfection, 195; relations, 224; value theory as, 294

Talent, axiological, 258
Tang (Chinese "ought"), 354
Technological ecology, 310
Technological value, 108
Technology, 310, 311
Teleological good, 62
Teleology, 36, 61, 243, 297, 298, 359
Teller, E., 297, 333
Temporal process, 127
Term: 194; applied value—, 180, 306; value—, and modalities, 179; axiomatic unit, 250; limit, 89–90; logarithmic nature of, 222; moral, 306; Ortega y Gasset on, 331; and system, 84–86; value—, and value proposition, 178. *See also* Value terms
Terminus, 332
Tertiary properties, 227
Tertiary value combinations, 278, 279
"The Good," 168, 349
Theologians, 114
Theologia perennis, 335
Theological space, 255
Theoretical axiology, 107
Theoretical constructs, 73
Theoretical specificity, 121
Theoretical specifics, 107, 108; of value, 112–15
Theory of types, 104, 157–60, 199–202, 221, 342, 343–44
Theory of value, and value, 5–9
Theory, value of, 293–303
Thinghood, 218
Thing variable, 181
Thinking: 118; abstract and formal, 25; dogmatic, 113; and doing, 11; empathic and emphatic, 114; modal, 171; orderly, 3–8, 14, 56; pragmatic, 113; schematic, 113
Thought, 117
Through the Looking Glass, 35, 303, 328
Thurber, J., 328
Thurstone, L. L., 307
Tillich, P., 323, 335
Tolstoy, L., 227, 328
Tompkins, 220, 321
Topological rule, 206
Total value, 216, 223
Total value product, 214
Toulmin, S. E., 327
Toynbee, A., 309
Toynbee, P., 322
T-propositions, 169–70, 228–48
Transcendentals, 101, 119, 164
Transfinite intensional structure, 224. *See also* Cardinalities
Transfinite ordinal, 352
Transfinite values, 224, 244, 248
Transition: from category to axiom, 42; from concept to term, 87; from moral philosophy to moral science, 15, 30, 36; from natural philosophy to natural science, 31; from philosophy to science, 14, 89

Transposition of values, 112, 244, 246, 268, 272–76, 278, 279, 280, 333, 343, 349. *See also* Composition
Transpositional concept, 271
Transpositional intension, 269, 271–72
True, 125
Truth: axiological, 186; and goodness, 164; subjective, 164, 308
Truth-value: 163; apophantic, 164; extrinsic, 164; and good-value, 164; intrinsic, 164; objective, 164;—pattern of value terms, 207; subjective, 164; systemic, 164; of value propositions, 170, 174, 178–92
Type differentiation, 205
Types: and definition, 199–202; theory of, 104, 157–60, 199–202, 221, 342, 343–44; of value, 221–23
Typical, 206
Typical constellation of properties, 261
Typical value terms, 206
Typhoon, 263

"Uncertainty relation" of moral science, 145. *See also* Indeterminacy
Underhill, E., 352
Unicorn, 198
Uniform, 273
"Unique," 107, 248, 295. *See also* "In a class by itself"
Uniqueness, 96, 113, 195, 199
Unique things, 210
Unit class, 182
Universality of value, 109, 296
Urmson, J. O., 99, 334
Use, axiological, 251

Vacation, 255
Valency, 228
Validity, 186, 346
Valuation: 107, 108; and class-membership, 179; emphatic and empathic, 114; essence of, 215–28; and exposition, 179, 197; and feeling, 129; process of reintegration, 219; refined, 68–69, 129; secret of, 214; sensitivity for, 68, 69, 129, 258, 279, 311; specific, 107; talent of, 258; and value theory, 5–7. *See also* Fallacy of Method
"Value," 4, 69, 107, 257. *See also* Formal value
Value: absoluteness of, 107, 109, 296; abstraction, 42; abundance—, 245; addition of, 228–29, 231–33, 242; adverbs, 344; aesthetic, 48, 108, 114–15, 180, 227, 309, 311, 324; —agreement, 110; —alchemy, 228; any "reduced" function, 160; awareness, 224; —argument, 123–24, 296; axiological, xi, 4, 17, 69, 156, 304, 329, 346;—axiom, 104, 110, 153, 154, 209; basic—terms, 162; and being, 108; better—, 254, 257; calculus, *see* Calculus; —capacity, 224; —cardinalities, *see* Cardinalities; —combinations, *see* Combinations; —complements 242; conceptual, 97; concreteness of, 98; a construction, 42, 53, 145; definition of, 18, 19, 52, 69, 92, 99, 107, 129, 223; determination of, 102; differences in, 142–43; —dimension, *see* Dimension; difference in—of things, 140; —disagreement, 110; disciplines, 227; division of, 231, 241–42; economic, 100, 108; elements of, 19, 226; empirical, 78; enrichment of, 98; ethical, 48, 108, 306, 308, 311; exponentiation, 243–48; expositional, 182; extrinsic, *see* Extrinsic value; and fact, 4, 19, 156–57, 209, 217–27; as fact, 223; of fact, 108, 273; formal, xi, 4, 17, 69, 75, 77, 143, 154, 156, 296; formal construction, 42, 53, 145; formal specifics of, 109–12; formal structure, 42, 145; formulae, *see* Calculus; fulfillment of intension, 254, *see also* Fulfillment; function of natural properties, 142; generic and specific, 95–101, 243; good-, 164; harmonies of, 258; hierarchy of, *see* Hierarchy; infinite—of person, 116–17; intensional richness, 98; intensional similarity, 18, 52; intrinsic, *see* Intrinsic value; and intrinsic nature, 139; intrinsicness of, 140–44; irrationality of, 107, 109, 296; "irreal," 145; juridical, 277, 310, 311; value, kinds of, 98; —language, *see* Language; —leaps, 223; —linguistics, 308, 309, 310; —logic, 30, 55, 62, 91, 107, 108; logical levels of, 97, 334; —logistic, 114; material specifics of, 115–20; meaning as—, *see*

Value (*continued*)
Meaning; measure, *see* Measure; —measurement, *see* Measurement; of mere fact, 246; metaphysical, 108; metaphysicians of, 12; methodological levels of, 4, 6, 69, 334; Moore's paradox of intrinsic, 131–49; multiplication of, 230, 238–41, 242; multiplier of a, 240; and nature of thing, 137, 138, 142; negative, 244, 271; no-good—, 244; nonnatural property, 104; non-systemic, 112; non-, extrinsic, 244; and nonvalue terms, 208–15; norm, 220; as normative ought-value, 243; and number, 17, 51, 52; objectivity of, 107, 110; —operations, 228–48, 305; orders of, 221–23; -ought, 240–42; ought-, 233–48; particular, 97, 98, 109; particularity of, 109, 296; value pattern, 155–56, 171, 172, 221; perfection as, 244; perfection of, 245; phenomenal, xi, 4, 17, 69, 143, 329; phenomenologists of, 12; —phenomenon, 4, 6, 7, 257, 303, 304, 329; —philosophy, 10, 13, 14, 77; —predicate, 103, 137, 144; primary—properties, 17, 219; principle of, 7; product, 214; —property, 104, 209, 226, *see also* Description; Properties; —of a property, 202–6; —propositions, 168–92; of proposition, 163; psychological, 108; rationality of, 107, 109, 296; —reality, 304; reality of, 220; realm of, 305, 306; "reduced" function, 157–60, 344; relations, 162–68; relativity of, 107, 220–25, 296; —science, 3, 4, 9, 77, 146, 149, 301, 304; —sciences, 107, 112, 114–15, 311; scientific knowledge of, 8; scientific, 17, 69, *see also* Axiological value; secondary—properties, 17, 227; secondary—product, 215; secret of, 158; self-fulfillment, 243–44; sense of, 258; sensitivity for, 68, 69, 129, 258, 279, 311; similarity of intensions, 18, 52, 75, 250, *see also* Measurement; singular, 98; *situational*, 17, *see also* Phenomenal value; situations of, 107, 297; specific and generic, 95–101; specific, 107, 108, 243; —specification, 104–6; stagnant, 244; structure of, 3, 249–51; subjectivity of, 107, 110, 296; subset of given set, 19, 223; subtraction of, 229–30, 233–38, 242, 246; sum of, 213, 232; symbolism, 260; synthetic concept, 257; systemic, *see* Systemic value; —talent, 258; technological, 108; as term, x, *see also* Value terms; —texture, 258; theoretical specifics of, 112–15; —theory, 41, 77, 293–302, *see also* Value theory; of thing, 132; topological determination of, 202; total, 216; total-product, 214; transfinite, 224, 244; transposition of, *see* Transposition; —type, 221–23; universality of, 109, 296; and "value", *see* Formal value; of valuation, 203; of value, 254; of value theory, 108, 293–302; a variable, 19, 99, 156; —world, 68–69, 225, 258, 296, 305; of world, 108, 111–12

Value terms: 51, 52, 103–4, 154–62, 182; arithmetical pattern, 208–15; formal specifics of value, 107; intensional pattern, 194–99; logical pattern, 199–208; logical and arithmetical nature, 353; and value propositions, 178; relationship of, 228–48; typical, 206

Value theory: philosophical and scientific, 41, 77; and valuing, 5–7. *See also* Fallacy of Method

Variable, 19, 85, 97, 99, 105, 181, 266, 278

Variation, combinatorial, 19, 265

Vasconcelos, J., 310

Velocity, 35, 300

Verum, 164

Vicious-circle principle, 135, 144, 326

Waddington, C. H., 339

Walter, B., 245

War, 254, 273

Weight of properties, 349

Weiss, P., 323, 335, 336, 356, 357, 358

Weissman, D., ix

Weitz, M., 343

Werfel, F., 328

Werkmeister, W. H., 330

Wesensschau, 90

White, Morton, 21

Whitehead, A. N.: analytic-synthetic, 91; axiom of mathematics, 105; extensive abstraction, 89–90; ideal of

Whitehead, A. N. (*continued*)
science, 4; "intensive abstraction," 92; on logic of value, 50, 324; mathematics and good, ix; paradox of knowledge, 2, 145, 301, 302, 326
Whitewashed house, 158, 159
Whyte, W. H., Jr., 355
Widersinn, 270
Wilkinson, J., 331
Williams, D., 319
Wittgenstein, L., 13, 145, 256, 310, 332, 356
Word counts, 310
World: of analytic concepts, 36, 64–68; aspect of, 123; badness of, 111–12; bestiality of, 124; of fact, 70, 220, 225; goodness of, 111–12; ideality of, 124; moral, 68–69; of primary properties, 220; rationality of, 110; of science and of philosophy; 64–68, 91; of synthetic concepts, 36, 64–68; of value, 68–69, 225, 258, 296, 305; value of, 108, 111–12
"Worse," 107, 108, 114, 163, 166, 184, 295
"Worse for," 107, 163, 166, 295
"Worst," 166

"Yellow," 135
"You"—"I," 355

Zen, 308
Ziff, P., 340
Zipf, G. K., 309
Zweig, Stefan, 328